I0131664

NX Nastran 9.0
for Designers

CADCIM Technologies

525 St. Andrews Drive
Schererville, IN 46375, USA
(www.cadcim.com)

Contributing Author

Sham Tickoo

Professor
Department of Mechanical Engineering Technology
Purdue University Northwest
Hammond, Indiana
USA

CADCIM Technologies

NX Nastran 9.0 for Designers
Sham Tickoo

CADCIM Technologies
525 St Andrews Drive
Schererville, Indiana 46375, USA
www.cadcim.com

Copyright © 2016 by CADCIM Technologies, USA. All rights reserved. Printed in the United States of America except as permitted under the United States Copyright Act of 1976.

No part of this publication may be reproduced or distributed in any form or by any means, or stored in the database or retrieval system without the prior permission of CADCIM Technologies.

ISBN   978-1-942689-16-4

NOTICE TO THE READER

Publisher does not warrant or guarantee any of the products described in the text or perform any independent analysis in connection with any of the product information contained in the text. Publisher does not assume, and expressly disclaims, any obligation to obtain and include information other than that provided to it by the manufacturer.

The reader is expressly warned to consider and adopt all safety precautions that might be indicated by the activities herein and to avoid all potential hazards. By following the instructions contained herein, the reader willingly assumes all risks in connection with such instructions.

The Publisher makes no representation or warranties of any kind, including but not limited to, the warranties of fitness for particular purpose or merchantability, nor are any such representations implied with respect to the material set forth herein, and the publisher takes no responsibility with respect to such material. The publisher shall not be liable for any special, consequential, or exemplary damages resulting, in whole or part, from the reader's use of, or reliance upon, this material.

www.cadcim.com

DEDICATION

*To teachers, who make it possible to disseminate knowledge
to enlighten the young and curious minds
of our future generations*

*To students, who are dedicated to learning new technologies
and making the world a better place to live in*

THANKS

*To the faculty and students of the MET department of
Purdue University Northwest for their cooperation*

To employees of CADCIM Technologies for their valuable help

Online Training Program Offered by CADCIM Technologies

CADCIM Technologies provides effective and affordable virtual online training on various software packages including Computer Aided Design, Manufacturing, and Engineering (CAD/CAM/CAE), computer programming languages, animation, architecture, and GIS. The training is delivered 'live' via Internet at any time, any place, and at any pace to individuals as well as the students of colleges, universities, and CAD/CAM /CAE training centers. The main features of this program are:

Training for Students and Companies in a Classroom Setting

Highly experienced instructors and qualified engineers at CADCIM Technologies conduct the classes under the guidance of Prof. Sham Tickoo of Purdue University Northwest, USA. This team has authored several textbooks that are rated "one of the best" in their categories and are used in various colleges, universities, and training centers in North America, Europe, and in other parts of the world.

Training for Individuals

CADCIM Technologies with its cost effective and time saving initiative strives to deliver the training in the comfort of your home or work place, thereby relieving you from the hassles of traveling to training centers.

Training Offered on Software Packages

CADCIM provides basic and advanced training on the following software packages:

CAD/CAM/CAE*: CATIA, Pro/ENGINEER Wildfire, PTC Creo Parametric, Creo Direct, SOLIDWORKS, Autodesk Inventor, Solid Edge, NX, AutoCAD, AutoCAD LT, AutoCAD Plant 3D, EdgeCAM, and ANSYS*

Architecture and GIS*: Autodesk Revit(Architecture/Structure/MEP), AutoCAD Civil 3D, AutoCAD Map 3D, Navisworks, Primavera, and Bentley STAAD Pro*

Animation and Styling*: Autodesk 3ds Max, Autodesk 3ds Max Design, Autodesk Maya, Autodesk Alias, The Foundry NukeX, MAXON CINEMA 4D, Adobe Flash, and Adobe Premiere*

Computer Programming*: C++, VB.NET, Oracle, AJAX, and Java*

*For more information, please visit the following link: **http://www.cadcim.com***

Note
If you are a faculty member, you can register by clicking on the following link to access the teaching resources: ***http://www.cadcim.com/Registration.aspx***. The student resources are available at ***http://www.cadcim.com***. We also provide **Live Virtual Online Training** on various software packages. For more information, write us at ***sales@cadcim.com***.

Table of Contents

Chapter 3: NX Nastran File Support

Chapter 4: Model Preparation for Analysis

Chapter 5: Meshing-I

Chapter 8: Connections and Contacts

Chapter 9: Defining Material and Boundary Conditions

Chapter 10: Solving and Post Processing

Chapter 11: Projects

Index

This page is intentionally left blank

Preface

NX Nastran 9.0

NX Nastran 9.0, a product of SIEMENS Corp., is one of the world's leading, widely distributed, and popular commercial CAE package. NX Nastran is being used by designers across a broad spectrum of industries such as aerospace, automotive, manufacturing, nuclear, electronics, biomedical, and many more to produce safe, reliable and optimized designs within increasingly shorter design cycles. It provides simulation solutions that enable designers to simulate design performance directly on the desktop. In this way, it dispenses fast, efficient, and cost-effective product development from design concept stage to performance validation stage of the product development cycle. NX Nastran package streamlines the product development process by resolving issues related to structural deformation, stress, vibration, buckling, heat transfer, acoustics and aeroelasticity, fluid flow, electromagnetic effects, a combination of these phenomena, and so on.

The **NX Nastran 9.0 for Designers** textbook has been written to assist engineers, engineering students, and practicing designers who are new to the field of Finite Element Analysis (FEA). The textbook covers FEA concepts, modeling, and analysis of engineering problems using NX Nastran. In addition, enough theoretical background is offered to allow individuals to use NX Nastran as simulation package. This textbook covers Structural Analysis, Buckling Analysis, and Response Analysis.

- **Tutorial Approach**

   The author has adopted the tutorial point-of-view and the learn-by-doing theme throughout the textbook. This approach guides the users through the process of creating the models in the tutorials.

- **Real-World Projects as Tutorials**

   The author has used about 18 real-world mechanical engineering projects as tutorials in this book. This enables the readers to relate the tutorials to the models in the mechanical engineering industry. In addition, there are about 10 exercises that are also based on the real-world mechanical engineering projects.

- **Tips and Notes**

   The additional information related to various topics is provided to the users in the form of tips and notes.

- **Learning Objectives**

   The first page of every chapter summarizes the topics that are covered in that chapter.

- **Self-Evaluation Test, Review Questions, and Exercises**

    Every chapter ends with Self-Evaluation Test so that the users can assess their knowledge of the chapter. The answers to the Self-Evaluation Test are given at the end of the chapter. Also, the Review Questions and Exercises are given at the end of each chapter and they can be used by the instructor as test questions and exercises.

Formatting Conventions Used in the Textbook

Please refer to the following list for the formatting conventions used in this textbook.

- Names of tools, buttons, options, groups, and toolbars, are written in boldface.

    Example: The **Sew** tool, the **OK** button, the **Feature** group, and so on.

- Names of dialog boxes, drop-downs, drop-down lists, list boxes, areas, edit boxes, check boxes, and radio buttons are written in boldface.

    Example: The **Model Display** dialog box, the **Change Window Drop-down** list in the **Context** group, the **Tool Option** drop-down list in the **Tool** rollout, **Merge Angle Tolerance** edit box, the **Create Idealized Part** check box, the **Create New** radio button, and so on.

- Values entered in edit boxes are written in boldface.

    Example: Enter **0.5** in the **Element Size Factor** edit box.

- Names and paths of the files saved are italicized.

    Example: *c03tut03.prt*, *C:\NX Nastran\c07\Tut03*, and so on.

- The methods of invoking a tool/option from the **Menu**, **Ribbon**, are enclosed in a shaded box.

Ribbon:	Home > Standard > New
Menu:	File > New

Naming Conventions Used in the Textbook

Dialog Box

In this textbook, different terms are used for referring to the components of a dialog box. Refer to Figure 1 for the terminology used.

Button

The item in a dialog box that has a 3D shape like a button is also termed as **Button**. For example, **OK** button, **Cancel** button, **Apply** button, and so on.

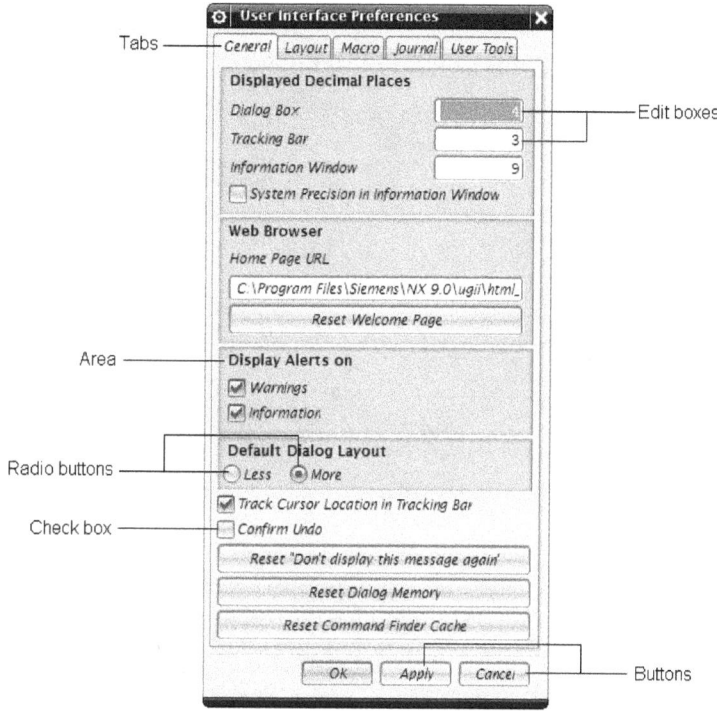

Figure 1 *The components in a dialog box*

Drop-down

A drop-down is the one in which a set of common tools are grouped together. You can identify a drop-down with a down arrow on it. These drop-downs are given a name based on the tools grouped in them. For example, **Load Type** drop-down, **Simulation Object Type** drop-down, and so on; refer to Figure 2.

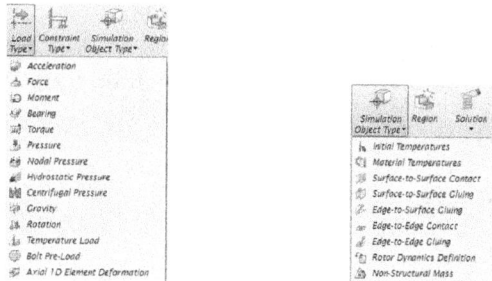

Figure 2 *The **Load Type** and **Simulation Object Type** drop-downs*

Gallery

A gallery is the one in which a set of common tools are grouped together. For example, **Face Operations** gallery of the **More** gallery; refer to Figure 3.

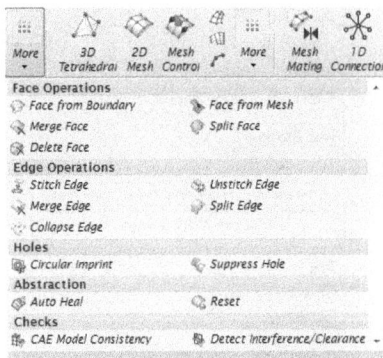

*Figure 3 The **Face Operations** gallery of the **More** gallery*

Drop-down List

A drop-down list is the one in which a set of options are grouped together. You can set various parameters using these options. You can identify a drop-down list with a down arrow on it. For example, **Type** drop-down list, **Geometry To Stitch** drop-down list, and so on; refer to Figure 4.

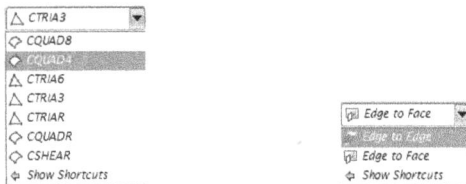

*Figure 4 The **Type** and **Geometry To Stitch** drop-down lists*

Options

Options are the items that are available in shortcut menu, drop-down list, dialog boxes, and so on. For example, choose the **Fit** option from the shortcut menu displayed on right-clicking in the drawing area; choose the **Manual** option from the **Method** drop-down list; refer to Figure 5.

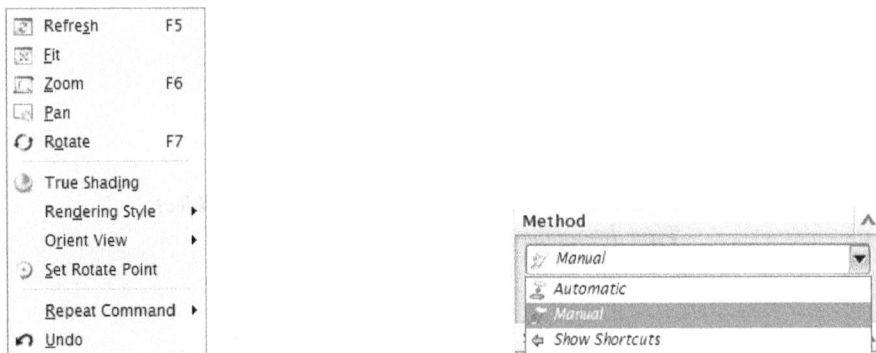

*Figure 5 Options in the shortcut menu and the **Method** drop-down list*

Symbols Used in the Textbook

Note
The author has provided additional information related to various topics in the form of notes.

Tip
The author has provided a lot of useful information to the users about the topic being discussed in the form of tips.

Free Companion Website

It has been our constant endeavor to provide you the best textbooks and services at affordable price. In this endeavor, we have come out with a Free companion website that will facilitate the process of teaching and learning of NX Nastran 9.0. If you purchase this textbook, you will get access to the companion website.

The following resources are available for the faculty and students in this website:

Faculty Resources
• **Technical Support**
The faculty can get online technical support by contacting *techsupport@cadcim.com*.

• **Instructor Guide**
Solutions to all review questions and exercises in the textbook are provided in this guide to help faculty members test the skills of the students.

• **PowerPoint Presentations**
The contents of the book are arranged in customizable PowerPoint slides that can be used by the faculty for their lectures.

• **Part Files**
The part files used in illustrations, tutorials, and exercises are available for free download.

Student Resources
• **Technical Support**
The students can get online technical support by contacting *techsupport@cadcim.com*.

• **Part Files**
The part files used in illustrations and tutorials are available for free download.

Stay Connected

You can now stay connected with us through Facebook and Twitter to get the latest information about our textbooks, videos, and teaching/learning resources. To stay informed of such updates, follow us on Facebook (*www.facebook.com/cadcim*) and Twitter (*@cadcimtech*). You can also subscribe to our YouTube channel (*www.youtube.com/cadcimtech*) to get the information about our latest video tutorials.

If you face any problem in accessing these files, please contact the publisher at *sales@cadcim.com* or the author at *stickoo@pnw.edu* or *tickoo525@gmail.com*.

Chapter 1

Introduction to FEA

Learning Objectives

After completing this chapter, you will be able to:
- *Understand basic concepts and the general working of FEA*
- *Understand advantages and limitations of FEA*
- *Understand the type of analysis*
- *Understand important terms and definitions used in FEA*
- *Understand theories of failure in FEA*

INTRODUCTION TO FEA

The finite element analysis (FEA) is a computing technique that is used to obtain approximate solutions to the boundary value problems in engineering. In this analysis, a numerical technique called the finite element method (FEM) is used to solve boundary value problems. FEA involves a computer model of a design that is loaded and analyzed for specific results. The FEM was first introduced by Richard Courant in 1943. He used the Ritz method of numerical analysis and minimization of variational calculus for getting approximate solutions to vibration systems. Later, the academic and industrial researchers created the finite element method for structural analysis.

The concept of FEA can be explained through a simple example of measuring the perimeter of a circle. To measure the perimeter of a circle without using the conventional formula, divide the circle into equal segments, as shown in Figure 1-1. Next, join the start point and endpoint of each of these segments by a straight line. Now, you can easily measure the length of the straight line, and thus the perimeter of the circle.

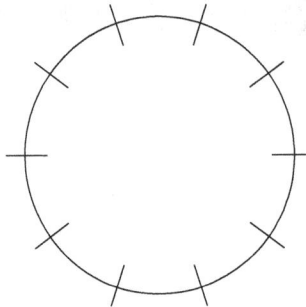

Figure 1-1 *The circle divided into equal segments*

If the number of segments into which the circle is divided is less, as shown in Figure 1-1, you will not get accurate results. For accuracy, divide the circle into more number of segments. In short, the more the number of segments, the more accurate the result will be. However, with more segments, the effort required is more. The same concept applies to FEA also, and therefore, there is always a compromise between accuracy and speed while using this method which makes it an approximation method.

The FEA was first developed to be used in the aerospace and nuclear industries where the safety of structures is crucial. Nowadays, the simplest of the products rely on the FEA for their design evaluation.

The FEA simulates the loading conditions of a design and determines the design response in those conditions. The design is modeled using the discrete building blocks called elements. Each element has some equations that describe its response to certain loads. The sum of the responses of all the elements in a model gives the total response of the design.

General Working of FEA

Better knowledge of FEA will help you build more accurate models. It will also help you understand the backend working of NX Nastran. A simple model is discussed here to give you a brief overview of FEA.

Figure 1-2 shows a spring assembly that represents a simple two-spring element model. These two springs are connected in series and one of the springs is fixed at the left endpoint, refer to Figure 1-2. The stiffness of the springs is represented by spring constants K_1 and K_2. The endpoints of each spring is restricted to the displacement or the translation in the X direction only. The change in position from the undeformed state of each endpoint can be defined by the variables X_1 and X_2. The forces acting on each endpoint of the springs are represented by F_1 and F_2.

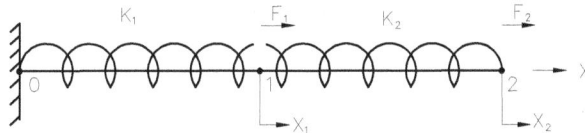

Figure 1-2 *Representation of a two-spring assembly*

To develop a model that can predict the state of this spring assembly, you can use the linear spring equation given below:

$$F = KX$$

If you use the spring parameters defined above and assume a state of equilibrium, the following equations can be written for the state of each endpoint:

$$F_1 - X_1 K_1 + (X_2 - X_1) K_2 = 0$$
$$F_2 - (X_2 - X_1) K_2 = 0$$

Therefore,

$$F_1 = (K_1 + K_2) X_1 + (-K_2) X_2$$
$$F_2 = (-K_2) X_1 + K_2 X_2$$

If the above set of equation is written in matrix form, it will be modified as follows:

$$\begin{bmatrix} F_1 \\ F_2 \end{bmatrix} = \begin{bmatrix} K_1 + K_2 & -K_2 \\ -K_2 & K_2 \end{bmatrix} \begin{bmatrix} X_1 \\ X_2 \end{bmatrix}$$

In the above mathematical model, if the spring constants (K_1 and K_2) are known and the forces (F_1 and F_2) are defined, then you can determine the resulting deformed shape (X_1 and X_2). Alternatively, if the spring constants (K_1 and K_2) are known and the deformed shapes (X_1 and X_2) are defined, then the resulting forces (F_1 and F_2) can be determined.

This type of spring system may be complicated to define, but it involves most of the key terminologies used in FEA. These FEA terminologies are discussed next.

1. Stiffness Matrix
2. Degree of Freedom
3. Boundary Conditions

Stiffness Matrix

The stiffness matrix represents the resistance offered by a body to withstand the load applied. In the previous equation, the following part represents the stiffness matrix (K):

$$\begin{bmatrix} K_1 + K_2 & -K_2 \\ -K_2 & K_2 \end{bmatrix}$$

This matrix is relatively simple because it comprises only one pair of spring, but it turns complex when the number of springs increases.

Degree of Freedom

The Degree of freedom is defined as the ability of a node to translate or transmit the load. In the previous example, you are only concerned with the displacement and forces. By making one endpoint fixed, one degree of freedom for displacement is removed. So, now the model has two degrees of freedom. The number of degrees of freedom in a model determines the number of equations required to solve the mathematical model.

Boundary Conditions

The boundary conditions are used to eliminate the unknowns in a system. A set of equations that is solvable is meaningless without the input. In the previous example, the boundary condition was $X_0 = 0$, and the input forces were F1 and F2. In either ways, the displacements could have been specified in place of forces as boundary conditions and the mathematical model could have been solved for the forces. In other words, the boundary conditions help you reduce or eliminate unknowns in the system.

The FEA technique needs the finite element model (FEM) for its final solution as it does not use the solid model. FEM consists of nodes, keypoints, elements, material properties, loading, and boundary conditions.

Nodes, Elements, and Element Types

Before proceeding further, you must be familiar with commonly used terms such as nodes, elements, and element types. These terms are discussed next.

Nodes

An independent entity in space is called a node. Nodes are similar to the points in geometry and represent the corner points of an element. You can change the shape of an element by moving the nodes in space. The shape of a node is shown in Figure 1-3.

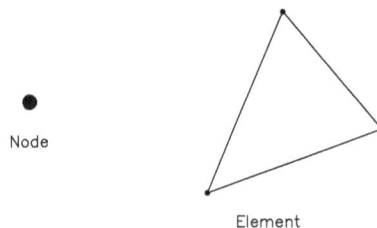

Node

Element

Figure 1-3 *A node and an element*

Elements

An element is an entity into which the system under study is divided. The shape (area, length, and volume) of an element is specified by nodes. Figure 1-3 shows a triangular shaped element.

Element Types

The following are the basic types of the elements:

Point Element

A point element is in the form of a point and therefore has only one node.

1D Element

A 1D element has the shape of a line or curve, therefore a minimum of two nodes are required to define it. There can be higher order elements that have additional nodes (at the middle of the edge of the element). The element that does not have a node at the middle of the edge of the element is called a linear element. The elements with node at the mid of the edges are called quadratic or second order elements. Figure 1-4 shows some line elements.

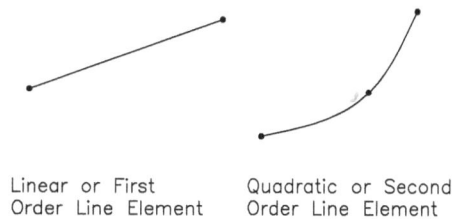

Linear or First
Order Line Element

Quadratic or Second
Order Line Element

Figure 1-4 *The 1D elements*

2D Element

An 2D element has the shape of a quadrilateral or a triangle, therefore it requires a minimum of three or four nodes to define it. Some of the 2D elements are shown in Figure 1-5.

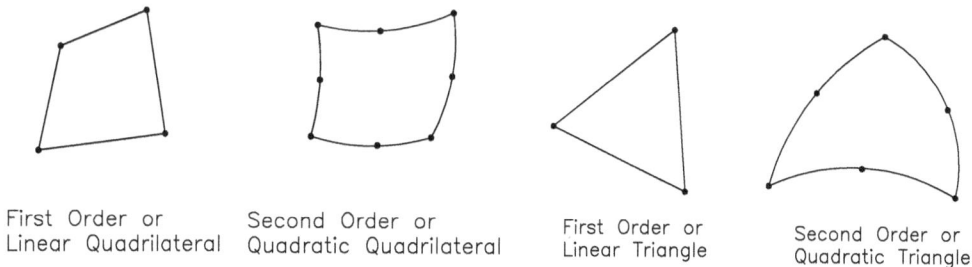

First Order or
Linear Quadrilateral

Second Order or
Quadratic Quadrilateral

First Order or
Linear Triangle

Second Order or
Quadratic Triangle

Figure 1-5 *The 2D elements*

Note

In this chapter, only the basic introduction of element types has been covered.

3D Element

A 3D element has the shape of a hexahedron (8 nodes), wedge (6 nodes), tetrahedron (4 nodes), or a pyramid (5 nodes). Some of the 3D elements are shown in Figure 1-6.

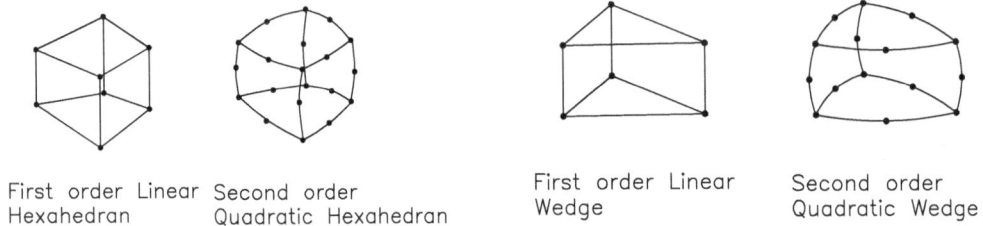

First order Linear   Second order   First order Linear   Second order
Hexahedran   Quadratic Hexahedran   Wedge   Quadratic Wedge

Figure 1-6 The 3D elements

Areas for Application of FEA

FEA is a very important tool for designing. It is used in the following areas:

1. Structural strength design
2. Structural interaction with fluid flows
3. Shock analysis
4. Acoustics
5. Thermal analysis
6. Vibrations
7. Crash simulations
8. Fluid flows
9. Electrical analysis
10. Mass diffusion
11. Buckling problems
12. Dynamic analysis
13. Electromagnetic analysis
14. Coupled analysis, and so on.

General Procedure of Conducting Finite Element Analysis

To conduct the finite element analysis, you need to follow certain steps. These steps are given next.

1. Set the type of analysis to be carried out.
2. Create or import the model.
3. Define the element type.
4. Divide the given problem into nodes and elements (generate a mesh).
5. Apply material properties and boundary conditions.
6. Solve the unknown quantities at nodes.
7. Interpret the results.

FEA through NX Nastran

In NX Nastran, the general process of finite element analysis is divided into three main phases, namely preprocessor, solution, and postprocessor, refer to Figure 1-7.

```
                    ┌──────────────────┐
                    │ Physical Problem │
                    └──────────────────┘
                             │
        ┌────────────────────────────────────┐
        │                FEM                  │
        │   (Generate  nodes,  elements,      │      Preprocessor
        │      boundary  conditions,          │
        │   material  properties,  loads,     │
        │         and  data  file)            │
        └────────────────────────────────────┘
                             │
        ┌────────────────────────────────────┐
        │                FEA                  │
        │   (Generate  elements               │
        │   matrices,  compute  nodal         │      Solution
        │   values,  derivatives,  and        │
        │         store  results)             │
        └────────────────────────────────────┘
                             │
        ┌────────────────────────────────────┐
        │         Analyze  Results            │
        │ (Display curves, counters,          │      Postprocessor
        │      deformed  shapes)              │
        └────────────────────────────────────┘
```

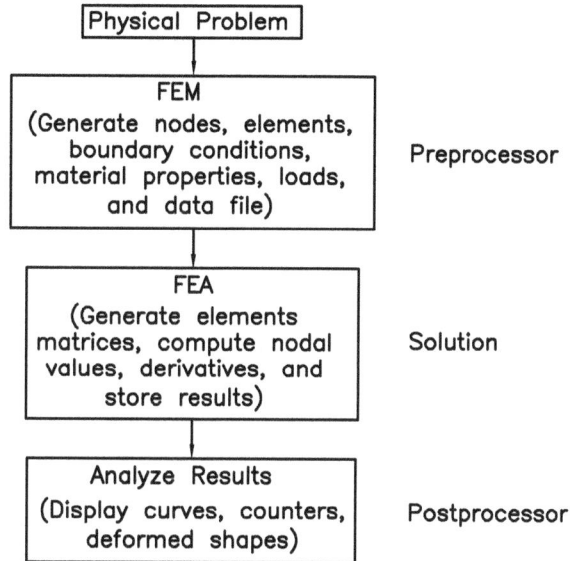

Figure 1-7   *FEA through NX Nastran*

Preprocessor

The preprocessor is a program that processes the input data to produce the output that is used as input in the subsequent phase (solution). Following are the input data that need to be given to the preprocessor:

1. Analysis type (structural or thermal, static or dynamic, and linear or nonlinear, etc.)
2. Element type
3. Real constants
4. Material properties
5. Geometric model
6. Meshed model
7. Loadings and boundary conditions

The input data will be preprocessed for the output data and preprocessor will generate the data files automatically with the help of users. These data files will be used by the subsequent phase (solution), refer to Figure 1-7.

Solution

Solution phase is completely automatic. The FEA software generates the element matrices, computes nodal values and derivatives, and stores the resulting data in files. These files are further used by the subsequent phase (postprocessor) to review and analyze the results through graphic display and tabular listings, refer to Figure 1-7.

Postprocessor

The output of the solution phase (result data files) is in numerical form and consists of nodal values of the field variable and its derivatives. For example, in structural analysis, the output is nodal displacement and stress in the elements. The postprocessor processes the result data and displays it in graphical form to check or analyze the result. The graphical output gives detailed information about the resultant data. The postprocessor phase is automatic and generates the graphical output in the form specified by the user, refer to Figure 1-7.

Effective Utilization of FEA

Some prerequisites for effective utilization of FEA from the perspective of engineers and FEA software are discussed next.

Engineers

An engineer who wants to work with this tool must have sound knowledge of Strength of Materials (for structural analysis), Heat Transfer, Thermodynamics (for thermal analysis), and a good analytical/designing skill. Besides this, the engineer must have a fair knowledge of advantages and limitations of the FEA software being used.

Software

The FEA software should be selected based on the following considerations:

1. Analysis type to be performed.
2. Flexibility and accuracy of the tool.
3. Hardware configuration of your system.

Nowadays, the CAE / FEA software can simulate the performance of most of the systems. In other words, anything that can be converted into a mathematical equation can be simulated using the FEA techniques. Usually, the most popular principle of GIGO (Garbage In Garbage Out) applies to FEA. Therefore, you should be very careful while giving/accepting the inputs for analysis. The careful planning is the key to a successful analysis.

Advantages and Limitations of FEA Software

Following are the advantages and limitations of FEA software:

Advantages

1. It reduces the amount of prototype testing; thereby saving the cost and time involved in performing design testing.
2. It gives graphical representation of the result of analysis.
3. The finite element modeling and analysis are performed in the preprocessor phase and the solution phase, which if done manually, will consume a lot of time and in some cases, may be impossible to carry out.
4. Variables such as stress, temperature can be measured at any desired point in the model.
5. It helps optimize the design.
6. It is used to simulate the designs that are not suitable for prototype testing such as surgical implants (artificial knees).
7. It helps you create more reliable, high quality, and competitive designs.

Limitations

1. It provides approximate solutions.
2. FEA packages are costly.
3. Qualified personnel are required to perform the analysis.

4. The results give solutions but not remedies.
5. Features such as bolts and welded joints cannot be accommodated in the model. This may lead to approximation and errors in the result obtained.
6. For more accurate result, more computer space and time are required.

KEY ASSUMPTIONS IN FEA

There are four basic assumptions that affect the quality of the solution and must be considered before carrying out finite element analysis. These assumptions are not comprehensive, but cover a wide variety of situations applicable to the problem. Make sure to use only those assumptions that apply to the analysis under consideration.

Assumptions Related to Geometry

1. When the displacement is small, the linear solution can be considered.
2. Stress behavior outside the area of interest is not important so the geometric simplifications in those areas will not affect the outcome.
3. Only internal fillets in the area of interest will be included in the solution.
4. Local behavior at the corners, joints, and intersection of geometries is of primary interest therefore no special modeling of these areas is required.
5. Decorative external features will be assumed insignificant for the stiffness and performance of the part, and will be omitted from the model.
6. The variation in mass due to the suppressed features is negligible.

Assumptions Related to Material Properties

1. Material properties will remain in the linear region and nonlinear behavior of the material property cannot be accepted. For example, it is understood that either the stress levels exceeding the yield point or the excessive displacement will cause a component failure.
2. Material properties are not affected by the load rate.
3. The component is free from surface imperfections that can produce stress risers.
4. All simulations will assume room temperature unless specified otherwise.
5. The effect of relative humidity or water absorption on the material used will be neglected.
6. No compensation will be made to account for the effect of chemicals, corrosives, wears or other factors that may have an impact on the long term structural integrity.

Assumptions Related to Boundary Conditions

1. Displacements will be small so that the magnitude, orientation, and distribution of the load remains constant throughout the process of deformation.
2. Frictional loss in the system is considered to be negligible.
3. All interfacing components will be assumed rigid.
4. The portion of the structure being studied is assumed a part separate from the rest of the system. As a result, the reaction or input from the adjacent features is neglected.

Assumptions Related to Fasteners

1. Residual stresses due to fabrication, preloading on bolts, welding, or other manufacturing or assembly processes will be neglected.
2. All the welds between the components will be considered ideal and continuous.

3. The failure of fasteners will not be considered.
4. Loads on the threaded portion of the parts is supposed to be evenly distributed among the engaged threads.
5. Stiffness of bearings, radially or axially, will be considered infinite or rigid.

TYPES OF ANALYSIS

The following types of analysis can be performed using FEA software:

1. Structural analysis
2. Thermal analysis
3. Fluid flow analysis
4. Coupled field analysis

Structural Analysis

In structural analysis, first the nodal degrees of freedom (displacement) are calculated and then the stress, strains, and reaction forces are calculated from the nodal displacements. The classification of the structural analysis is shown in Figure 1-8.

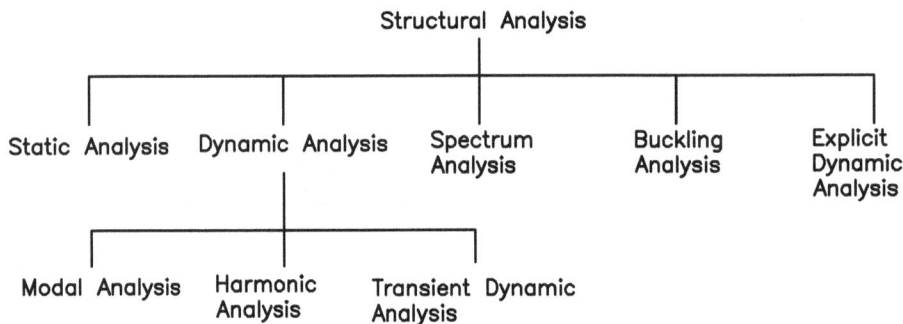

Figure 1-8 *Types of structural analysis*

Static Analysis

In static analysis, the load or field conditions do not vary with respect to time and therefore, it is assumed that the load or field conditions are applied gradually not suddenly. The system under analysis can be linear or nonlinear. Inertia and damping effects are ignored in static structural analysis. In static structural analysis, the following matrices are solved:

$$[K] \times [X] = [F]$$

Where,
$$K = \text{Stiffness Matrix}$$
$$X = \text{Displacement Matrix}$$
$$F = \text{Load Matrix}$$

The above equation is called as the force balance equation for a linear system. If the elements of matrix [K] are a function of [X], the system is known as a nonlinear system. Nonlinear systems include large deformation, plasticity, creep, and so on. The loadings that can be applied in a static analysis include:

1. Externally applied forces and pressures
2. Steady-state inertial forces (such as gravity or rotational velocity)
3. Imposed (non-zero) displacements
4. Temperatures (for thermal strain)
5. Fluences (for nuclear swelling)

The outputs that can be expected from a FEA software are given next.

1. Displacements
2. Strains
3. Stresses
4. Reaction forces etc.

Dynamic Analysis

In dynamic analysis, the load or field conditions do vary with time. The assumption here is that the load or field conditions are applied suddenly. The system can be linear or nonlinear. The dynamic load includes oscillating loads, impacts, collisions, and random loads. The dynamic analysis is classified into the following three main categories:

Modal Analysis
It is used to calculate the natural frequency and mode shape of a structure.

Harmonic Analysis
It is used to calculate the response of the structure to harmonically time varying loads.

Transient Dynamic Analysis
It is used to calculate the response of the structure to arbitrary time varying loads.

In dynamic analysis, the following matrices are solved:

For the system without any external load:

[M] x Double Derivative of [X] + [K] x [X]= 0

where

M = Mass Matrix
K = Stiffness Matrix
X = Displacement Matrix

For the system with external load:

[M] x Double Derivative of [X] + [K] x [X]= [F]

where

K = Stiffness Matrix
X = Displacement Matrix
F = Load Matrix

These equations are called as force balance equations for a dynamic system. By solving this set of equations, you will be able to extract the natural frequencies of a system. The load types applied in the dynamic analysis are the same as that in the static analysis. The outputs that can be expected from a software are:

1. Natural frequencies
2. Mode shapes
3. Displacements
4. Strains
5. Stresses
6. Reaction forces

All the outputs mentioned above can be obtained with respect to time.

Spectrum Analysis
This is an extension of the modal analysis and is used to calculate the stress and strain due to the response of the spectrum (random vibrations).

Buckling Analysis
This type of analysis is used to calculate the buckling load and the buckling mode shape. The slender structures and the structures with slender part loaded in the axial direction buckle under relatively small loads. For such structures, the buckling load becomes a critical design factor.

Explicit Dynamic Analysis
This type of structural analysis is used to calculate fast solutions for large deformation dynamics and complex contact problems.

Thermal Analysis
Thermal analysis is used to determine the temperature distribution and related thermal quantities such as:

1. Thermal distribution
2. Amount of heat loss or gain
3. Thermal gradients
4. Thermal fluxes

All the primary heat transfer modes such as conduction, convection, and radiation can be simulated. You can perform two types of thermal analysis, steady-state and transient.

Steady State Thermal Analysis
In this analysis, the system is studied under steady thermal loads with respect to time.

Transient Thermal Analysis
In this analysis, the system is studied under varying thermal loads with respect to time.

Fluid Flow Analysis
This analysis is used to determine the flow distribution and temperature of a fluid. It simulate the

laminar and turbulent flow, compressible and electronic packaging, automotive design, and so on. The outputs that can be expected from the fluid flow analysis are:

1. Velocities
2. Pressures
3. Temperatures
4. Film coefficients

Coupled Field Analysis

This type of analysis considers the mutual interaction between two or more fields. It is impossible to solve the fields separately because they are interdependent. Therefore, you need a program that can solve both the physical problems by combining them.

For example, if a component is exposed to heat, you may first require to study the thermal characteristics of the component and then the effect of the thermal heating on the structural stability.

Alternatively, if a component is bent into different shapes using one of the metal forming processes and then subjected to heating, the thermal characteristics of the component will depend on the new shape of the component and therefore the shape of the component has to be predicted through structural simulations first. This is called as coupled field analysis.

IMPORTANT TERMS AND DEFINITIONS

Some of the important terms and definitions used in FEA are discussed next.

Strength

When a material is subjected to an external load, deformation occurs and the resistance offered by the material against this deformation is by the virtue of its strength.

Load

The external force acting on a body is called the load.

Stress

The force of resistance offered by a body per unit area against the deformation is called stress. The stress is induced in the body while the load is applied on it. Stress is calculated as load per unit area.

$$p = F/A$$

where

p = Stress in N/mm^2
F = Applied Force in Newton
A = Cross-Sectional Area in mm^2

The material can undergo various types of stresses which are discussed next.

Tensile Stress

If the resistance offered by a body is against the increase in length, the body is said to be under tensile stress.

Compressive Stress

If the resistance offered by a body is against the decrease in length, the body is said to be under compressive stress. Compressive stress is just the reverse of tensile stress.

Shear Stress

Shear stress exists when two materials tend to slide over each other in opposite directions. Note that in any typical plane of shear, the force parallel to the plane is usually known as shear stress.

Shear Stress = Shear resistance (R) / Shear area (A)

Strain

When a body is subjected to a load (force), its length changes. The ratio of the change in length to the original length of the member is called strain. If the body returns to its original shape on removing the load, the strain is called as elastic strain. If the metal remains distorted, the strain is called as plastic strain. The strain can be of three types, tensile, compressive, and shear strain.

Strain (e) = Change in Length (*dl*) / Original Length (*l*)

Elastic Limit

The maximum stress that can be applied to a material without producing permanent deformation is known as the elastic limit of the material. If the stress applied is within the elastic limit then on the removal of the stress, the material will return to its original shape and dimension.

Hooke's Law

It states that the stress is directly proportional to the strain within the elastic limit.

Stress / Strain = Constant (within the elastic limit)

Young's Modulus or Modulus of Elasticity

In case of axial loading, the ratio of intensity of tensile or compressive stress to the corresponding strain is constant. This ratio is called Young's modulus and it is denoted by E.

E = p/e

Where,

p = Stress
e = strain

Shear Modulus or Modulus of Rigidity

In case of shear loading, the ratio of shear stress to the corresponding shear strain is constant. This ratio is called shear modulus, and it is denoted by C, N, or G.

Ultimate Strength

The maximum stress that a material can withstand when load is applied is called its ultimate strength.

Factor of Safety

The ratio of the ultimate strength to the estimated maximum stress (design stress) is known as factor of safety. It is necessary that the design stress should be well below the elastic limit and to achieve this condition, the ultimate stress should be divided by a 'factor of safety'.

Lateral Strain

If a cylindrical rod is subjected to an axial tensile load, the length (l) of the rod will increase (dl) and the diameter (ϕ) of the rod will decrease ($d\phi$). In short, the longitudinal stress will not only produce a strain in its own direction, but will also produce a lateral strain. The ratio dl/l is called the longitudinal strain or linear strain, and the ratio $d\phi/\phi$ is called the lateral strain.

Poisson's Ratio

The ratio of lateral strain to the longitudinal strain is constant within the elastic limit. This ratio is called as Poisson's ratio and is denoted by $1/m$. For most of the metals, the value of the 'm' lies between 3 and 4.

$$\text{Poisson's ratio} = \text{Lateral Strain} / \text{Longitudinal Strain} = 1/m$$

Bulk Modulus

If a body is subjected to equal stresses along three mutually perpendicular directions, the ratio of the direct stresses to the corresponding volumetric strain is found to be constant for a given material when the deformation is within a certain limit. This ratio is called the bulk modulus and is denoted by K.

Creep

At elevated temperatures and constant stress or load, many materials continue to deform but at a slow rate. This behavior of materials is called creep. At a constant stress and temperature, the rate of creep is approximately constant for a long period of time. After this period and after a certain amount of deformation, the rate of creep increases thereby causing fracture in the material. The rate of creep is highly dependent on both stress and temperature.

Classification of Materials

Materials are classified into three main categories: elastic, plastic, and rigid. In case of elastic materials, the deformation disappears on the removal of load. In plastic materials, the deformation is permanent. A rigid material does not undergo any deformation when subjected to an external load. However, in actual practice, no material is perfectly elastic, plastic, or rigid. The structural members are designed such that they remain in elastic conditions under the action of working loads. All engineering materials are grouped into three categories that are discussed next.

Isotropic Material

In case of Isotropic materials, the material properties do not vary with direction, which means that they have same material properties in all directions. The material properties are defined by Young's modulus and Poisson's ratio.

Orthotropic Material

In case of Orthotropic material, the material properties vary with the change in direction. They have three mutually perpendicular planes of material symmetry. The material properties are defined by three separate Young's modulus and Poisson's ratios.

Anisotropic Material

In case of Anisotropic material, the material properties vary with the change in direction, but in this case, there is no plane of material symmetry.

THEORIES OF FAILURE

The following are the theories of failure.

Von Mises Stress Failure Criterion

The von Mises stress criterion is also called Maximum Distortion Energy theory. The theory states that a ductile material starts yielding at a location when the von Mises stress becomes equal to the stress limit. In most cases, the yield strength is used as the stress limit.

Maximum Shear Stress Failure Criterion

The Maximum Shear Stress failure criterion is based on the Maximum Shear Stress theory. This theory predicts failure of a material when the absolute maximum shear stress reaches the stress limit that causes the material to yield in a simple tension test. The Maximum Shear Stress criterion is used for ductile materials.

Maximum Normal Stress Failure Criterion

This criterion is used for brittle materials. It assumes that the ultimate strength of the material in tension and compression is the same. This assumption is not valid in all the cases. For example, cracks considerably decrease the strength of the material in tension while their effect is not significant in compression because the cracks tend to close. Brittle materials do not have a specific yield point and hence it is not recommended to use the yield strength to define the stress limit for this criterion.

Self-Evaluation Test

Answer the following questions and then compare them to those given at the end of this chapter:

1. The _____ are used to eliminate the unknowns in a system.

2. The shape of an element is specified by _____.

3. In NX Nastran, the general process of finite element analysis is divided into three main phases: _____ , _____ , and _____.

4. In _____ analysis, the load or field conditions do not vary with respect to time.

5. FEA is a computing technique that is used to obtain approximate solutions to the boundary value problems in engineering. (T/F)

6. The Model Analysis is used to calculate the natural frequency and mode shape of a structure. (T/F)

7. The degree of freedom is defined as the ability of a node to translate or transmit the load. (T/F)

8. A 1D element has the shape of a line or curve. (T/F)

Review Questions

Answer the following questions:

1. When the stiffness of material is the function of displacement then the behavior of material is known as _____.

2. In _____ analysis, the load or field conditions vary with respect to time.

3. The ratio of the change in length to the original length of the member is called _____.

4. The _____ states that the stress is directly proportional to the strain within the elastic limit.

5. A 3D element can have the shape of hexahedron (8 nodes), _____ , _____ , or _____ .

6. The stiffness matrix represents the resistance offered by a body to withstand the load applied. (T/F)

Answers to Self-Evaluation Test

1. Boundary conditions, **2.** nodes, **3.** preprocessor, solution, postprocessor, **4.** Static, **5.** T, **6.** T, **7.** T, **8.** T

Chapter 2

Introduction to NX Nastran

Learning Objectives

After completing this chapter, you will be able to:

- *Understand basic concepts of NX Nastran*
- *Understand different environments in NX Nastran*
- *Understand the system requirements for NX Nastran*
- *Start NX Nastran*
- *Understand important terms and definitions in NX Nastran*
- *Understand the functions of mouse buttons*
- *Understand default tabs in different environments*
- *Use dialog boxes in NX Nastran*
- *Use the NX Nastran Help*

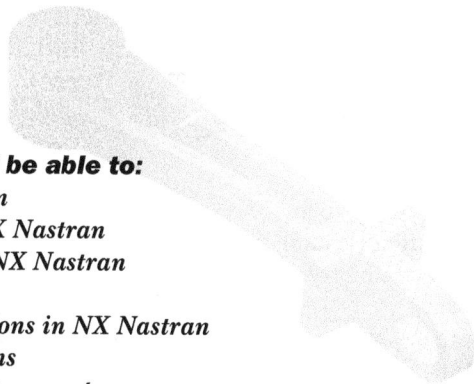

INTRODUCTION TO NX Nastran

NX Nastran is a premium computer-aided engineering (CAE) tool that manufacturers use worldwide for their critical engineering computing needs. For over 30 years, NX Nastran has been the analysis solution in almost every major industry including aerospace, defense, automotive, shipbuilding, heavy machinery, medical, and consumer products.

In CAE, often you can work with data imported from different CAD softwares. You can easily modify a design using NX CAD software and then use the modification as a design variable to arrive at an optimal design.

NX Nastran has the following analysis capabilities:

1. Linear statics (including inertia relief)
2. Normal modes and buckling
3. Heat transfer (steady-state and transient)
4. Transient response
5. Frequency response
6. Response spectrum and random response
7. Geometric nonlinear static and transient response
8. Material nonlinear static and transient response
9. Design optimization and sensitivity (including dynamic and shape optimization)
10. Composite materials
11. Acoustic response
12. Aeroelasticity
13. Superelements
14. Complex eigen analysis
15. Axisymmetric analysis
16. Cyclic symmetry

In NX Nastran you can analyze the basic and as well as complex geometries using different environments. Each environment consists of a set of tools that allows the user to perform specific analysis tasks in a particular area. After starting a new part file, you can invoke the required environment of NX.

BASIC ENVIRONMENTS IN NX

The basic environments in NX are Modeling, Shape Studio, Drafting, Assembly, Sheet metal, and the Manufacturing. Some of these environments are discussed next.

Modeling Environment

The Modeling environment is a parametric and feature-based environment in which you can create solid models. The basic requirement for creating solid models in this environment is a sketch which can be drawn directly in this environment by using the tools available in the **Direct Sketch** group of the **Home** tab. The sketch can also be drawn in the Sketching environment. The Sketching environment can be invoked by choosing the **Sketch** tool from the **Direct Sketch** group of the **Home** tab or by choosing the **Sketch in Task Environment** tool from the **Curve** tab. When a sketch is drawn, various applicable constraints and dimensions get automatically applied to it. Additional constraints and dimensions can also be manually applied. After drawing the sketch, you need to convert the sketch into a feature. The tools to convert a sketch into a feature are available in the Modeling environment. You can also create features such as fillets, chamfers, taper, and so on by using the other tools available in this environment. These features are called placed features. You can also assign materials to the model in the Modeling environment.

Shape Studio Environment

The Shape Studio environment is also a parametric and feature-based environment in which you can create surface models. The tools in this environment are similar to those in the Modeling environment with the difference that the tools in this environment are used to create surfaces, both basic and advanced. This environment also provides surface editing tools that are used to manipulate surfaces to obtain required shape. This environment is useful for conceptual and industrial designs.

Assembly Environment

The Assembly environment is used to assemble components using the assembly constraints available in this environment. There are two types of assembly design approaches in NX, Bottom-up and Top-down.

In the bottom-up approach, you can assemble the components that were created earlier whereas in the top-down approach, components are created in the Assembly environment.

In the Assembly environment, you can also assemble an existing assembly with the current assembly. The performed analysis in this environment provides the facility to check the interference and clearance between the components in an assembly.

SYSTEM REQUIREMENTS

The following are the minimum system requirements to ensure smooth functioning of NX Nastran on your system:

* Operating System: Windows 64-bit (Windows XP 64 SP2, Windows Vista 64 SP1, Windows 7, Windows 8, Windows HPC Server 2008 R2)
* Platform: Intel Pentium class, Intel 64 or AMD 64
* Memory: 4 GB or more
* DVD drive: For installing the software
* Graphics adapter: should be capable of supporting 1024x768 High Color (16-bit)
* Microsoft Internet Explorer 6.0 or higher

STARTING NX NASTRAN 9.0

To start NX Nastran 9.0, double-click on the shortcut icon of NX 9.0 displayed on the desktop of your computer. After the necessary files are loaded and the licenses are verified, the initial screen of NX 9 will be displayed.

The default initial screen of NX 9.0 is shown in Figure 2-1. This screen displays information about NX 9.0 which helps you learn more about NX 9.0. You can also view other related information by moving the cursor over the topics displayed on the left of the NX 9.0 screen.

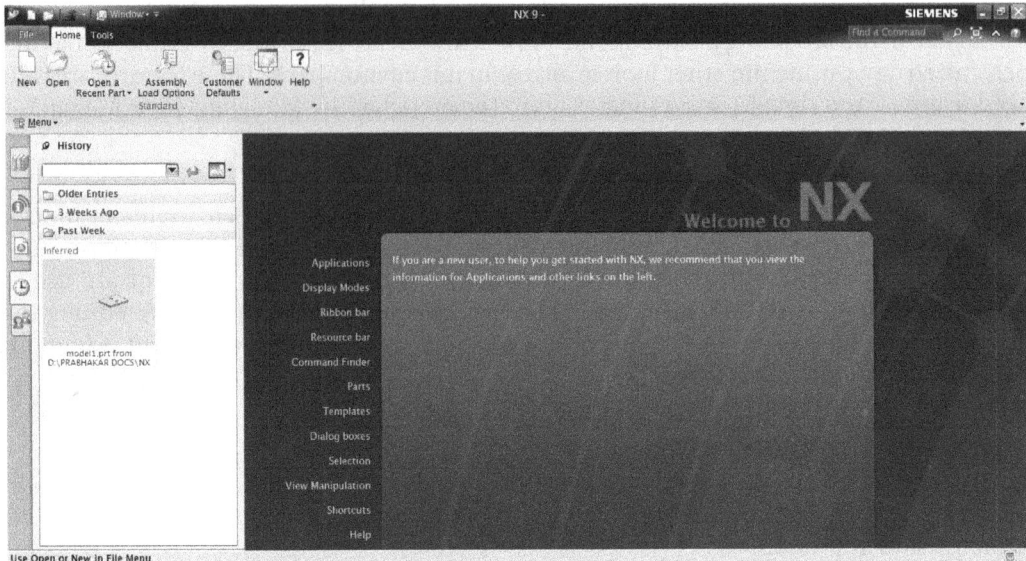

Figure 2-1 *The initial screen that appears after starting NX 9.0*

STARTING A NEW DOCUMENT IN NX NASTRAN 9.0

Ribbon: Home > Standard > New
Menu: File > New

To invoke the Advanced Simulation environment of NX Nastran, choose **Menu > File > New** from the **Top Border Bar**; the **New** dialog box will be displayed with the **Model** tab chosen, as shown in Figure 2-2. The **New** dialog box contains tabs to switch between different working environments like Modeling, Drawing, Simulation, Manufacturing, and so on.

To enter in NX Nastran, choose the **Simulation** tab and select **NX Nastran** from the **Name** column for the **FEM** type from the **Template** rollout, if not selected by default, refer to Figure 2-3.

The rollouts available in the **Simulation** tab for invoking NX Nastran are discussed next.

Templates Rollout
In this rollout, different FEM and SIM type templates are available for different solvers. You can also select the **Millimeters** or **Inches** as a unit from the **Units** drop-down list available in the **Filters** sub-rollout.

New File Name Rollout
In this rollout, the **Name** and **Folder** edit boxes are available. You can enter the name of FEM or SIM in the **Name** edit box and the location to save the file in the **Folder** edit box.

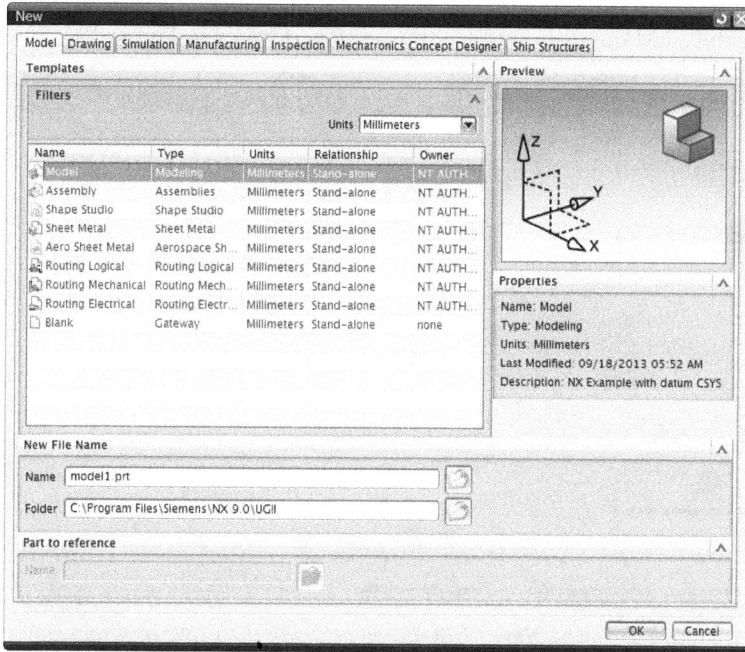

Figure 2-2 The *New* dialog box with the *Model* tab chosen

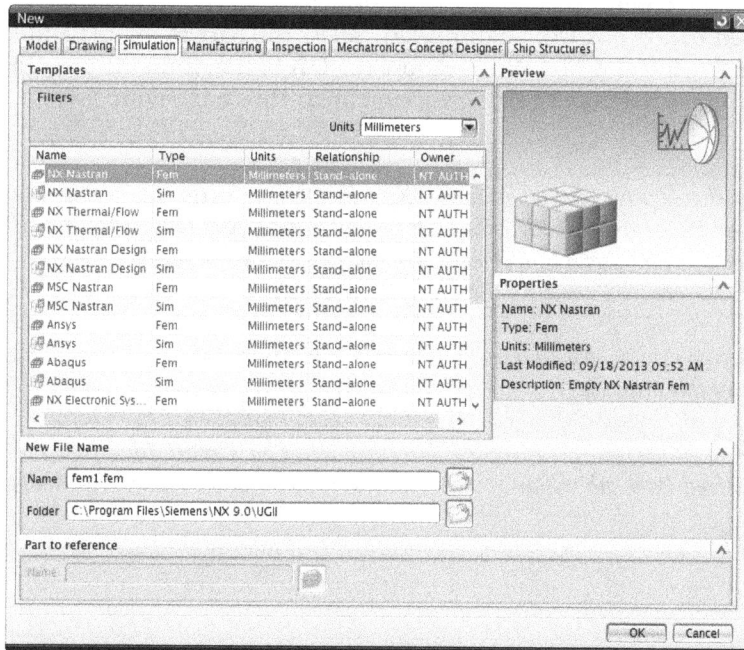

Figure 2-3 The *New* dialog box with the *Simulation* tab chosen

After defining the required parameters in these rollouts, choose the **OK** button; the **New FEM** dialog box with the FEM environment will be displayed, refer to Figure 2-4.

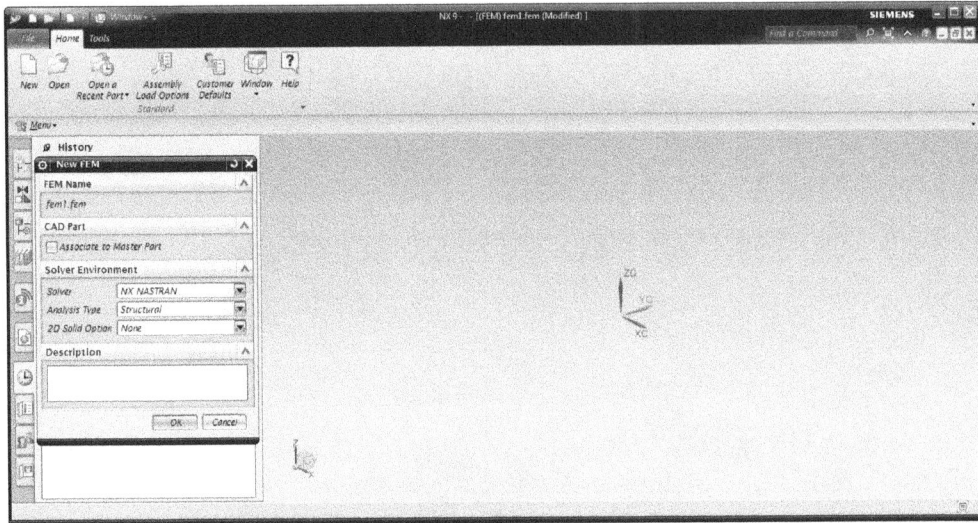

*Figure 2-4 The **New FEM** dialog box with the FEM environment*

For importing any NX model in the FEM environment, select the **Associate to Master Part** check box from the **CAD Part** rollout of the **New FEM** dialog box; the **CAD Part** rollout will be modified and the **Open Part** button will be displayed, as shown in Figure 2-5. Choose this button; the **Open** dialog box will be displayed. Select the part to import and choose the **OK** button from the **Open** dialog box. Next, choose the **OK** button from the **New FEM** dialog box; the part will get imported in the FEM environment, as shown in Figure 2-6.

Note
*1. If you choose a file type other than NX such as ***.igs***, ***.stp***, you need to leave the **Associate to Master Part** check box cleared and then choose the **OK** button from the dialog box; the FEM environment window will be displayed, refer to Figure 2-6. Now, open the part from this window to import the non-NX model in the FEM environment.*

You will learn more about the importing and exporting of the CAD model in NX in the later chapters of this textbook.

*2. If you want to create an Idealized Part, then select the **Create Idealized Part** check box from the **Idealized Part** sub-rollout in the **CAD Part** rollout of the **New FEM** dialog box.*

You will learn more about creating an Idealized Part later in the textbook.

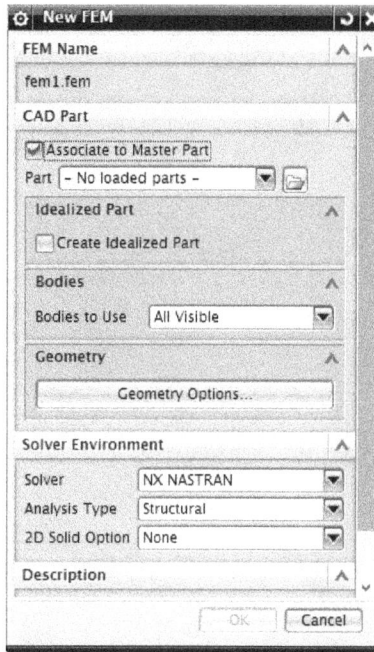

Figure 2-5 The modified **New FEM** dialog box

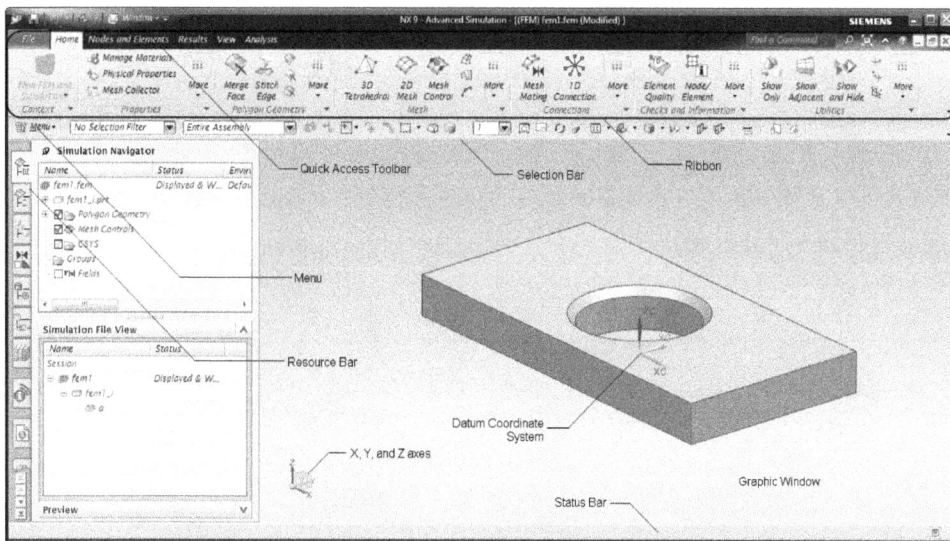

Figure 2-6 The part imported in the FEM environment

In the FEM environment, you can define materials, meshing, and other physical properties of the model. The model after defining materials and applying mesh is shown in Figure 2-7.

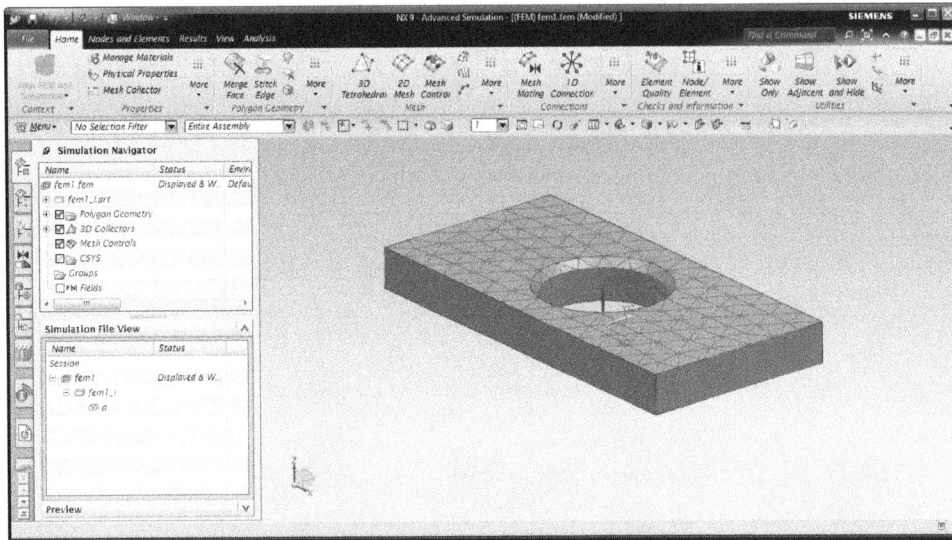

Figure 2-7 Meshed model in the FEM environment

For applying loads, constraints, and other boundary conditions, you need to switch to the
Simulation environment. To do so, choose the **New Simulation** button from the **Change
Window Drop-down** of the **Home** tab; the **New Part File** window will be displayed. Select the
NX Nastran template with the **Sim** type from the **Template** rollout if not selected by default,
as shown in Figure 2-8.

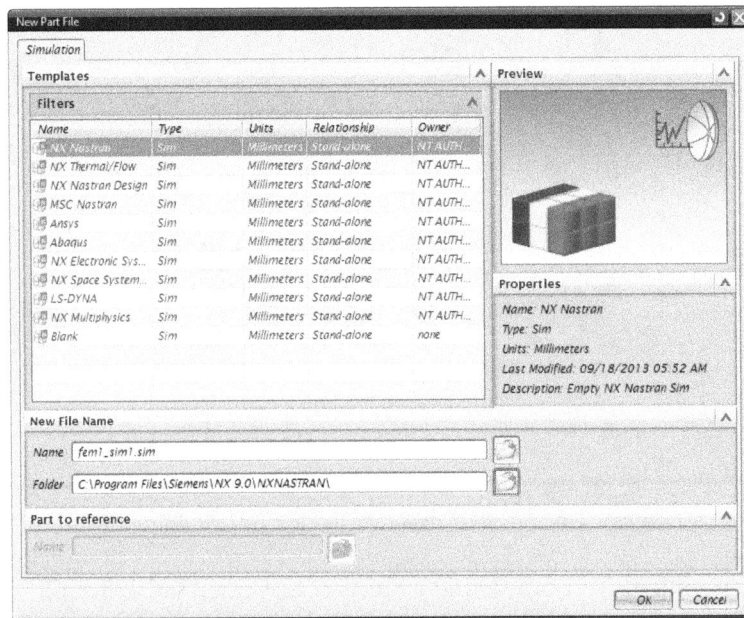

Figure 2-8 The New Part File window

Choose the **OK** button from the **New Part File** window; the **New Simulation** dialog box with
the Simulation environment will be displayed. Choose the **OK** button from the **New Simulation**

dialog box; the **Solution** dialog box will be displayed. In this dialog box, you can describe the analysis details. Choose the **OK** button from the dialog box; you will enter in the Simulation environment. The Simulation environment with the meshed model is shown in Figure 2-9. In this environment, you can apply boundary conditions, loads, and solve the analysis.

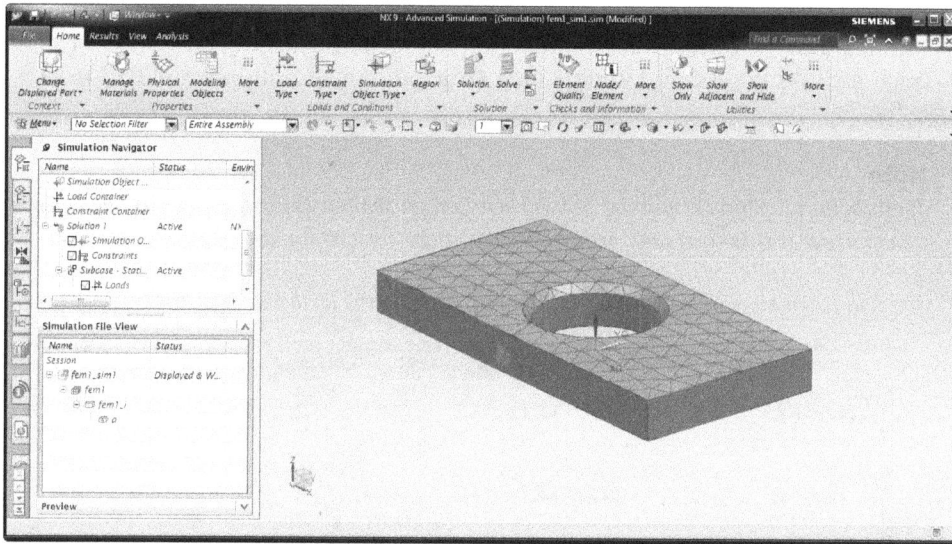

Figure 2-9 *The Simulation environment with meshed model*

Note
You will learn more about the Simulation environment and its dialog boxes later in the textbook.

IMPORTANT TERMS AND DEFINITIONS
Some important terms used in NX Nastran are discussed next.

Advanced Simulation
The Advanced Simulation is a comprehensive finite element modeling and results visualization application. It includes pre- and post-processing and supports a broad range of product performance evaluation solutions. Advanced Simulation provides numerous additional features that support advanced analysis processes.

1.   Advanced Simulation provides separate simulation and FEM files that help in the development of FE models across a distributed work environment.

2.   Advanced Simulation supports a complete range of element types (0D, 1D, 2D, and 3D) to control specific part meshing.

3.   Advanced Simulation includes a number of geometry modification tools to manipulate any CAD geometry for making its analysis easier and faster.

fem
The *fem* file used in NX Nastran contains information about the nodes and elements in the

finite element model. It also includes physical properties, materials, and beam element sections of the model. The file contains the idealized (modified) model that you create from the master CAD part.

sim

The *sim* (Simulation) file contains the simulation data such as loads, constraints, and all motion objects (such as links, joints, and connectors). You can create multiple simulation files for a single *fem* file.

> **Note**
> *If any of the properties is missing in the fem or sim file while solving the analysis, the **Information** window will be displayed with errors and warnings, refer to Figure 2-10.*

You will learn more about solving an analysis later in the textbook.

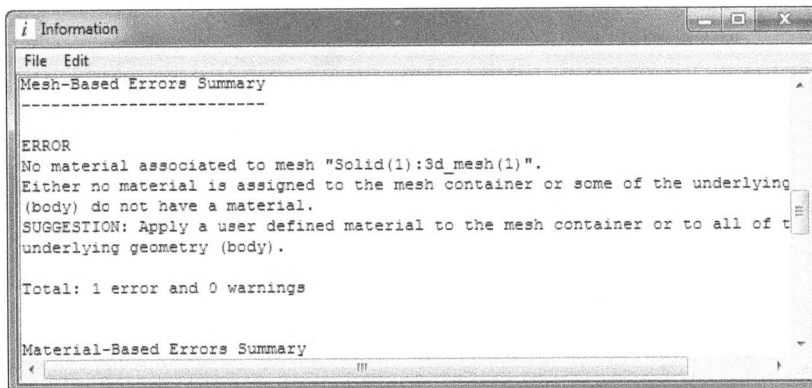

*Figure 2-10 The **Information** window*

Resource Bar

The Resource Bar combines all navigation windows, a history palette, an integrated web browser, and a part template in one common place for a better user interface. By default, the Resource Bar is located on the left side of the interface window.

Simulation Navigator

The **Simulation Navigator** is located at top of the Resource Bar. It displays the following files and steps associated with analysis.

1. FEM file
2. Associated CAD geometry
3. Mesh collectors and meshes
4. Loads, constraints, and objects that define boundary conditions
5. Solution and results

Using the **Simulation Navigator**, you can review the structure, content, and status of your analysis. You can also show and hide the geometry, meshes, and boundary conditions of the model. A partial view of the **Simulation Navigator** is shown in Figure 2-11.

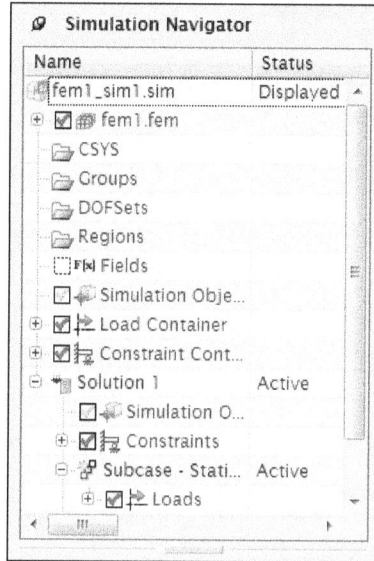

Figure 2-11 *Partial view of the* ***Simulation Navigator***

Simulation File View

The **Simulation File View** is located below the **Simulation Navigator**. You can use the **Simulation File View** in the following ways:

1. View all the loaded parts as well as all FEM and Simulation files in sequential order, as shown in Figure 2-12.

2. Create Idealized part without manipulating the master part.

3. Create multiple FEM and Simulation files for any design or idealized part.

You can switch to any simulation environment by choosing any option from the **Simulation File View**. To do so, right-click on a simulation step; a shortcut menu will be displayed, refer to Figure 2-13. Choose the **Make Displayed Part** option from the shortcut menu; the simulation environment will be invoked. You can also switch to simulation environment by double-clicking.

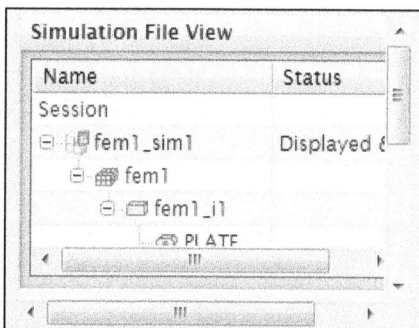

Figure 2-12 *The* ***Simulation File View***

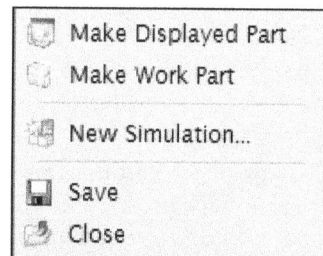

Figure 2-13 *Shortcut menu displayed*

Post Processing Navigator

The **Post Processing Navigator** is available in the **Resource Bar** below the **Simulation Navigator**. The **Post Processing Navigator** is used to manage, view, and interrogate the results of an analysis. Using the **Post Processing Navigator**, you can:

1. Create nodal and elemental contour plots of results on the model and also import results.

2. Create cross sections in post views to show internal variation in the larger portion of part.

3. Annotate results to show maximum and minimum variation points.

4. Identify the values of selected nodes and elements and export them for further analysis.

5. Create and display graphs of result data across selected nodes.

6. Quickly generate a contour plot of the selected results or result components.

7. Animate post views.

8. Control the display of results based on meshes or groups.

You can open the **Post Processing Navigator** to get the current simulation results. To do so, right-click on the **Results** step in the **Simulation Navigator**; a shortcut menu will be displayed, refer to Figure 2-14. From this shortcut menu, choose the **Open** option from the menu; the **Post Processing Navigator** will be displayed with the current simulation results, refer to Figure 2-15.

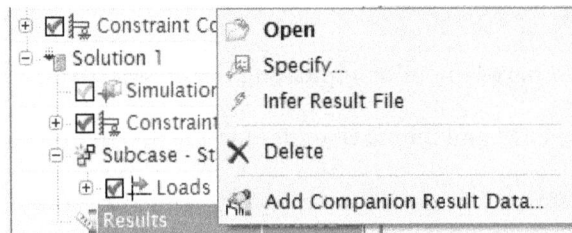

*Figure 2-14 Shortcut menu displayed on choosing the **Results** step of **Simulation Navigator***

Note
*You can also import an existing results file in the **Post Processing Navigator** using the **Imported Results** option shown in the navigator. You will learn more about importing results in the **Post Processing Navigator** later in the textbook.*

Part Navigator

The **Part Navigator** located in the Resource Bar keeps a track of all the operations related to model preparation. Figure 2-16 shows the part navigator that appears when you choose the **Part Navigator** tab in the Resource Bar.

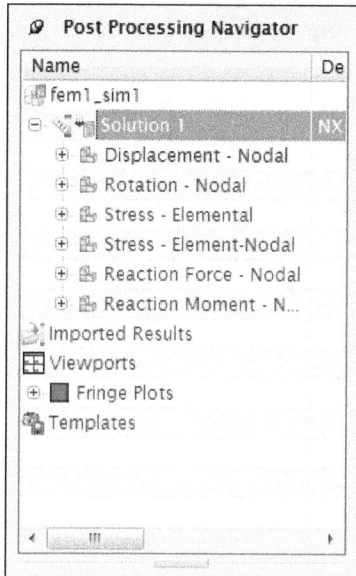

*Figure 2-15 The **Post Processing Navigator***

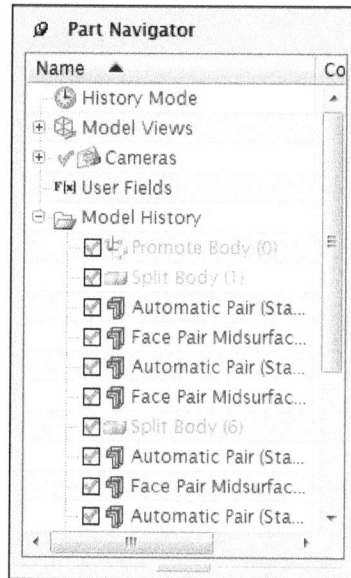

*Figure 2-16 The **Part Navigator***

UNDERSTANDING THE FUNCTIONS OF THE MOUSE BUTTONS

To work in NX Nastran environments, it is necessary that you understand the functions of mouse buttons. The efficient use of the mouse buttons along with the CTRL key can reduce the time required to complete any design task. Different combinations of the CTRL key and the mouse buttons are described next:

1. The left mouse button helps you make a selection by simply selecting a face, surface, loads, or constraints from the geometry area or from the Navigators. For multiple selections, press the CTRL key along with the left mouse button.

2. The right mouse button is used to invoke shortcut menu which has different options and tools.

3. Press and hold the middle and right mouse buttons to invoke the **Pan** tool. Next, drag the mouse to pan the model. You can also invoke the **Pan** tool by first pressing and holding the SHIFT key and then the middle mouse button. Figure 2-17 shows the use of a three buttons mouse in performing the pan functions.

4. Press and hold the middle mouse button to invoke the **Rotate** tool. Next, drag the mouse to dynamically rotate the view of the model in the geometry area and view it from different directions. Figure 2-17 shows the use of the three buttons mouse in performing the rotate operation.

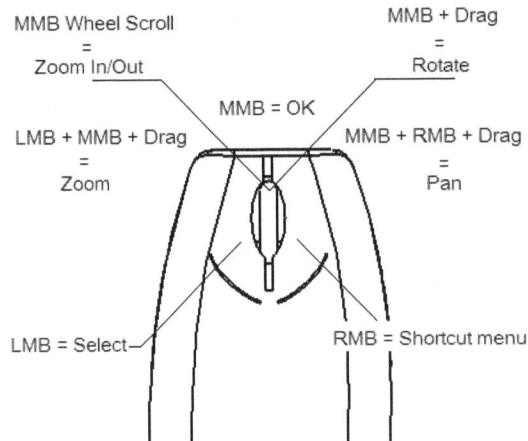

Figure 2-17 *Functions of the mouse buttons*

5. Press and hold the CTRL key and then the middle mouse button to invoke the **Zoom** tool. Alternatively, press and hold the left mouse button and then the middle mouse button to invoke the **Zoom** tool. Next, drag the mouse dynamically to zoom in or out the model in the geometry area. Figure 2-17 shows the use of three mouse buttons in performing the zoom functions.

QUICK ACCESS TOOLBAR

The **Quick Access Toolbar** toolbar is common to all environments of NX Nastran, refer to Figure 2-18. The buttons in this toolbar are used to start a new file, open an existing file, save a file of the current document, cut and place the selection on a temporary clipboard, copy a selection, paste content from the clipboard to some selected location, undo, redo, search a tool, and invoke the help topics.

Figure 2-18 *The* ***Quick Access Toolbar***

RIBBON

The Ribbon in the NX Nastran interface comprises of a series of tabs which group various tools and options based on their functionality. These tabs and groups are displayed depending upon the environment invoked. The different environments and some of their respective tabs and groups are discussed next.

FEM Environment

In the FEM environment, you can define the type of material, mesh, and other physical properties of the model from the **Home** tab. This tab is discussed next.

Home Tab

The tools in the **Home** tab, as shown in Figure 2-19, are used in the FEM environment to:

1. Entering into the Simulation environment
2. Assign material to the given model
3. Mesh the given model
4. Define connection between two or more parts
5. Refining the model for better mesh quality
6. Check element or mesh quality

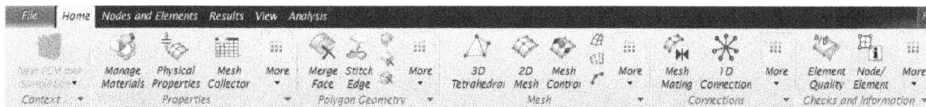

*Figure 2-19 Partial view of the **Home** tab in the FEM environment*

The **Context** group of the **Home** tab is discussed next.

Context Group

This group available in the **Home** tab and is used to enter in the Simulation environment and to create a new simulation file, refer to Figure 2-20. The tools in the **Context** group will be modified in the Simulation environment.

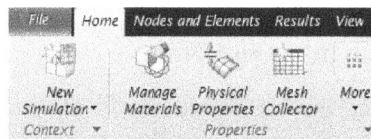

*Figure 2-20 The **Context** group*

Simulation Environment

The Simulation environment is used for defining loads, constraints, and other boundary conditions. The **Home** tab of this environment is discussed next.

Home Tab

The **Home** tab in the Simulation environment is shown in Figure 2-21. The tools in the **Home** tab are used in the Simulation environment to:

1. Activating the **Simulation** tool to re-enter in the Simulation environment from the FEM environment
2. Managing materials in the model
3. Managing modeling objects
4. Defining loads and constraints
5. Defining simulation object type and its operational region
6. Creating solutions and verify the model analysis steps
7. Solving the analysis and create the analysis report
8. Checking the element or mesh quality

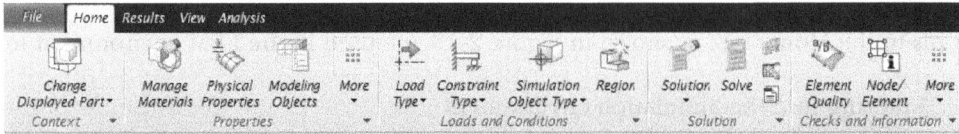

*Figure 2-21 Partial view of the **Home** tab in the simulation environment*

The **Context** group of the **Home** tab is discussed next.

Context Group

After entering in the Simulation environment, the **Context** group gets modified, as shown in Figure 2-22. The tools in the **Context** group are used to activate meshing and to again enter in the FEM environment from the Simulation environment.

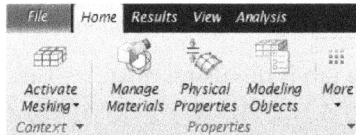

*Figure 2-22 The **Context** group*

Results Tab

After the analysis of model is performed and results are generated the **Post Processing Navigator** is displayed and the **Results** tab gets available along with the **Post Processing**, **Animation**, **Manipulation** and other groups in the Simulation environment, as shown in Figure 2-23.

*Figure 2-23 The **Results** tab*

The tools in the **Results** tab are used to:

1. Return to the Simulation environment
2. Edit the given view and manipulate the result
3. Render the result
4. Create the annotation
5. Animating results

View Tab

The **View** tab is common to all environments of NX Nastran. The tools in the **View** tab, as shown in Figure 2-24, are used for manipulating the views of the model.

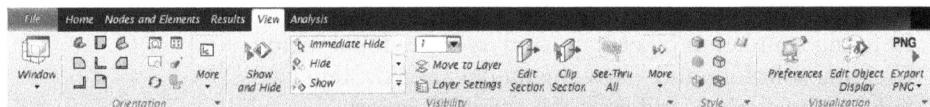

*Figure 2-24 The **View** tab*

STATUS BAR

The Status Bar appears at the bottom of the drawing window and comprises of two areas and a button, as shown in Figure 2-25. These areas and button are discussed next.

Cue line area | Status area | Enters or exits full screen mode button

Create a new FEM and Simulation | Double-click the WCS to activate WCS Dynamics

Figure 2-25 The Status Bar

Cue Line Area

The cue line area is the prompt area where prompts appear for completing the tool task.

Status Area

This area provides information about the operations to be carried out in the interface.

Enters or exits full screen mode

If you choose this button, the graphics window will be maximized and you can get a full screen display. For getting the default screen display, you need to choose this button again.

Selection Bar

The **Selection** bar, as shown in Figure 2-26, appears below the **Ribbon**. This bar is used for easy selection of the model features like faces or edges.

Figure 2-26 The Selection bar

Tip

By default, all the tools are not available in a group. Therefore, you may need to customize the group to add those tools that are not available by default. To customize a group, click on the down arrow at the bottom right corner of the group; a drop-down list will be displayed. Click on the tool to be added or removed from the group. Note that the tick mark on the left of a tool indicates that it is already added to the group.

*Similarly, you can add or remove groups from the **Ribbon** by using the **Ribbon Options** arrow available at its bottom right corner.*

DIALOG BOXES IN NX Nastran

To perform a simulation, you need to follow certain processes in an order. The steps to perform these processes are placed in a top-down order in the corresponding dialog boxes, refer to Figure 2-27. In the dialog box, the current selection step is highlighted in orange. The required steps are marked with asterisks and the completed steps are marked with green check marks.

The **Reset** button is used to reset the dialog box to its initial settings. The **Close** button is used to exit the dialog box.

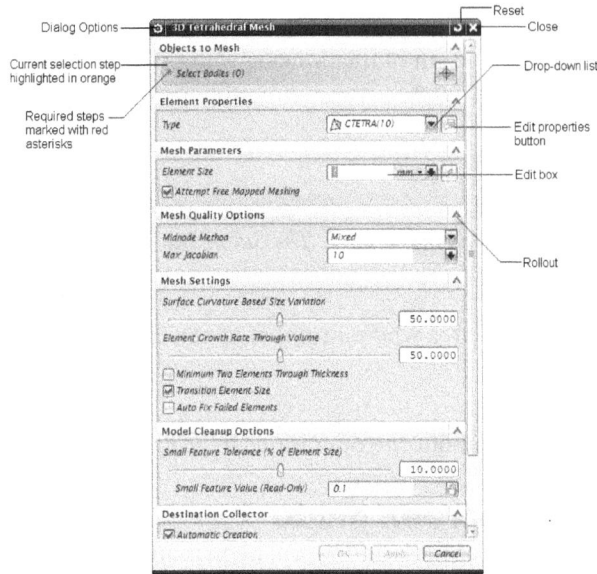

*Figure 2-27 The **3D Tetrahedral Mesh** dialog box*

USING HELP IN NX Nastran

In NX Nastran, there are different ways to access help. These methods are discussed next.

Note
*For using the NX Nastran help file, you must install the **Documentation** file in your system.*

NX Help

To access help in NX Nastran, choose **Menu > Help > NX Help** from the **Top Border Bar**; the **NX 9.0 Help** menu will open in a browser window, as shown in Figure 2-28.

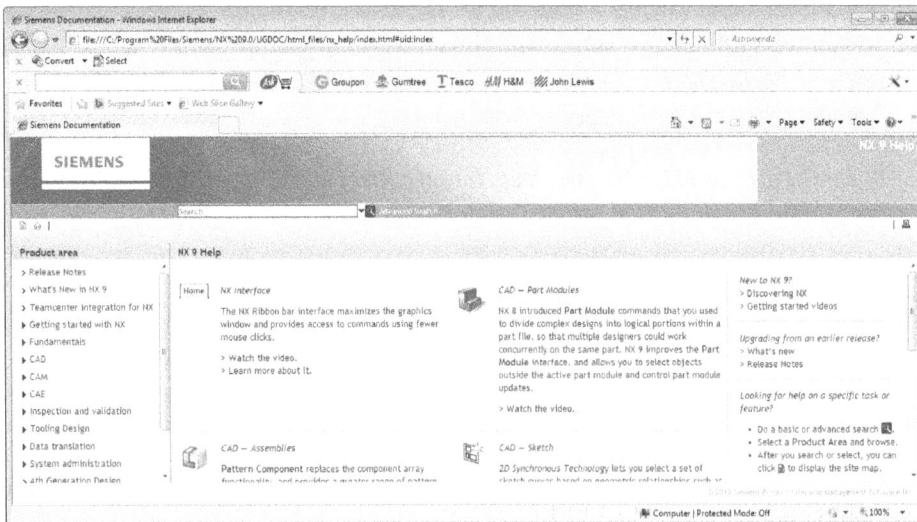

*Figure 2-28 The **NX 9.0 Help** opened in the browser window*

The help window is divided into two parts: **Product area** and **NX 9 Help**. The **Product area** provide with hyperlinks of various topics and the **NX 9 Help** provide product description with summary. Through this browser window, NX provides description of each module provided in the software package. You can browse as well as search any description of any related topic. NX also provides video tutorials for the description of topic being searched as well as the whole package.

Apart from the **NX 9.0 Help** browser window, NX offers more ways to access help which are discussed next.

On Context Help

The **On Context Help** option also provides description of the invoked environment. For example, if you are in the Simulation environment and choose **Menu > Help > On Context** from the **Top Border Bar**, the **Advanced Simulation Overview** will open in the browser window, as shown in Figure 2-29.

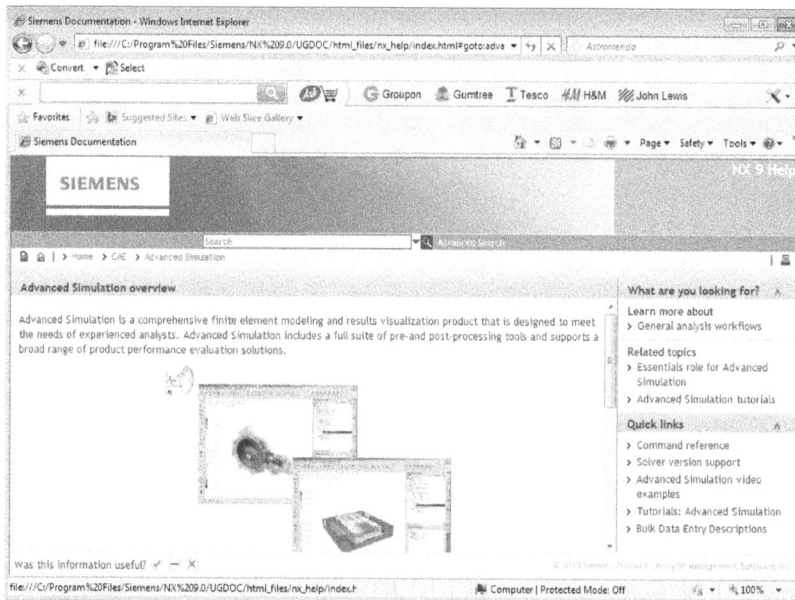

Figure 2-29 The *Advanced Simulation Overview* in the browser window

Command Finder

The **Command Finder** option available in NX Nastran is used to find a particular tool or command. Choose **Menu > Help > Command Finder** from the **Top Border Bar**; the **Command Finder** dialog box will be displayed, refer to Figure 2-30. You can also invoke this dialog box from the **Ribbon**.

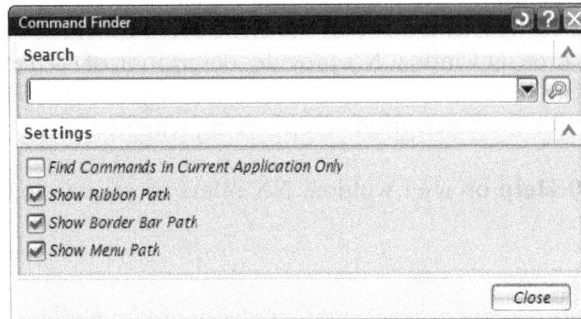

*Figure 2-30 The **Command Finder** dialog box*

To search any tool from the **Command Finder**, enter the name of tool in the **Search** box and choose the **Find Command** button adjacent to the **Search** box. The available tools will be displayed in the dialog box. When you hover the mouse on any option, the software will locate the particular option in the window, as shown in Figure 2-31.

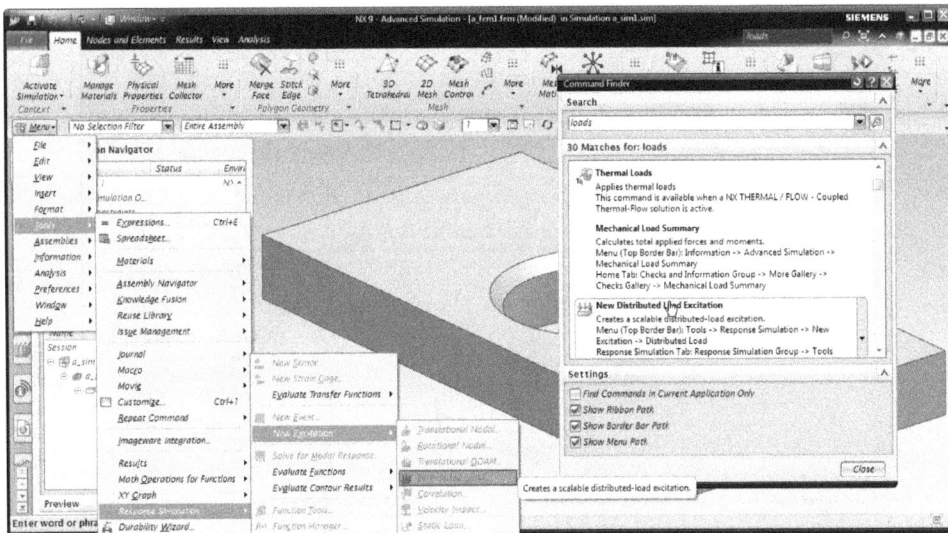

*Figure 2-31 The **Command Finder** dialog box with the located tool*

Note
*In the **Command Finder** dialog box, as you hover the mouse on the tool which is not related to the existing environment, the dialog box will not display the location of the tool. But it will only display the description of the tool.*

Self-Evaluation Test

Answer the following questions and then compare them to those given at the end of this chapter:

1. The _____ environment is used to define boundary conditions of the model.

2. The _____ environment is used to define material for a model.

3. The tools in the _____ tab are used for manipulating the views of the model.

4. In NX Nastran, the FEM environment is used to create mesh in a model. (T/F)

5. The **Command Finder** locates only the tools available in the current working environment. (T/F)

6. The **On Context** is used to provide the description of the current working environment in the browser window. (T/F)

Review Questions

Answer the following questions:

1. Which of the following analysis steps cannot be performed in the FEM environment?

    (a) Meshing   (b) Material applying
    (c) Defining loads   (d) None of these

2. Which of the following processes cannot be performed using the **Post Processing Navigator**?

    (a) Plotting results   (b) Applying boundary conditions
    (c) Creating graphs   (d) None of these

3. The tools available in _____ are used to animate the results generated through analysis.

4. Using the **Simulation File View**, you can switch between working environments. (T/F)

5. The **Post Processing Navigator** is used to view and manipulate results. (T/F)

Answers to Self-Evaluation Test

1. Simulation, **2.** FEM, **3. View**, **4.** T, **5.** F, **6.** T

Chapter 3

NX Nastran
File Support

Learning Objectives

After completing this chapter, you will be able to:
- *Open NX and non-NX CAD files*
- *Understand unit support in NX Nastran*
- *Import files in NX Nastran*
- *Export files from NX Nastran*
- *Understand different solver environments*

INTRODUCTION

For conducting a Finite Element Analysis (FEA), NX Nastran supports a very large range of files. NX Nastran also supports software environment file systems of the other software. In this chapter, you will learn different files and software environments that NX Nastran supports.

OPENING NX CAD FILE

To open an NX CAD file, start NX by double-clicking on the shortcut icon; the NX window will be displayed. Choose the **New** button from the **Home** tab of the **Ribbon**; the **New** dialog box will be displayed. In this dialog box, choose the **Simulation** tab and select the **NX Nastran** template with the **Fem** type from the **Template** rollout. Enter the name of the file in the **Name** text box and specify the location from the **Folder** edit box to save the file. Next, choose the **OK** button; the **New FEM** dialog box will be displayed, as shown in Figure 3-1. The options in this dialog box are discussed next.

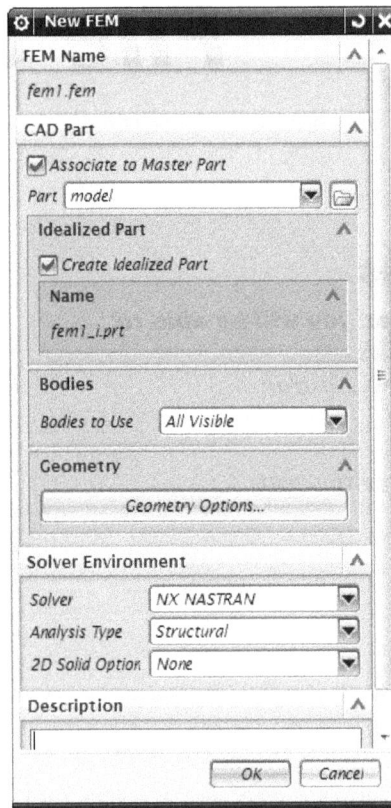

*Figure 3-1 The **New FEM** dialog box*

FEM Name

The **FEM Name** rollout displays the name of the FEM file created in the **New** dialog box.

CAD Part

The **CAD Part** rollout is used to open an NX CAD model in the FEM environment. When you select the **Associate to Master Part** check box, the other options in this area are also displayed, refer to Figure 3-1. The options in this area are discussed next.

Associate to Master Part
This check box is used to create a link between the FEM and Simulation files and the master part. This check box remains selected when you switch to the Advanced Simulation environment from the Modeling environment.

Note
You will learn to switch from the Modeling to Advanced Simulation environment later in this chapter.

Part
The **Part** drop-down list displays the part name are associated with the FEM file. To select a part for associating it with the FEM file, choose the **Open Part** button next to the **Part** drop-down list and select the file to be associated. The name of the selected file will be displayed in the **Part** drop-down list, refer to Figure 3-1. Also, the preview of the part will displayed in a separate **CAD Part** preview window, as shown in Figure 3-2.

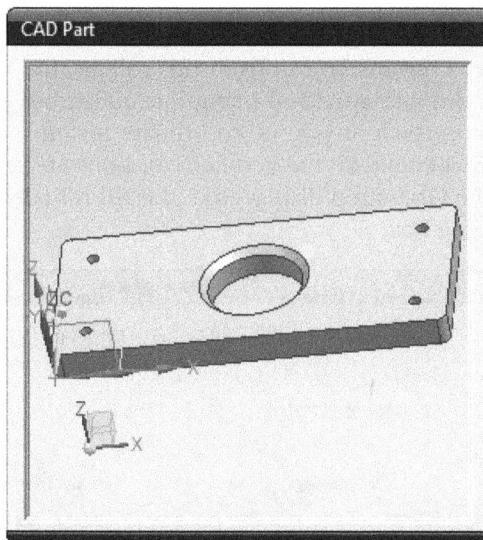

*Figure 3-2 The **CAD Part** preview window*

Idealized Part
The options in this sub-rollout are used to create an idealized part file. When you select the **Create Idealized Part** check box in this sub-rollout, a file with the extension i.prt (for example **(name)_i.prt**) will be created. You will use this file for creating the idealized (editing) part.

Bodies
The **Bodies** sub-rollout allows you to select the part of the body which will be included in the idealized part, FEM, or in Simulation. This sub-rollout is used when the master part contains multiple parts such as assembly. In this sub-rollout, the **Bodies to Use** drop-down list is used

to select the part from the preview window. The options available in this drop-down list are discussed next.

All Visible
The **All Visible** option is used to include only the currently visible part of the master part file.

All
The **All** option is used to include all the parts of the master file even if they are hidden or invisible in the preview.

Select
The **Select** option is used to select the parts one by one from the currently visible parts of the master part file.

None
If you select the **None** option, the FEM environment will get invoked with the master part file name but no part will be displayed in it.

Geometry
When choose the **Geometry Options** button from this rollout, the **Geometry Options** dialog box will be displayed, as shown in Figure 3-3. Using this dialog box, you can import geometry as well as non-geometry parts such as points, coordinate systems, lines, arcs/circles, splines, conics, and sketch curves. By default, all the geometry options are set to **Off**. You can select all the options at once using the **On** button displayed in the **All** rollout. Choose the **OK** button to return to the **New FEM** dialog box.

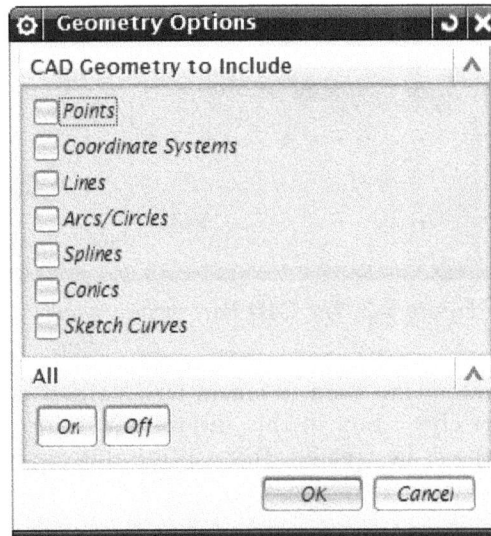

Figure 3-3 *The* **Geometry Options** *dialog box*

Solver Environment

In this rollout, you can define the analysis type and the solver environment type using the **Analysis Type** and **Solver** drop-down lists, respectively.

Description

Under the **Description** rollout, you can provide the description of the project.

After defining all the settings, choose the **OK** button from the **New FEM** dialog box; the model will open in the FEM environment, as shown in Figure 3-4. You can see the FEM, idealize, and master part file names in the **Simulation File View** area.

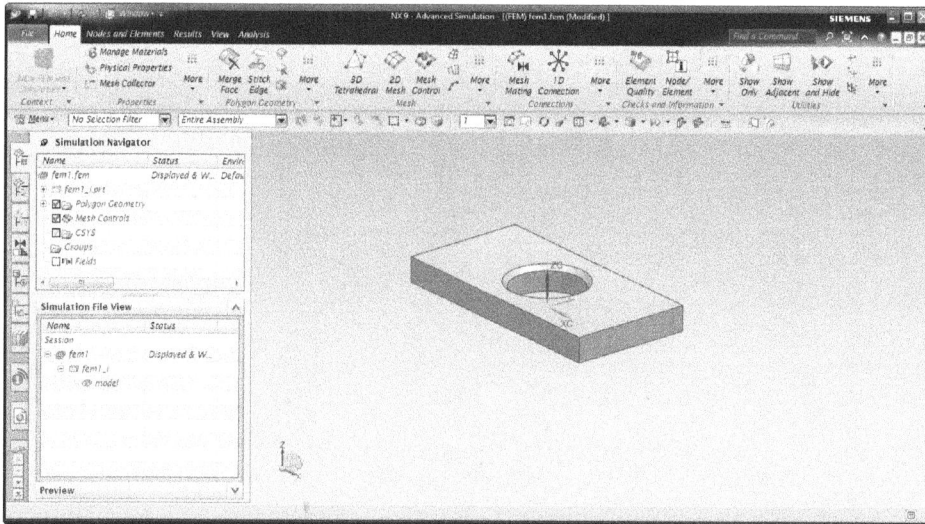

Figure 3-4 *Model in FEM environment*

SWITCHING FROM MODELING TO ADVANCED SIMULATION ENVIRONMENT

NX provides multi-file working environment for modeling as well as for analysis. If you are in the Modeling environment of NX, you can switch to the Advanced Simulation environment. To invoke the Advanced Simulation environment, choose **File > Applications > Advanced Simulation** from the **Ribbon**, as shown in Figure 3-5. Figure 3-6 shows the Advanced Simulation environment. From the Advanced Simulation environment, you can easily perform the analysis through FEM and the Simulation environment. Similarly, you can switch back to the Modeling environment and can make changes in your model.

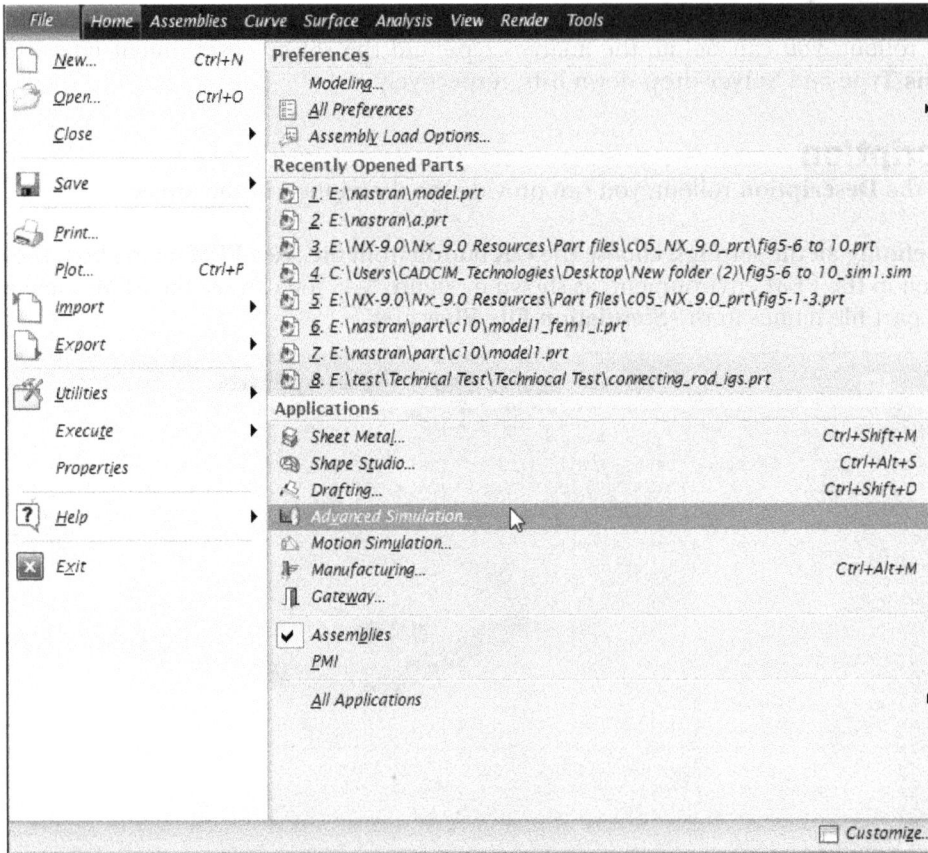

Figure 3-5 The **File** menu

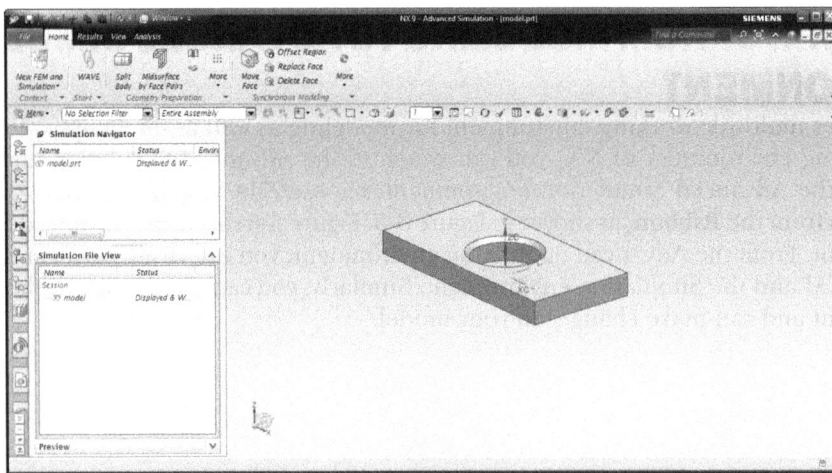

Figure 3-6 The Advanced Simulation environment

OPENING A NON NX CAD MODEL

In NX Nastran, it is not necessary that the model to be analysed is NX CAD models. You can import and analyze models created in other software as well, such as SolidWorks, Solid Edge, or CATIA models and also the models in standard file format like *.iges, or *.step. To import such models in NX Nastran, choose the **Open** button from the **Home** tab of the **Ribbon**; the **Open** window will be displayed, as shown in Figure 3-7. Choose the file type from the **Files of type** drop-down list; the available files will be displayed in the respective folder. Select the file and choose the **OK** button from the window; the respective file will open in the NX Gateway environment, as shown in Figure 3-8. The NX Gateway environment provides an expanded set of capabilities to work within the multi-workflow files. It also provides general tools which are common to all the environment. Using this environment, you can edit the model displayed and also edit the work section in case of large models or assemblies.

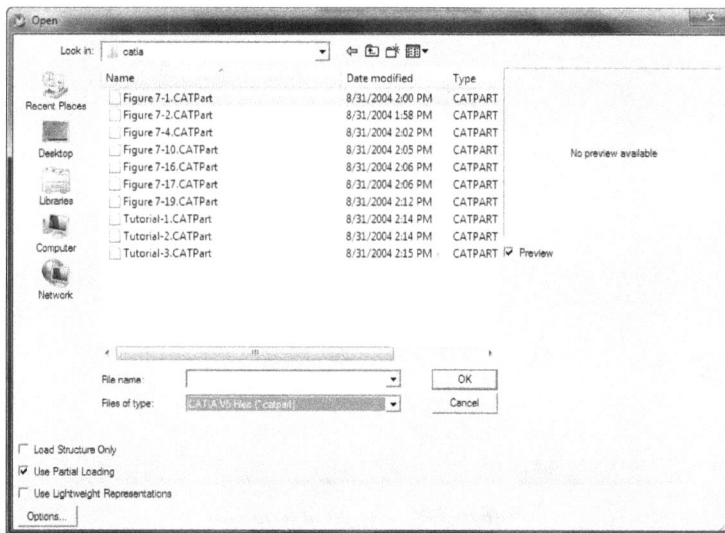

Figure 3-7 The **Open** window

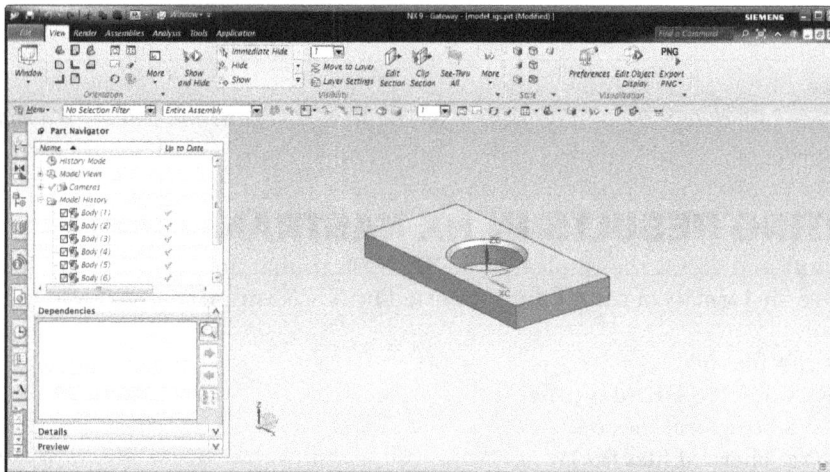

Figure 3-8 The NX Gateway with model

You can switch to the Advanced Simulation environment from NX Gateway by choosing **File > Applications > Advanced Simulation** from the **Ribbon** to start your analysis. You can also switch to the Modeling environment of NX by choosing **File > Applications > Modeling** from the **Ribbon** to edit or modify the model.

UNITS SUPPORT IN NX NASTRAN

NX Nastran supports two main unit formats, Milimeters and Inches. When you choose the **New** button to open a NX Nastran session, both units are displayed in the **Units** drop-down list of the **Templates** rollout in the **New** dialog box, refer to Figure 3-9. Choose the unit from the drop-down list and start your session.

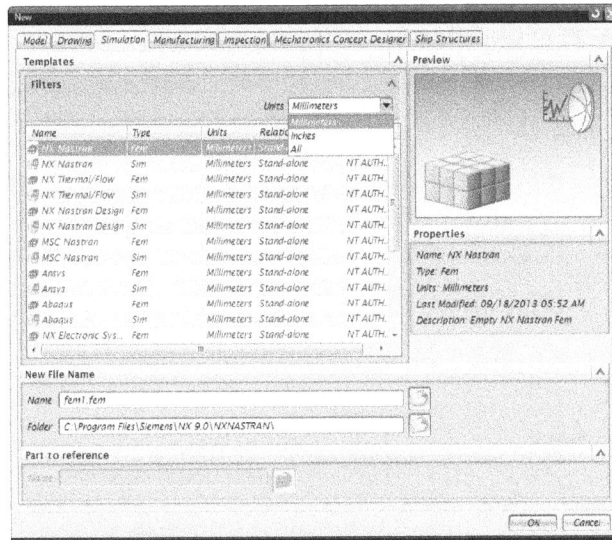

*Figure 3-9 The **New** window*

As NX Nastran supports only millimeter and inch as unit formats so when you import a model in other unit such as centimeter and foot, the software converts the model units into Milimeter and Inch.

Note
You can also change the unit of results after analyzing the part. You will learn about it later in the book.

IMPORTING RESULTS IN NX NASTRAN

You can import and access the result files in NX Nastran that are solved in another software. The following file formats of results are supported by NX Nastran.

- Nastran files (*op2*)
- STRUCTURES P.E. files (*vdm*)
- ANSYS Structural files (*rst*)
- ANSYS Thermal files (*rth*)
- ABAQUS files (*fil*)
- I-DEAS files (*unv*)

- I-DEAS Bun files (*bun*)
- ABAQUS ODB files (*odb*)
- RS2 files (*rs2*)
- LS-DYNA state databases (*d3p*)

The results files in NX Nastran can be imported in two ways: using the **Post Processing Navigator** or using the simulation node in the **Simulation Navigator**. To import results from the **Post-Processing Navigator**, right-click on the **Imported Results** option available in the navigator; a shortcut menu will be displayed. Select the **Import Results** option from the shortcut menu; the **Import Results** dialog box will be displayed, as shown in Figure 3-10.

Using the Browse button available in the dialog box, you can open a result file. The file location and name will be displayed in the **File** and **Name** areas, respectively. You can also change the name of the file. In the **Units** area, two radio buttons are available: **Infer** and **User Defined**. If you choose the **Infer** option, the software will apply default unit to the results. But, if you choose the **User Defined** option; the **Change** button will be activated in this dialog box for changing the units of the results. To change the units, choose the **Change** button; the **Result Units** dialog box will be displayed, as shown in Figure 3-11. You can change the displayed units according to the requirement.

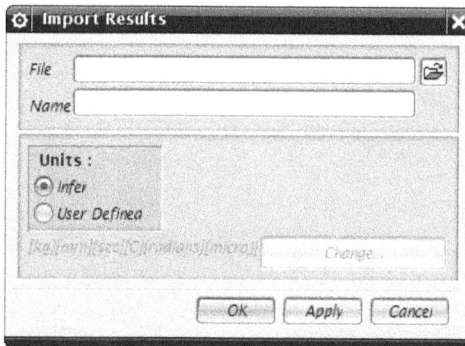

*Figure 3-10 The **Import Results** dialog box*   *Figure 3-11 The **Result Units** dialog box*

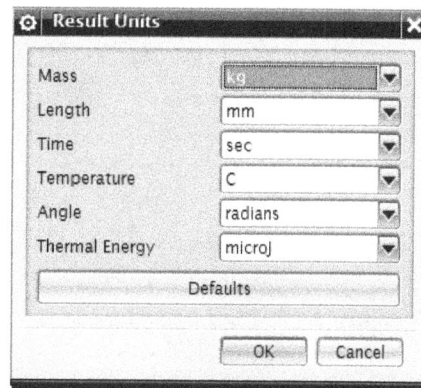

The process of importing of results using the simulation node in the **Simulation Navigator** is also similar to the one discussed previously. You can also import the results file using the **Import Simulation** dialog box discussed in the next section.

You will learn more about importing and manipulating the results in the later chapters.

IMPORTING SIMULATION FILES

NX also supports simulation files created in other software for analyzing. When you import the data, the software creates FEM and Simulation files of a corresponding solution type. The imported files may include the following attributes of analysis.

- Elements
- Material properties

- • Physical properties
- • Coordinate system definitions
- • Boundary conditions
- • Contact parameters
- • Solution attributes

There are different options available in Nastran for importing files. To import the simulation data, choose the **Simulation** option from **File > Import** of the **Ribbon**; the **Import** dialog box will be displayed, as shown in Figure 3-12. This dialog box displays the solver type for solving the simulation data. To solve the simulation data, select the solver type from the **Select solver** area and choose the **OK** button; the **Import Simulation** dialog box will be displayed for the selected solver. Figure 3-13 shows the **Import Simulation** dialog box for the NX Nastran solver. The options available in this dialog box are discussed next.

Figure 3-12 The **Import** dialog box

Figure 3-13 The **Import Simulation** dialog box for NX Nastran solver

Input File Units

In the **Input File Units** area, the **Default Units (Force)(Length)(Mass)** drop-down list is available. You can choose the unit type from this drop-down list. Below this option, the **Use solver file units if available** check box is displayed. If you select this check box, NX Nastran will use the units specified in the file.

In the **Input File Temperature Unit** area, select the **Specify Temperature Unit** check box; a drop-down list will be displayed. You can change the temperature unit from this drop-down list.

File Type

In the **File Type** area, two radio buttons are available, **ASCII** and **OP2 binary**. These radio buttons are used to specify the type of import file. Below the **File Type** area, the **File Name** edit box will be displayed along with the **Browse** button. In this edit box, choose the **Browse** button to select a file to import.

Note

*In Nastran, the file used in the **ASCII** option is in the dat format while the op2 format is used for the **OP2 binary** option. You will learn about the file formats to import in Advanced Simulation environment later in the chapter.*

Selective Import

Under the **Selective Import** rollout, the cards options for importing the cards are displayed. Using the **Filter Method** drop-down list in this rollout, you can change the card display options. The options available in this drop-down list are discussed next.

By card family

When you choose this option from the drop-down list, the check boxes corresponding to the cards and properties of the model will be displayed. Select the desired check boxes to import the cards.

By card name

When you choose this option from the drop-down list, three boxes will be displayed below the list, **Cards to import**, **Cards to exclude**, and **Cards to process as user-defined text**. The cards available for the selected input file will be displayed in the **Cards to import** box. You can exclude cards or process cards as user-defined text using the buttons below the boxes.

Note

In analysis software, the cards are collective form of the properties of the respective analysis. For example, there are cards related to Meshing, Material, Loads and so on.

General Options

Under the **General Options** rollout, the **Auxiliary Input File (ASCII)** option will be displayed along with the **Browse** button. This option is used to import an auxiliary ASCII input file other than the one selected in the **File Type** area. The parameters of this ASCII input file will be used in the main imported file. You can use this type of parameter file option whenever you import a *.dat format file. Note that the **Auxiliary Input File (ASCII)** option will be available only when the **ASCII** option is chosen.

Processing Options

This rollout contains two check boxes which are discussed next.

Enable extended data checking

This check box is used to ensure whether the software checks for the existence of each node (data) on a given element in the input file or not.

Output verbose messaging, up to

This check box allows you to check the verbose message used by the software during import process. You can also control the number of messaging lines reported.

Field Data Options

This rollout contains two check boxes which are discussed next.

Import 3D MOVs as field data

This option controls the import of material orientation vectors for solid elements like node ID tables (spatial fields). If you clear this check box, the material orientation vectors data will be imported for solid elements as physical property table data. Note that, if a coordinate system is used in a solver input file to provide orientation for the 3D material orientation vector, then the software does not import the coordinate system. This ensures that the software does not import multiple coordinate systems which are not required for the analysis.

Import selective loads as field data

This check box controls certain types of loads to be imported as field data. If you select this check box, NX Nastran will import certain loads such as Nastran PLOAD2 pressure loads as spatial field.

Round Trip Options

The check boxes in this rollout will be activated only if the **ASCII** option is selected. These check boxes are discussed next.

Create Round Trip Parameters modeling object

Select this check box to preserve the order and format of bulk data entries in the original file for later use.

Import unsupported cards

Select this check box to import unsupported cards as commented or uncommented user defined text.

Use comment cards for entity names

Select this check box to use comment lines in input file to determine the names of the imported entities, such as materials, physical properties, or meshes.

Note
*The options in the **Import Simulation** dialog box change based on the solver type chosen in the **Select Solver** area of the **Import** dialog box.*

After applying settings in the **Import Simulation** dialog box, choose the **OK** button; the **New FEM and Simulation** dialog box will be displayed, as shown in Figure 3-14. Choose the **OK** button from this dialog box; the FEM and Simulation files will be created for the imported file.

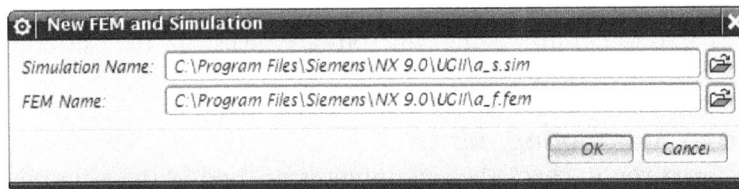

*Figure 3-14 The **New FEM and Simulation** dialog box*

After specifying the path of FEM and Simulation files in respective areas, choose the **OK** button from the **New FEM and Simulation** dialog box; the software imports the file or files in environment. After importing the file, the **Information** window will be displayed, as shown in Figure 3-15, showing the information of the imported file.

Note
Simulation files, similar to simulation entities, can also be imported in the model to be analyzed. The simulation entities may contain the bolts, welds, or any other type of connections. The model should be available in the environment. You will learn more about importing simulation entities in the later chapters of the book.

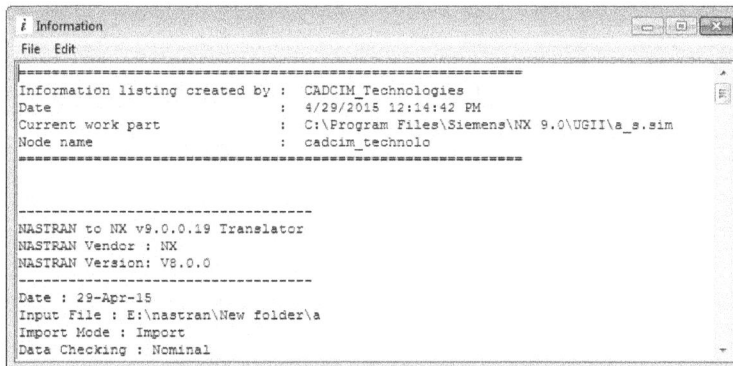

Figure 3-15 The Information window

EXPORTING SIMULATION FILES

You can also export the simulation files and analysis created in NX Nastran for solving in another solver. Using the options available in the **Export Simulation** dialog box, you can export a prepared model. The following things can be done in the export analysis data.

• Selected FEM or simulation files can be exported for further use.

• The location of the input file as well as the units for the file can be changed.

• Cards to be imported in the data file can be selected.

Note that when you export a simulation file using the **Export Simulation** dialog box, geometry or part data are not exported.

After preparing the model to export, choose the **Simulation** option from **File > Export** in the **Ribbon**; the **Export Simulation** dialog box will be displayed. The **Export Simulation** dialog box changes according to the solver chosen. Figure 3-16 shows the **Export Simulation** dialog box for Nastran solver and Figure 3-17 for the Abaqus solver.

Tip
*1. While creating the report of your analysis, you need to have snapshot of the result display. From the **Ribbon** you can create image files such as *.jpeg, *.gif, *.png, and so on. For example, the **JPEG** option from the **Ribbon** is used to create a *.jpeg image.*

2. You will learn more about creating the report of analysis in the later chapters of the book.

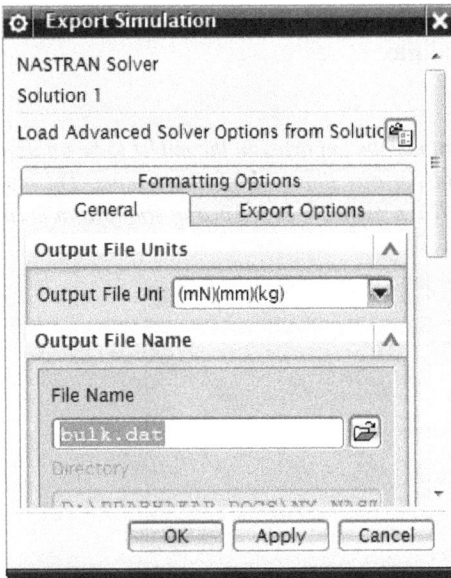

Figure 3-16 The **Export Simulation** *dialog box for Nastran solver*

Figure 3-17 The **Export Simulation** *dialog box for Abaqus solver*

SOLVER ENVIRONMENT

NX supports major software solvers. You can create elements, loads, boundary conditions, and solution options for these solvers. Major software solvers for which you can prepare a file in NX are as follows.

Nastran Environment

NX software provides similar file system for NX Nastran and MSC Nastran environments. You can create elements, loads, boundary conditions, and solution options in both the environments. Nastran environment supports various file formats. In Nastran environment you can:

- Import *dat* file of ASCII format
- Import *op2* file of Binary format
- Export *dat* file of ASCII format
- Post-process results as *op2*
- Use *f06* files of ASCII format as result

Beside these files, there are some other fields such as elements and analysis cards that are different in the Nastran environment and are discussed next.

Elements

Elements are small divisions of a part. Depending upon the part structure, there are four types of elements in the Nastran environment, namely 0D elements, 1D elements, 2D elements, and 3D elements. Different types of elements in the environment are as follows:

- 0D Elements: CBUSH, CBUSH1D, CDAMP2, CMASS2
- 1D Elements: CBAR, CBUSH, CELAS2, CONROD

- 2D Elements: CQUAD4, CQUAD8, CQUADX4, CSHEAR
- 3D Elements: CHEXA(8), CHEXA(20), CPENTA, CPYRAM

Analysis Cards

For doing analysis of a part, you need to define some cards in the environment. These cards contain the definitions of the analyses which you can use at the time of solving them. Some of the analysis cards and their functions in the Nastran are as follows:

- SOL 101: For linear statics
- SOL 103: For response simulation and for flexible body
- SOL 105: For linear buckling

Note

You will learn more about working on the files, elements, and cards in NX Nastran in the later chapters of the books.

Abaqus Environment

You can create files for the Abaqus environment. Abaqus environment supports different file formats. In Abaqus environment, you can:

- Import *inp* file of ASCII format
- Export *inp* file of ASCII format
- Post-process results as *fil* or *odb*

Beside these files, the elements and analysis cards in the Abaqus environment are as follows:

Elements

There are four types of elements in the Abaqus environment, 0D elements, 1D elements, 2D elements, and 3D elements. Some main type of elements in the Nastran environment are as follows:

- 0D Elements: Mass
- 1D Elements: B21, B31, DC1D2
- 2D Elements: S3, DS4, CAX3H
- 3D Elements: C3D4, DC3D6, DC3D10

Analysis Cards

Cards for the analysis in the Abaqus environment are as follows:

- Static stress or displacement analysis
- Static linear perturbation analysis
- Eigenvalue buckling prediction analysis

ANSYS Environment

You can create files for Ansys environment. Ansys environment supports the different file formats. In Ansys environment, you can:

- Import *prep7* or *cdb* file of ASCII format
- Import *rst* or *rth* file of Binary format
- Export *inp* file of ASCII format
- Post-process results as *rst* or *rth*
-

Beside these files, the elements and analysis cards in the Ansys environment are as follows:

Elements

There are five types of elements in the Ansys environment: 0D elements, 1D elements, 2D elements, 3D elements, and Contact elements. Some of the types of elements in the environment are as follows:

- 0D Elements: MASS21
- 1D Elements: BEAM44, COMBIN39, LINK34
- 2D Elements: INTER192, SHELL57 (4), SHELL181 (3)
- 3D Elements: INTER195, PLANE82 (8), SOLID92 (10), SURF154
- Contact elements: CONTA172, TARGE169, CONTA178

Analysis Cards

Cards for the analysis in the Ansys environment are as follows:

- Static
- Modal
- Linear and Nonlinear Buckling

LS-DYNA Environment

You can create file for the LS-DYNA environment. LS-DYNA environment supports different file formats. In LS-DYNA environment, you can:

- Import *k* file of ASCII format
- Export *k* file of ASCII format
- Post-process results as *d3p*

Beside these files, the elements and analysis cards in the Ansys environment are as follows:

Elements

There are four types of elements in LS-DYNA environment, 0D elements, 1D elements, 2D elements, and 3D elements. Some of the types of elements in the environment are as follows:

- 0D Elements: ELEMENT_MASS, ELEMENT_INERTIA
- 1D Elements: ELEMENT_BEAM, ELEMENT_BEAM_OFFSET,
- 2D Elements: ELEMENT_SHELL_MCID, ELEMENT_SHELL_OFFSET
- 3D Elements: ELEMENT_TSHELL (6), ELEMENT_SOLID(4),(6),(8),(10)

Analysis Cards

In NX, you can use the LS-DYNA environment to define basic options for the General Impact solution.

Note

From the FEM type column, you can choose a solver type environment to be displayed when NX Nastran is started. The same solver environment will have to be chosen when a Simulation file is created otherwise an error will be displayed during solving or exporting an analysis.

Self-Evaluation Test

Answer the following questions and then compare them to those given at the end of this chapter:

1. NX Nastran supports only _____ and inch units.

2. Using the _____ tool, you can export a model from one environment to another environment.

3. The _____ file is created in NX Nastran when a file is exported.

4. The **New FEM** dialog box is used to open a model in NX Nastran. (T/F)

5. You can refine a model using the Idealized part. (T/F)

6. You can not export an analysis file from one environment to another other solving environment. (T/F)

Review Questions

Answer the following questions:

1. Using the _____ tool, you can create images in jpeg, png, or other formats.

2. The _____ file is used in NX Nastran for previewing the results.

3. The solver environment used for creating the FEM and Simulation files should be same. (T/F)

4. In NX Nastran, only selected properties of a file can be imported. (T/F)

5. Same type of files can be created in all types of solver environments. (T/F)

Answers to Self-Evaluation Test

1. Milimeter, **2.** Export, **3.** *.dat, **4.** T, **5.** T, **6.** F

Chapter 4

Model Preparation for Analysis

Learning Objectives

After completing this chapter, you will be able to:

- *Prepare the model for analysis*
- *Idealize and defeature a selected geometry*
- *Create midsurfaces*
- *Split a body*
- *Trim and extend a midsurface model*
- *Sew a geometry*
- *Abstract a geometry using the Auto Heal Geometry tool*
- *Split an edge and a face*
- *Merge an edge and a face*
- *Stitch and unstitch an edge*
- *Collapse an edge*
- *Create a face from boundary*
- *Delete a face*
- *Reset the operations done previously*
- *Create circular imprint on a geometry*
- *Suppress a hole*

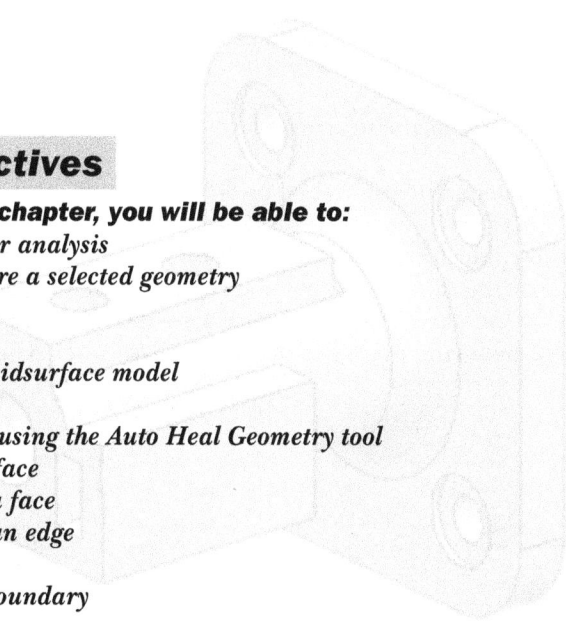

INTRODUCTION

In this chapter, you will learn about preparing a model for analysis. Before starting the analysis of a model, many times it is required to simplify or repair the model. For simplifying or repairing the model, two types of operations namely Geometry Idealization and Geometry Abstraction can be used. After these processes, meshing and feature selection becomes easier.

MODEL PREPARATION

Geometry Idealization and Geometry Abstraction allow you to manipulate some geometry according to the needs of the analysis. However, the two are fundamentally distinct processes that operate on different aspects and environments. After these processes, meshing and feature selection becomes easier.

GEOMETRY IDEALIZATION

Geometry idealization is the process which removes or suppresses features in a model for proper defining of mesh. It is used in the following ways for preparing the model.

• Features such as bosses, fillet, chamfers which are not significant for analysis can be removed.

• Faces can be moved to change the structure of a model without interrupting the master model.

• Midsurfaces can be created to apply shell meshing on uniform face parts.

Note that all the geometry idealization process are performed on the idealized part. To perform the geometry idealization operation, you first need to create an idealized part as no idealization is performed directly on the master model. Figure 4-1 shows the master part with all the small features and Figure 4-2 shows the prepared part for analysis after removing the small features.

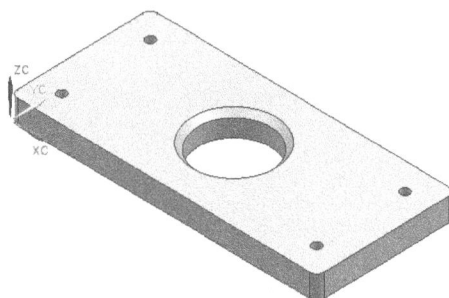

Figure 4-1 The master part   *Figure 4-2* The prepared part for analysis

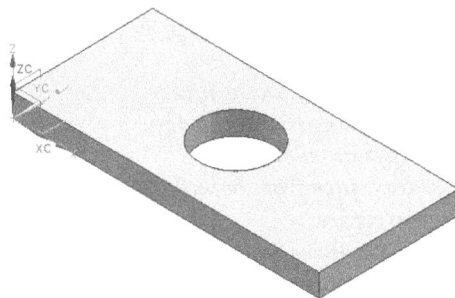

To create an idealized part, invoke the **New FEM** dialog box. Select the **Associate to Master Part** check box, refer to Figure 4-3 and open the part file using the **Open Part** button. Next, select the **Create Idealized Part** check box, refer to Figure 4-3, and choose the **OK** button from the dialog box; the idealized part will be created for the respective model and will be displayed in the **Simulation File View** area, as shown in Figure 4-4.

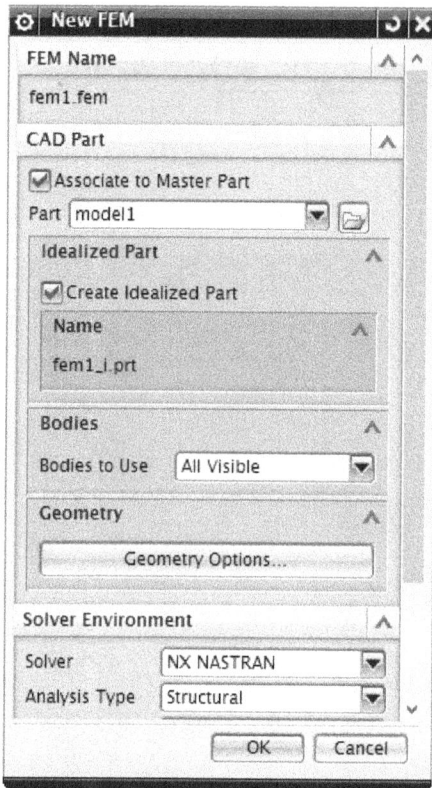

Figure 4-3 *The* **New FEM** *dialog box*

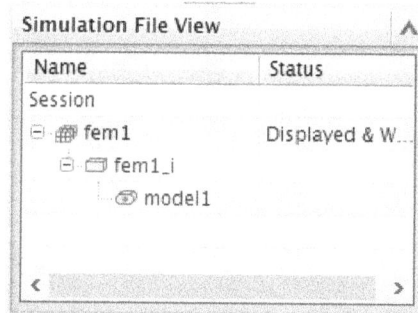

Figure 4-4 *The* **Simulation File View** *area*

After creating the idealized part, you need to switch to the Ideal environment for performing geometry idealization operations. To invoke the Ideal environment, right-click on the Idealize file name in the **Simulation File View** area; a shortcut menu will be displayed, as shown in Figure 4-5.

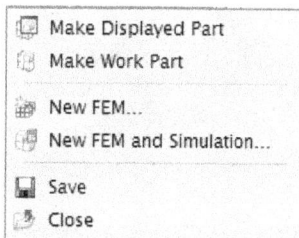

Figure 4-5 *Shortcut menu displayed on right-clicking a file name*

Choose the **Make Displayed Part** option from the shortcut menu; the **Idealized Part Warning** message box will be displayed with the warning that an associative copy of the geometry must be created, refer to Figure 4-6. Choose the **OK** button from the message box; the Ideal environment window will be displayed along with the model, refer to Figure 4-7.

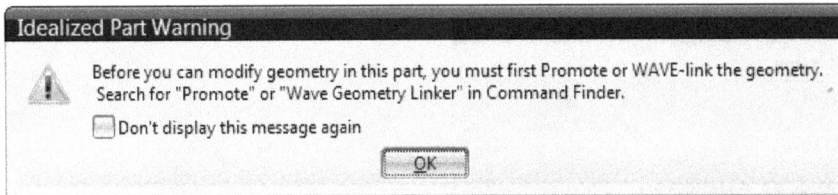

Figure 4-6 The **Idealized Part Warning** *message box*

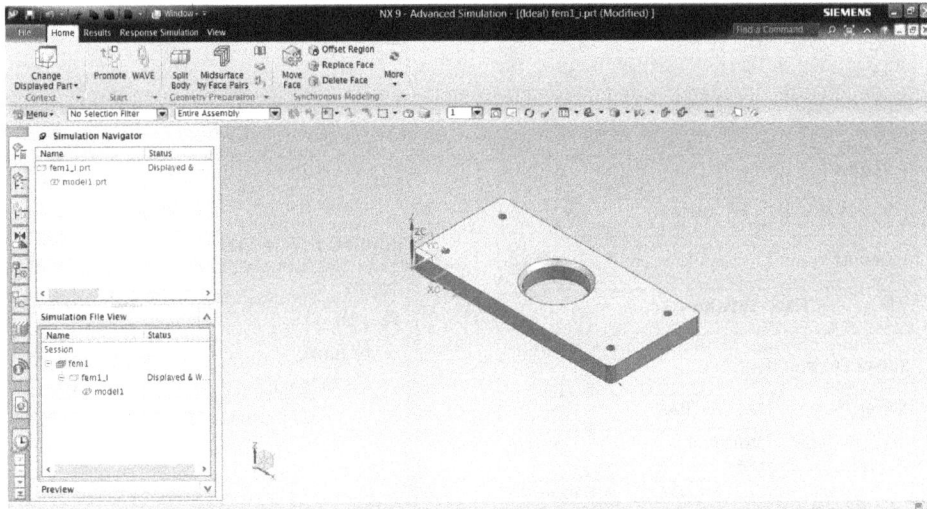

Figure 4-7 The part to idealize displayed in the Ideal environment

Before idealizing the part, first you must create an associative copy of the part. To do so, choose the **Promote** tool from the **Start** group of the **Home** tab; the **Promote Body** dialog box will be displayed, as shown in Figure 4-8. Select the part from the Ideal environment and choose the **OK** button; an associative copy of the part selected will be created.

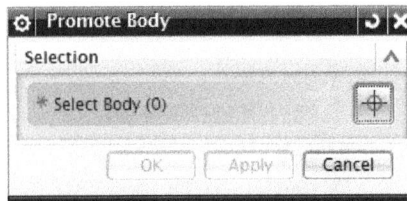

Figure 4-8 The **Promote Body** *dialog box*

Note
*You can use the **Promote** tool in assembly component to promote individual parts to be idealized.*

Now, you can perform the geometry idealization operations such as trimming, extending, splitting, defeaturing on the associative copy. Note that the tools used for the idealiztion process can be used only after promoting the part. These tools are available in the Ideal environment and are discussed next.

Idealize Geometry

Ribbon: Home > Geometry Preparation > Idealize Geometry
Menu: Insert > Model Preparation > Idealize

In any model, generally there are some small geometric features that do not play any major role in the functioning of the model but take long time in getting solved. You can use the **Idealize Geometry** tool to simplify model geometry by removing such features from a part or from a region of a part by applying certain criteria. In Figure 4-9, the master part is shown with small holes and blends. Figure 4-10 shows the model after idealizing the part.

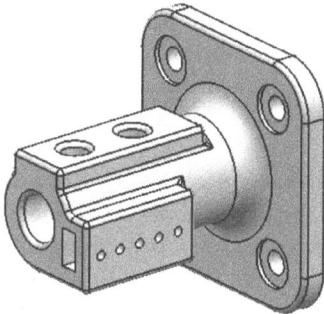

Figure 4-9 *The master part*

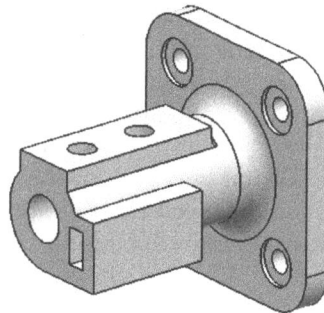

Figure 4-10 *The idealized model*

To create an idealized geometry, choose the **Idealize** tool from **Menu > Insert > Model Preparation** of the **Top Border Bar**; the **Idealize Geometry** dialog box will be displayed, as shown in Figure 4-11. The options in this dialog box are discussed next.

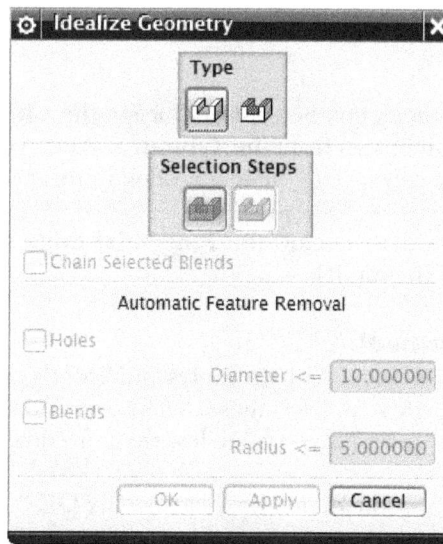

Figure 4-11 *The **Idealize Geometry** dialog box*

Type

The **Type** area is used to define the type of geometry to be used to idealize. There are two buttons in this area: **Body** and **Region**. The **Body** button is used to select model and the **Region** button is used to select a particular region or face of the model.

Selection Steps

The **Selection Steps** area is used to define the selection on the basis of the button chosen in the **Type** area. The buttons in this area are displayed based upon the button chosen in the **Type** area.

If the **Body** button is chosen in the **Type** area, the **Body (Required)** and **Removed Faces (Optional)** buttons will be displayed in the **Selection Steps** area. The **Body (Required)** button is selected by default and is used to perform the idealization process on the whole model. The **Removed Faces (Optional)** button is used to remove specific features. This button will get activated only where the model has been selected.

Choose the **Region** button from the **Type** area, the **Seed Face (Required)**, **Boundary Faces (Optional)**, and **Removed Faces (Optional)** buttons will be displayed in the **Selection Steps** area. The **Seed Face (Required)** button is used to start the features removal process by selecting the first face of the model for defining region. The **Boundary Faces (Optional)** button is used to define outer boundary of the region by selecting the face or a set of faces. The **Removed Faces (Optional)** button is used to remove specific faces.

Chain Selected Blends

The **Chain Selected Blends** check box is displayed below the **Selection Steps** area. On selecting this check box, the adjacent blends of same radius are automatically selected. Once you select a blend face, the software creates a chain of highlighted adjacent blend faces having the same radius as the selected blend face.

Tangential Edge Angle

The **Tangential Edge Angle** check box is displayed below the **Chain Selected Blends** check box when you select the **Region** button from the **Type** area. When you select this check box, an edit box will be displayed. You can specify the angle value in this edit box.

The **Preview Region** button available below the **Tangential Edge Angle** check box is used to preview the selected region of the model.

Automatic Feature Removal

Under the **Automatic Feature Removal** area, the **Holes** and **Blends** check boxes are available. To remove holes and blends in a single step, select these check boxes and enter values in the respective edit boxes to remove all the features having value less than or equal to the specified value.

After specifying all the options in the dialog box, choose the **OK** button to idealize the model and exit the dialog box.

Note
*For using the **Idealize Geometry** tool, you must have to promote and then create an idealized part for creating an associative copy.*

Defeature Geometry

Ribbon: Home > Geometry Preparation > More > Edit and Defeature > Defeature Geometry
Menu: Insert > Model Preparation > Defeature

The **Defeature Geometry** tool works similar to the **Idealize Geometry** tool. This tool removes specific features from the model. To defeature a geometry, choose the **Defeature Geometry** tool from **Menu > Insert > Model Preparation** of the **Top Border Bar**; the **Defeature Geometry** dialog box will be displayed, as shown in Figure 4-12.

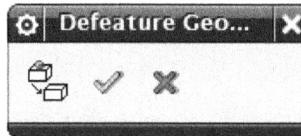

*Figure 4-12 The **Defeature Geometry** dialog box*

For removing any feature or features from the model, select the face or faces of that feature and choose the **Defeature Geometry** button from the dialog box; the modified **Defeature Geometry** dialog box will be displayed. Next, choose the **Apply** button from this dialog box to remove another feature of the model or click **OK** to exit the tool. Figure 4-13 shows the master part with all the features. Figure 4-14 shows the defeatured model.

Figure 4-13 The master part

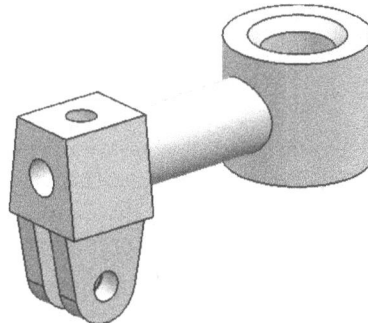

Figure 4-14 The defeatured model

Note
*For using the **Defeature Geometry** tool, you have to promote and then create the idealized part for creating an associative copy.*

In some models, you can simplify the geometry for analysis by creating a midsurface. The midsurface tool generates a surface at the middle of the model between pairs of surfaces. Using a midsurface, you can directly perform analysis and apply uniform shell meshing. You can also ignore fillets, rounds, or bosses that are not important for the analysis. Face Pairs, User Defined, and Offset are the tools used to create a midsurface and are discussed next.

Note

For creating midsurface, you have to promote and then create the idealized part for creating an associative copy.

Face Pairs

Ribbon:	Home > Geometry Preparation > Midsurface by Face Pairs
Menu:	Insert > Model Preparation > Midsurface > Face Pairs

Using the **Face Pairs** tool, you can create a midsurface of a part. The size, shape, and thickness of the midsurface is automatically determined by the software. Figure 4-15 shows a master part of default thickness. Figure 4-16 shows the idealized part after creating the midsurface.

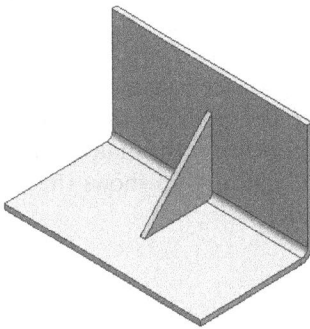

Figure 4-15 The master part

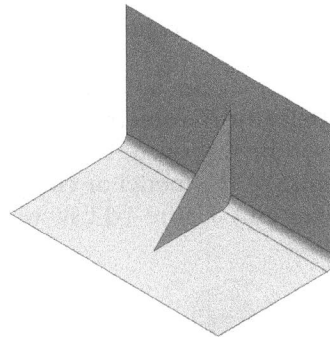

Figure 4-16 The midsurface part

To create a midsurface, choose **Menu > Insert > Model Preparation > Midsurface > Face Pairs** from the **Top Border Bar**; the **Midsurface by Face Pairs** dialog box will be displayed, as shown in Figure 4-17. You can also invoke the **Face Pairs** tool from the **Geometry Preparation** group of the **Ribbon**. The options and areas available in the **Midsurface by Face Pairs** dialog box are discussed next.

Solid Body
In this rollout, the **Select Solid Body (0)** area is highlighted implying that you need to select the geometry for creating the midsurface.

Faces to Exclude from Pairing
The options in this rollout are used for specifying the faces that you want to exclude from auto-midsurfacing. For excluding the faces, choose the **Face** button from the rollout; you will be prompted to select the faces to be excluded from pairing. Now, you can select the faces to be excluded. Figure 4-18 shows a model with faces to be excluded and Figure 4-19 shows the preview of the midsurface created after excluding the specified faces.

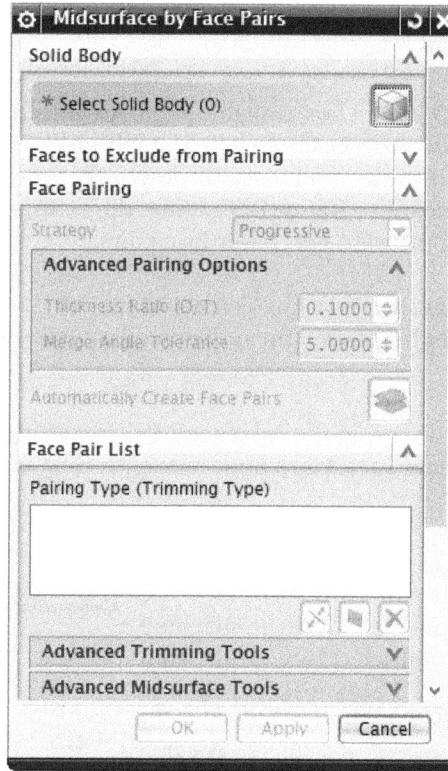

Figure 4-17 The **Midsurface by Face Pairs** *dialog box*

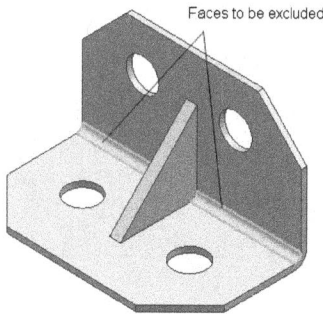

Figure 4-18 The faces to be excluded from midsurfacing

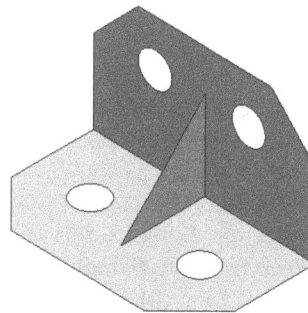

Figure 4-19 The preview of the midsurface after excluding the faces

Face Pairing

The options available in the **Face Pairing** rollout get activated only when a part is selected. In the **Strategy** drop-down list, three options are available: **Progressive**, **Thickness**, and **Manual**. These options are discussed next.

Progressive

When this option is selected, the midsurface is created by automatic pairing method. This option is best for the parts whose the thickness does not change randomly. On selecting this option, the **Advanced Pairing Options** sub-rollout and the **Automatically Create Face Pairs** option is displayed.

The **Thickness Ratio (D/T)** edit box available in the **Advanced Pairing Options** sub-rollout is used to define the thickness of the face pairs. The software divides the smallest characteristic length of a face (D) by the maximum local thickness (T) between the two faces to be paired. If you decrease the **Thickness Ratio (D/T)** value, the software finds more face pairs and if you increase this value, software gets fewer face pairs.

The **Merge Angle Tolerance** edit box is used to specify the angle tolerance between two face pairs. When the angle between the individual face pairs is less than the specified value the faces will merge.

After specifying all the values in the **Advanced Pairing Options** sub-rollout, choose the **Automatically Create Face Pairs** button; the preview of the midsurface will be displayed in the part.

Thickness

The **Thickness** option pairs two surfaces if the thickness between them is less than the specified thickness value. When you select this option from the drop-down list, the **Thickness** edit box is displayed below the **Strategy** drop-down list. In this edit box, you can specify the thickness value of the part along which you want to create the midsurface. The remaining options in this dialog box function the same way as in the **Progressive** option.

Manual

When you select the **Manual** option from the **Strategy** drop-down list, the options in the dialog box will be modified, as shown in Figure 4-20. In the **Face Pairing** rollout, two options are available for face selection: **Select Side 1 Face (0)** and **Select Side 2 Face (0)**. The **Select Side 1 Face (0)** and **Select Side 2 Face (0)** options are used to select two different sides of the surface between which the midsurface is to be created. Figure 4-21 shows the selected part and surface to be selected. After selecting the two surfaces the **Create Face Pair** button gets activated. By choosing this button, the midsurface will be created in the part, as shown in Figure 4-22.

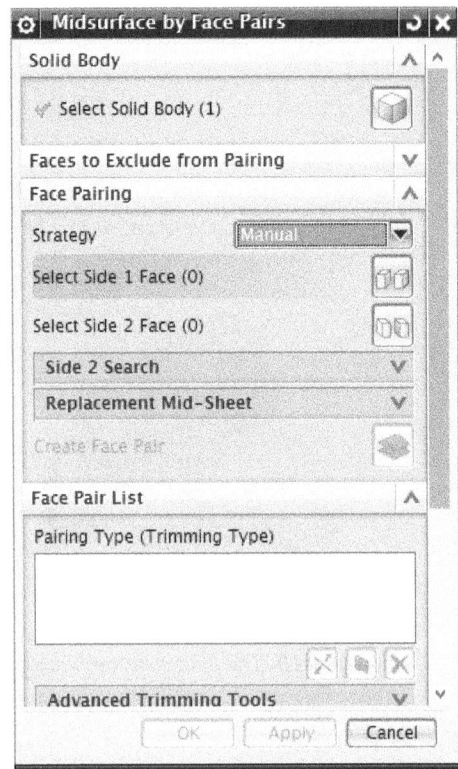

*Figure 4-20 The **Midsurface by Face Pairs** dialog box displayed on selecting the **Manual** option*

The options in the **Side 2 Search** sub-rollout are used to search side 2 of the given surface. You can use this option after selecting the first face. Specify the search distance for the second face, in the **Search Distance** edit box displayed under the **Side 2 Search** sub-rollout. Choose the **Find Side 2 Face Candidates** button; software will find the second face within the specified range. If you select the **Dynamic Search** check box, the software will automatically find the second face after selecting the first face. The options in the **Replacement Mid-Sheet** sub-rollout are used to define a face or mid-sheet as midsurface for the selected face pairs.

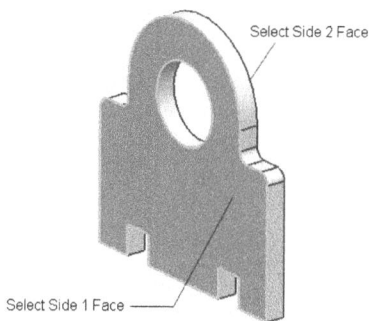

Figure 4-21 The surfaces to be selected for creating midsurface

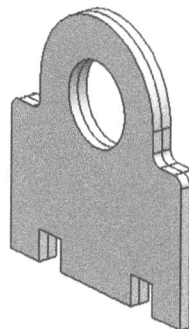

Figure 4-22 The preview of the midsurface

Face Pair List

In this rollout, the options for modification of midsurface are available. The options are discussed next.

Pairing Type (Trimming Type)

After creating the midsurface by choosing the **Automatically Create Face Pairs** button, the default name of the midsurface will be displayed in this area. Three buttons are displayed below this area; **Reverse Face Pair**, **Merge Face Pair**, and **Delete Face Pair**. When you choose any created midsurface from this area, the **Reverse Face Pair** and **Delete Face Pair** buttons become activated. The **Reverse Face Pair** button is used to reverse the side of the face pair in the midsurface. The **Delete Face Pair** button is used to remove the selected face pair from the midsurface.

Advanced Trimming Tools

The options in the **Advanced Trimming Tools** sub-rollout are used to control the trim operation of the midsurface for a selected face pair. When you choose the name of any midsurface from the **Pairing Type (Trimming Type)** area, the buttons available for trimming will get activated. The usage of these buttons are as follows.

The **Trim to Side 1 Faces** tool trims the midsurface using the boundaries of side 1 in the face pair.

The **Trim to Solid Body** tool trims the midsurface using the solid body selected for creating the midsurface.

The **Standard Trimming** tool trims the midsurface based on the side 1 and side 2 surfaces in the pair. This is the default trimming method for all the pairs. You can also use this tool to reset the surface that was trimmed using other tools.

The **No Trimming** tool will only be available when you use the **Replacement Mid-Sheet** option. This tool is used to skip the trimming of the replacement mid-sheet against the solid body, the faces in the pair, or other mid-sheets for avoiding any modifications in its size or shape.

Advanced Midsurface Tools

The options in this rollout are used to manipulate the midsurface. When you select a midsurface name from the **Pairing Type (Trimming Type)** box, the options available for manipulating the midsurface will get activated.

The **Use Side 1** tool creates the mid-sheet directly on the faces in side 1 of the pair. You can choose the **Reverse Face Pair** button available below the **Pairing Type (Trimming Type)** box to change the side of the faces.

Using the **Offset** button, you can create the mid-sheet by offsetting a surface from side 1 of the face pair.

The **Cloud of Points** tool creates midsurface from random generated points that software creates between sides 1 and side 2 faces.

The **Use Largest Face** tool creates midsurface using the largest surface area of the side 1 face.

Display Options

In this rollout, the **Hide Solid Body Upon Apply**, **Show Pairing Faces As Transparent**, and **Show Midsheets As Transparent** check boxes are available. These options are used for manipulating the midsurface. The **Refresh Display** button is used to refresh the settings.

Preview

The **Show Result** button in this rollout is used to preview the midsurface.

User Defined

Ribbon:	Home > Geometry Preparation > More > Midsurface > User Defined
Menu:	Insert > Model Preparation > Midsurface > User Defined

You can create a midsurface by using the **User Defined** tool as well. In this method, an existing surface of the solid body is defined as the midsurface.

To create the midsurface, choose **Menu > Insert > Model Preparation > Midsurface > User Defined** from the **Top Border Bar**; the **User Defined Midsurface** dialog box will be displayed, as shown in Figure 4-23. The options available in this dialog box are discussed next.

Associated Solid Body

In this rollout, the **Select Solid Body (0)** option is selected by default and therefore you need to select the part for which the midsurface is to be created.

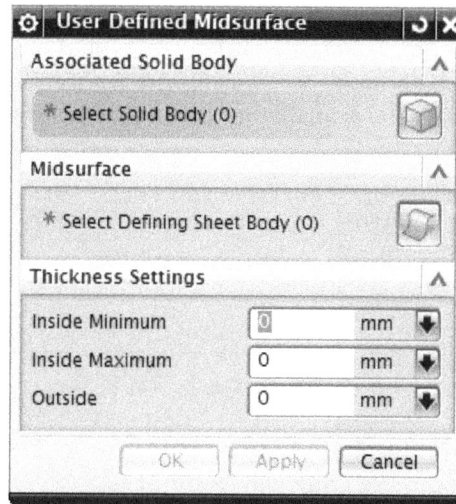

Figure 4-23 *The* **User Defined Midsurface** *dialog box*

Midsurface

In this rollout, the **Select Defining Sheet Body (0)** option is selected by default and you are prompted to select the sheet part that will behave as the midsurface of the solid body. You can select any surface within the solid body as the midsurface.

Thickness Settings

In this rollout, three edit boxes are available: **Inside Minimum**, **Inside Maximum**, and **Outside**. The **Inside Minimum** option is used to assign the minimum thickness value at a point that lies within the solid body. The **Inside Maximum** option is used to assign maximum thickness value at a point that lies within the solid body. The **Outside** option is used to assign the outside thickness (distance) value at a point that lies outside the solid body.

Figure 4-24 shows the solid body for which the midsurface is to be created and the sheet body which will behave as the midsurface. After defining the midsurface, the edges of the sheet body will be highlighted, as shown in Figure 4-25.

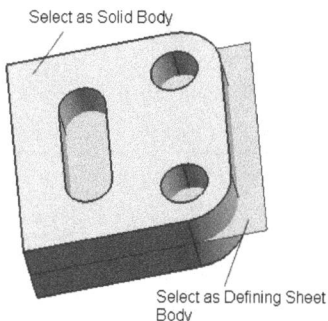

Figure 4-24 *Selection of the solid body and sheet body*

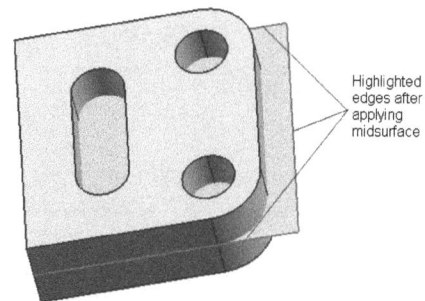

Figure 4-25 *Highlighted edges of the sheet body showing midsurface*

After defining the sheet body as midsurface, switch to FEM environment. In the **Simulation Navigator**, you can view that the sheet body node is converted into a midsurface, as shown in Figure 4-26. Hide the solid body node by clearing the respective check box; the remaining midsurface (sheet body) part will be available in the window, refer to Figure 4-27.

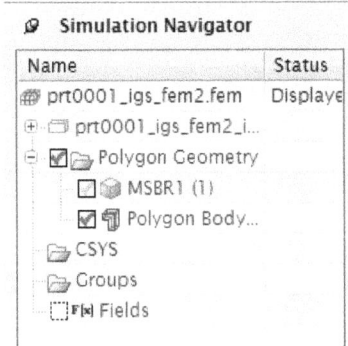

Figure 4-26 *Midsurface node in* **Simulation Navigator**

Figure 4-27 *Final converted midsurface*

Offset

Ribbon:	Home > Geometry Preparation > More > Midsurface > Offset
Menu:	Insert > Model Preparation > Midsurface > Offset

You can also use the offset method to create midsurface. To create a midsurface using the offset method, choose **Menu > Insert > Model Preparation > Midsurface > Offset** from the **Top Border Bar**; the **Offset Midsurface** dialog box will be displayed, as shown in Figure 4-28. You can use the options available in the dialog box to create midsurface. These options are discussed next.

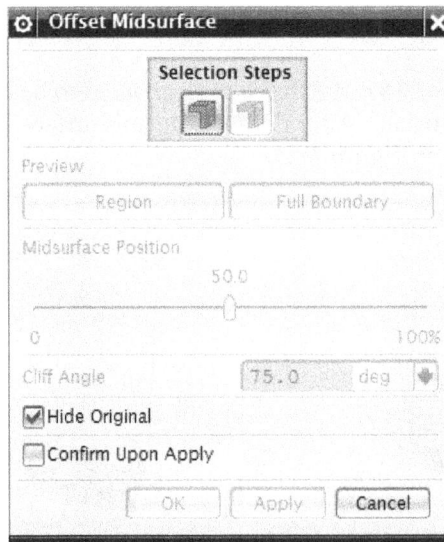

Figure 4-28 *The* **Offset Midsurface** *dialog box*

Selection Steps

The options in this area are used for selecting the model and for specifying the face for midsurfacing. The buttons available in this area are discussed next.

Target Body

This button allows you to select the part from the display environment.

Seed Face

After selecting the part using the **Target Body** button, the **Seed Face** button gets activated. Choose this button to select the face that will be used as a reference to create the midsurface.

Preview

The **Preview** area is used preview any area of a part to be midsurfaced. Two buttons available in this area are discussed next.

Region

Using the **Region** button, you can highlight face of the given part to define the region along which you want to generate the midsurface.

Full Boundary

This button is used to highlight the full boundary faces for generating the midsurface.

Midsurface Position

Using the slider bar in this area, you can define the depth of the midsurface relative to the seed face. The following position of the slider bar define the midsurface creation.

0% position — midsurface created at the seed face
50% position (default) — midsurface created between the seed face and opposing face
100% — midsurface created at the opposite face

Cliff Angle

This edit box is used to specify angle value to determine the boundary faces. You can adjust the cliff angle value to ensure that the correct face is selected.

Hide Original

If the **Hide Original** check box is selected, the original model will be hidden and only the sheet body will be displayed.

Confirm Upon Apply

If you select the **Confirm Upon Apply** check box, another dialog box will be displayed asking to confirm the settings specified, as shown in Figure 4-29. Using this dialog box, you can preview the results as well as examine the created midsurface.

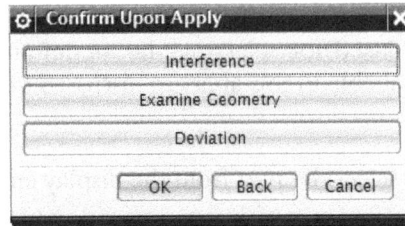

Figure 4-29 The **Confirm Upon Apply** *dialog box*

Split Body

Ribbon:	Home > Geometry Preparation > Split Body
Menu:	Insert > Model Preparation > Trim > Split Body

Proper meshed model is necessary for any analysis. You can create 3D mesh on a model, but for more specific results you can use swept meshing on the model. For swept meshing, you need to divide the model into multiple uniform geometries and create swept meshing on all of them. The **Split Body** tool is used to divide the model into multiple uniform sheet or solid bodies. Note that after splitting the model, you can edit it using the **Part Navigator** in the Ideal environment.

Note
You will learn about swept meshing and creating it on a split body in later chapters.

You can use the **Split Body** tool in the following ways for dividing selected geometries.
• Using an existing face or datum plane.
• By creating a new datum plane.
• By creating a revolved or extruded feature.

In Figure 4-30, the master part is shown without splitting. Figure 4-31 shows the part after splitting.

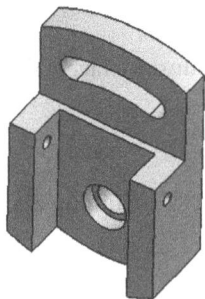

Figure 4-30 The master part

Figure 4-31 The split part

For splitting a model using the **Split Body** tool, choose **Menu > Insert > Model Preparation > Trim > Split Body** from the **Top Border Bar**; the **Split Body** dialog box will be displayed, as shown in Figure 4-32. You can use the options available in the dialog box for dividing the model. The options available in the dialog box are discussed next.

Figure 4-32 The **Split Body** *dialog box*

Target

The button in this rollout is used to select the solid body that you need to split.

Tool

In the **Tool** rollout, the options for dividing the model are displayed. In the **Tool Option** drop-down list, four options for dividing model are available. These options are **Face or Plane**, **New Plane**, **Extrude**, and **Revolve**. Other options in this rollout will be updated on the basis of option chosen from the **Tool Option** drop-down list. The options available in the drop-down list are discussed next.

Face or Plane

Using the **Face or Plane** option, you can divide the model with an existing plane or face of the part as the splitting plane. After selecting this option, the **Select Face or Plane(0)** option gets highlighted. Choose the face or the plane to divide the model.

New Plane

The **New Plane** option is used to create new splitting plane. After selecting this option, the **Specify Plane** option gets highlighted.

There is the **Plane Dialog** button along the **Specify Plane** option. When you choose this button, the **Plane** dialog box will be displayed. Using this dialog box, you can create the datum plane. You can also create the plane using the edges or faces of the model and the options available in the plane creation drop-down list. These options are available on the right of the **Plane Dialog** button. Figure 4-33 shows the model with plane created using the edges of the model and plane creation tool.

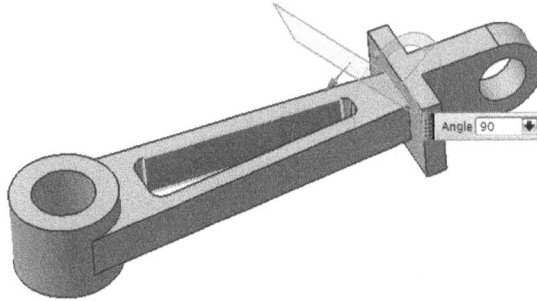

Figure 4-33 *The part with plane preview*

Extrude

The **Extrude** option is used to extrude selected curves to split the body. On selecting this option, the **Section** and **Direction** areas will be displayed. You can use these areas in the following manner:

Section: The **Section** area is used to define the curve to be extruded. The **Sketch Section** button available in this area is used to create a new section sketch. On choosing this button, the **Create Sketch** dialog box will be displayed. Choose the plane along which you want to create the section sketch. Using the **Curve** button, you can also use the edges in the model for extruding.

Direction: The **Direction** area is used to define the direction of extrusion. The **Vector Dialog** button available in this area is used to define the vector direction. On choosing this button, the **Vector** dialog box will be displayed. Using this dialog box, you can define the vector direction. You can define the direction also by using the vector defining drop-down list.

Figure 4-34 shows the model with an edge selected for extrusion to split the model. The arrow shows the direction of extrusion.

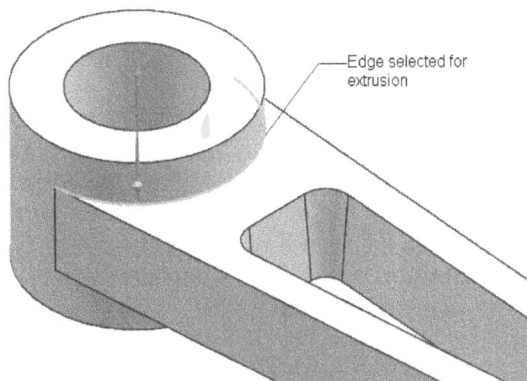

Figure 4-34 *The edge selected for extrusion*

Revolve

The **Revolve** option is used to revolve the selected curves to split the body. On selecting this option, the **Section** and **Axis** areas will be displayed. The options available in these areas are similar to those available in the **Extrude** option.

Simulation Settings

In the **Simulation Settings** rollout, two check boxes are available: **Create Mesh Mating Conditions** and **Check for Sweepable Body**. These check boxes are discussed next.

Create Mesh Mating Conditions

When you select the **Create Mesh Mating Conditions** check box, a connection is established between the resulting bodies. When you switch from the Ideal environment to the FEM environment, NX Nastran automatically creates a glue coincident mesh mating condition between the splitted bodies. Using mesh mating conditions while applying swept meshing, the mesh can be connected to each other in the bodies. In Figure 4-35, the part shows the swept mesh without applying the mesh mating condition. Figure 4-36 shows the part with mesh mating condition applied.

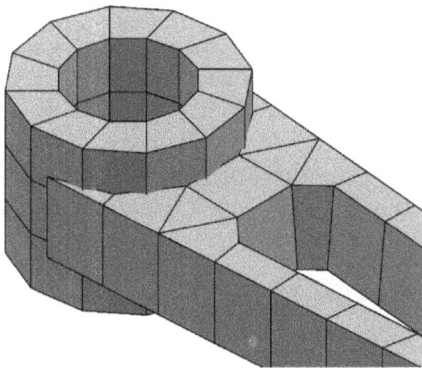

Figure 4-35 The part without applying the mesh mating condition

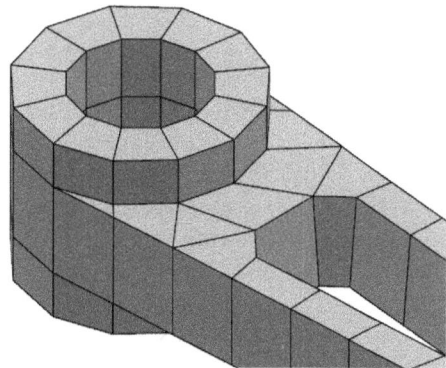

Figure 4-36 The part with mesh mating condition applied

Check for Sweepable Body

When you select the **Check for Sweepable Body** check box, the part temporarily changes the color. While splitting the part, their color changes to red, yellow, green. These colors indicate whether the 3D swept mesh can be generated on the splitted part or not. The detailed descriptions of the displayed colors are as follows:

- Red color indicates that you cannot currently generate a swept mesh on the part. For generating a swept mesh, you must split the part.

- Yellow color indicates ambiguity condition. It indicates you may be able to generate a swept mesh on the part but the software will show warning about the mesh. However, you still need to divide the part further to successfully generate a swept mesh.

- Green color indicates that you are able to generate a swept mesh on the part without further splitting it.

Settings

In the **Settings** rollout, the **Tolerance** edit box and the **Keep Imprinted Edges** check box are available. The **Tolerance** edit box is used to specify the range within which the part will split. When you select the **Create Mesh Mating Conditions** check box, the **Keep Imprinted Edges** check box gets selected by default. This helps to make connection between the edges.

Note
*For using the **Split Body** tool, you need to create the idealized part and promote it for creating an associative copy.*

Trim and Extend

Ribbon: Home > Geometry Preparation > More > Trim and Divide > Trim and Extend
Menu: Insert > Model Preparation > Trim > Trim and Extend

Most of the time after using the **Midsurface** tool, the mid-sheets are created at a gap from their respective sheets. The **Trim and Extend** tool is used to remove this gap by doing trimming and extending operation on the mid-sheets.

To use the **Trim and Extend** tool, choose **Menu > Insert > Model Preparation > Trim** from the **Top Border Bar**; the **Trim and Extend** dialog box will be displayed, as shown in Figure 4-37. Based on the options available in the drop-down list of the **Type** rollout, you can trim or extend the mid-sheets. Four options are available in this drop-down list: **By Distance**, **Percentage of Measured**, **Until Selected**, and **Make Corner**. The other options displayed in the dialog box change as per the options chosen from the **Type** rollout. You can trim and extend a midsurface model using the following steps:

*Figure 4-37 The **Trim and Extend** dialog box*

1. Invoke the **Trim and Extend** dialog box from **Menu > Insert > Model Preparation > Trim > Trim and Extend** of the **Top Border Bar**.
2. Select required option from the drop-down list available in the **Type** rollout.
    If you select the **By Distance** option, you can define the length of the surface extension by specifying the distance value.

If you select **Percentage of Measured**, you can extend the surface upto the specified percentage length of the selected edge that you choose using the **Extension** rollout options. If you select **Until Selected** option, you can extend the surface up to some selected reference object or surface.

If you select the **Make Corner** option, you can create a corner by intersecting the extended surface with the selected surface.

3.   After specifying all the options, choose the **OK** button to trim and extend the sheet.

Figure 4-38 image shows the default created mid-sheets. Figure 4-39 shows the mid-sheets after performing trim and extend operation using the **Make Corner** and **Until Selected** options from the **Type** rollout.

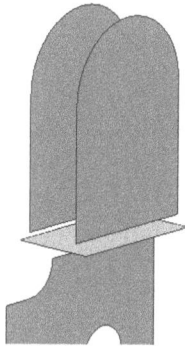

Figure 4-38 *The default created mid-sheets*

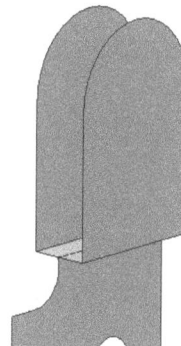

Figure 4-39 *The mid-sheets after performing the trim and extend operation*

Sew

Ribbon:	Home > Geometry Preparation > Sew
Menu:	Insert > Model Preparation > Sew

Most of the time when you import a Non-NX model in the NX environment, the faces of the model do not combine together. In such a case, the **Sew** tool is used to sew the free edges of the model. You can also sew the faces of two solid bodies to combine them.

While doing the CAE operations, the import model must fulfill the following requirements:

•   The model must be fully stitched or watertight. There should not be any free edges, internal cracks, or voids in it.

•   All edges must have precise edges. The large tolerance edges are not valid for CAE operations.

Although a CAD model may appear to be fully stitched on inspecting visually but still there may be several issues related to free edges.

To sew the free edges, choose **Menu > Insert > Model Preparation > Sew** from the **Top Border Bar**; the **Sew** dialog box will be displayed, as shown in Figure 4-40. Two options are available in the drop-down list of the **Type** rollout: **Sheet** and **Solid**. For sewing surface, you need to use the **Sheet** option. You can use the following steps to sew two or more surfaces.

1. Invoke the **Sew** dialog box.
2. Select the **Sheet** option from the drop-down list available in the **Type** rollout; you will be prompted to select the target sheet body. Select the target sheet body; you will be prompted to select the tool sheet bodies to sew. Select the tool sheet bodies; the target sheet will be sewed with the tool sheet bodies.
3. Choose the **OK** button to sew the sheet and exit from the dialog box.

*Figure 4-40 The **Sew** dialog box*

Note that if the selected sheet edges are placed at a distance in the model, a warning window will appear to apply a looser tolerance to sew them, as shown in Figure 4-41. Change the tolerance value from the **Tolerance** edit box available in the **Settings** rollout and then choose the **OK** button to sew.

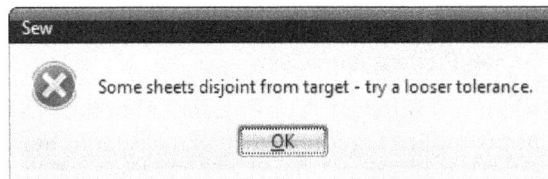

Figure 4-41 The tolerance warning window

Figure 4-42 image shows two sheets at some gap. Figure 4-43 shows the sheets after sewing operation.

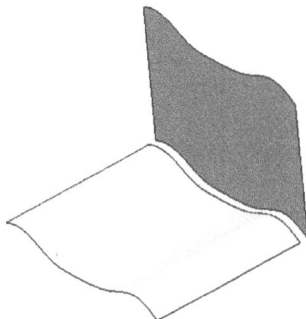

Figure 4-42 Two sheets at some gap

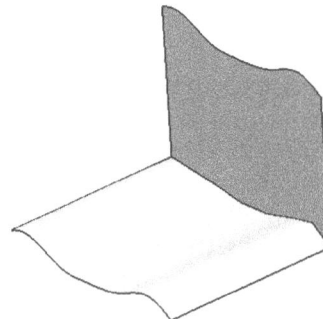

Figure 4-43 The sheets after sewing

Tip

*You can check the free edges of the import model any time while idealizing. For checking this, you must be in the FEM environment. Use the **Simulation File View** to switch to the FEM environment. To check the free edges, choose **Menu > Preferences > Model Display** from the **Top Border Bar** of the FEM environment; the **Model Display** dialog box will be displayed. Choose the **Polygon Geometry** tab and select the **Display Free Edges** check box, as shown in Figure 4-44. Choose the **Apply** button; all free edges will be highlighted.*

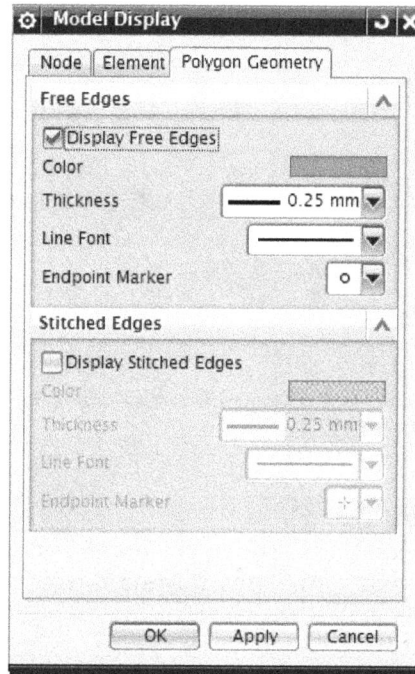

Figure 4-44 The **Model Display** *dialog box with the* **Polygon Geometry** *tab chosen*

Note

*For using the **Sew** tool, you must create the idealized part and promote it for creating an associative copy. For promoting the model, you can choose faces of the model one by one or drag a window around the model while holding the left mouse button; all faces will be selected at once.*

Tip

*After completing the idealizing process, you can examine whether your geometry is proper or not. To examine the geometry, choose **Menu > Analysis > Examine Geometry** from the **Top Border Bar** of the Ideal environment; the **Examine Geometry** dialog box will be displayed, as shown in Figure 4-45. Select the model from the environment and select the respective check boxes from the dialog box to examine it as per your requirement. Next, from the **Actions** rollout, choose the **Examine Geometry** button to display the results.*

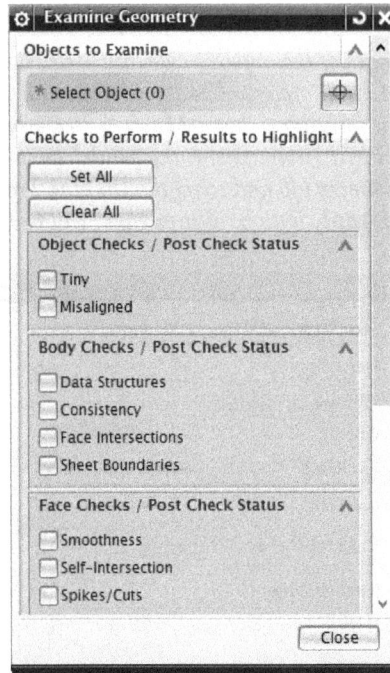

*Figure 4-45 Partial view of the **Examine Geometry** dialog box*

GEOMETRY ABSTRACTION

The geometry abstraction operations can be performed on a geometry within the FEM environment. Using the geometry abstraction operation, you can eliminate issues from the geometry that can cause undesirable results while meshing. This process is used in the following ways to prepare a model.

- The process is used to remove small fillets as well as small holes from a model.

- You can create a face and also delete an unwanted face from the boundary curve.

- You can split, merge, or stitch two faces of a model.

The tools that are used in the geometry abstraction are discussed next. Note that all the tools used for abstraction operations will work only in the FEM environment.

Auto Heal Geometry

Ribbon:	Home > Polygon Geometry > More > Abstraction > Auto Heal
Menu:	Insert > Model Cleanup > Auto Heal Geometry

The **Auto Heal Geometry** tool is used to abstract fillets from the model that may be problematic for meshing. You can specify the model tolerance to abstract the fillets. To heal the geometry, choose **Menu > Insert > Model Cleanup > Auto Heal Geometry** from the **Top Border Bar**; the **Auto Heal Geometry** dialog box will be displayed, as shown in Figure 4-46. The options in this dialog box are discussed next.

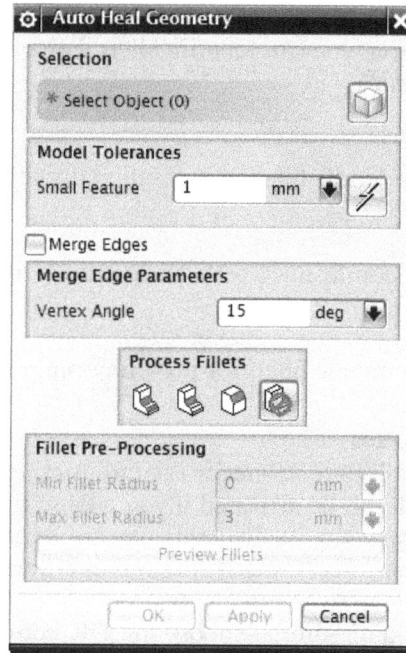

Figure 4-46 *The Auto Heal Geometry* dialog box

Selection Area

In the **Selection** area of the dialog box, the **Select Object (0)** button is activated by default. As a result, you are prompted to select the object or the model for healing the geometry.

Model Tolerances Area

In this area, the **Small Feature** edit box is available. This edit box is used to specify the feature dimension value. If the dimension of the feature is smaller than this value, then the **Auto Heal Geometry** tool will abstract that feature.

Next to the **Small Feature** edit box, the **Auto Element Size** button is available. This button is used to examine the selected geometry and calculate the appropriate tolerance value for the selected model.

Merge Edge Parameters Area

The **Merge Edges** check box is available above the **Merge Edge Parameters** area. When you select this check box, the two selected edges will merge and unnecessary vertices will be removed. In the **Merge Edge Parameters** area, the **Vertex Angle** edit box is available. Using this edit box, you can specify an angle value between the two angled edges to be merged.

Process Fillets Area

You can specify the feature parameters using the buttons available in the **Process Fillets** area. In this area, four buttons are available: **All Fillets**, **Inside Radius Fillets**, **Outside Radius Fillets**, and **No Fillets**. These buttons can be used as follows:

All Fillets

When you choose this button, you can identify fillets with either inside or outside radii within the specified value.

Inside Radius Fillets

Using this button, you can identify fillets with inside radii within the specified value.

Outside Radius Fillets

Using this button, you can identify fillets with outside radii within the specified value.

No Fillets

Using this button, you can disable fillets for pre-processing.

Fillet Pre-Processing Area

The **Fillet Pre-Processing** area is used to specify the radius criteria for identification of fillets. The edit boxes in this area will not be activated if you choose the **No Fillets** button from the **Process Fillets** area. Two edit boxes available in this area are: **Min Fillet Radius** and **Max Fillet Radius**. Using these edit boxes, you can specify the radius criteria.

After specifying all the values and criteria, choose the **OK** button to heal geometry and exit the dialog box.

Split Edge

Ribbon:	Home > Polygon Geometry > More > Edge Operations > Split Edge
Menu:	Insert > Model Cleanup > Split Edge

Using the **Split Edge** tool, you can divide a single edge into two separate edges at some specified location. The split edges can be used in the following ways:

• For defining separate boundary conditions on different portions of an edge.

• For splitting a face.

To split the edge, choose **Menu > Insert > Model Cleanup > Split Edge** from the **Top Border Bar**; the **Split Edge** dialog box will be displayed, as shown in Figure 4-47. In the drop-down list of the **Type** rollout, three options are available: **Location on Edge**, **Project Point on Edge**, and **Project Point Along a Vector**. These options can be used in the following ways:

Figure 4-47 The Split Edge dialog box

Location on Edge

Using this option, you can divide the edge at any point selected on the edge. When you select this option, the **Split Location** rollout will be displayed in the dialog box for picking a point on the edge.

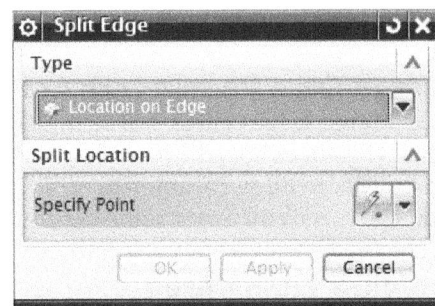

Project Point on Edge

Using this option, you can divide the edge by projecting a point onto the edge. When you select this option, two rollouts will be displayed: **Edge to Split** and **Point to Project**. Using the option available in the **Edge to Split** rollout, you can select the edge to be projected and using the option in the **Point to Project** rollout, you can select the point to be projected.

Project Point Along a Vector

Using this option, you can divide the edge by projecting a point onto the edge along a specified vector. When you select this option, three rollouts will be displayed: **Edge to Split**, **Point to Project**, and **Line of Projection**. Using the options available in the **Line of Projection** rollout, you can specify the vector along which you want to project the point. The options in the remaining rollouts will function in similar manner as explained above.

Figure 4-48 shows a split edge created by using the **Project Point on Edge** option of the **Split Edge** tool.

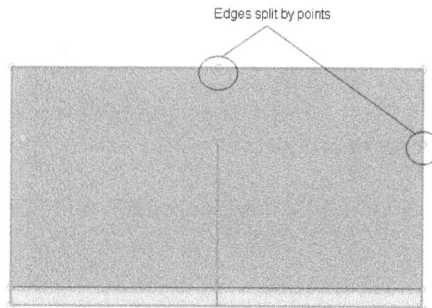

*Figure 4-48 Split edge created by using the **Project Point on Edge** tool*

Note
*You can check the split edges by using the **Model Display** dialog box. To do so, invoke the **Model Display** dialog box by choosing **Menu > Preferences > Model Display** from the **Top Border Bar** in the FEM environment and then choose the **Polygon Geometry** tab. Select the **Display Free Edges** check box and choose the **Apply** button to display split edges.*

Split Face

Ribbon:	Home > Polygon Geometry > Split Face
Menu:	Insert > Model Cleanup > Split Face

Using the **Split Face** tool, you can divide a selected face into two separate faces. The split face can be used in the following ways:

• For dividing a face into several smaller faces to define individual meshes.

• For creating edges on a face to define needed boundary conditions.

To split the face, choose **Menu > Insert > Model Cleanup > Split Face** from the **Top Border Bar**; the **Split Face** dialog box will be displayed, as shown in Figure 4-49. In the dialog box, two options are available in the drop-down list of the **Type** rollout: **Split face by points** and **Split face by suppressed edges**. These options are discussed next.

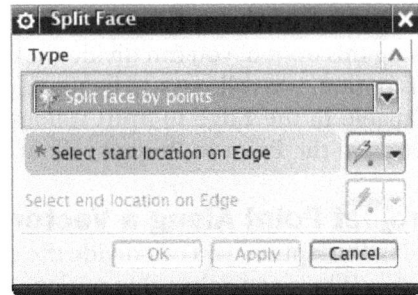

Split face by points

Using the **Split face by points** option, you can split a face by selecting two points on the respective face edges. After selecting this option from the drop-down list, you are prompted to select start and end point location of the face.

*Figure 4-49 The **Split Face** dialog box*

> **Note**
> *You can also use split points for creating a split face. These points were created while splitting an edge using the **Split Edge** tool, as discussed earlier.*

Split face by suppressed edges

Using this option, you can divide a face by restoring a suppressed edge that was previously removed by another abstraction operation.

Figure 4-50 shows a split face created by using the **Split face by points** option of the **Split Face** tool.

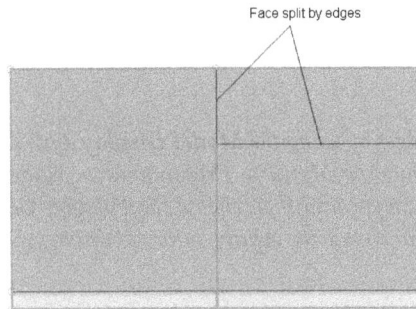

*Figure 4-50 Split face created by using the **Split face by points** tool*

Merge Edge

Ribbon:	Home > Polygon Geometry > Merge Edge
Menu:	Insert > Model Cleanup > Merge Edge

Using the **Merge Edge** tool, you can merge two edges into a single entity. To merge two edges, choose **Menu > Insert > Model Cleanup > Merge Edge** from the **Top Border Bar**; the **Edge Merge** dialog box will be displayed, as shown in Figure 4-51. Also, the split points of the edges will be highlighted in the model. Now, select the point which is common in the desired edges and then choose the **Apply** button from the dialog box; the edges will be merged.

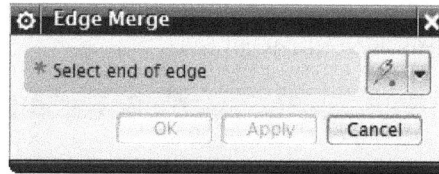

Figure 4-51 *The Edge Merge dialog box*

Note
If you select an edge which was already used by the Split Face tool to split a face then that edge will not be merged.

Merge Face

Ribbon: Home > Polygon Geometry > Merge Face
Menu: Insert > Model Cleanup > Merge Face

Using the **Merge Face** tool, you can merge multiple faces into a single face. To merge faces, choose **Menu > Insert > Model Cleanup > Merge Face** from the **Top Border Bar**; the **Face Merge** dialog box will be displayed, as shown in Figure 4-52. In the dialog box, two buttons are displayed: **Merge** and **Auto Merge**.

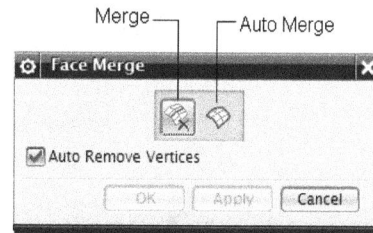

Figure 4-52 *The Face Merge dialog box*

When the **Merge** button is chosen, you have to select individual edges to remove and merge required faces. And when the **Auto Merge** button is chosen, two edit boxes will be displayed in the dialog box: **Edge Angle** and **Vertex Angle**. You can describe the angle value for tolerance between edges and vertices by using these boxes. After specifying the tolerance values, select all the faces of the model and choose the **Apply** button. The faces along the specified tolerances will be merged.

The **Auto Remove Vertices** check box available in the dialog box is used to automatically remove the associated vertices of the merge edges.

Stitch Edge

Ribbon: Home > Polygon Geometry > Stitch Edge
Menu: Insert > Model Cleanup > Stitch Edge

You can use the **Stitch Edge** tool to join two separate edges into a single edge or to stitch two separate faces. The **Stitch Edge** tool is particularly used to eliminate free edges of the created midsurface or imported non-NX CAD geometry.

To stitch edges, choose **Menu > Insert > Model Cleanup > Stitch Edge** from the **Top Border Bar**; the **Stitch Edge** dialog box will be displayed, as shown in Figure 4-53. You can stitch the geometry by using the options available in the **Geometry To Stitch** drop-down list in the **Method** rollout of the dialog box. You can invoke these options by selecting the **Automatic** or

Manual option from the drop-down list available below the **Method** rollout. The method to stitch geometry using these options is discussed next.

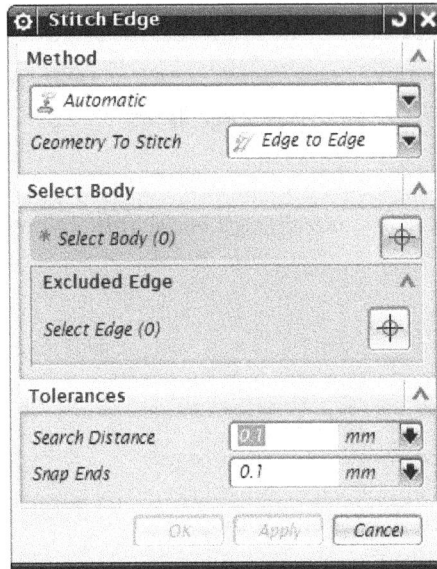

*Figure 4-53 The **Stitch Edge** dialog box*

When you select the **Automatic** option from the **Method** drop-down list, two rollouts will be displayed: **Select Body** and **Tolerances**. Under the **Select Body** rollout, the **Select Body (0)** is highlighted, and you are prompted to select the body to be stitched. Select the body to stitch. You can also exclude the edges from the stitching operation using the option in the **Excluded Edge** sub-rollout. After selecting the model for stitching, specify the tolerance value in the **Snap Ends** edit box available in the **Tolerance** rollout. Choose the **Apply** button to perform stitching. When you select the **Automatic** option, the **Both** option also gets enabled in the **Geometry To Stitch** drop-down list. Using this option, you can simultaneously perform edge to edge and edge to face stitching.

On selecting the **Manual** option from the **Method** drop-down list, three rollouts will be displayed: **Source Edge**, **Target Geometry**, and **Tolerances**. Using the option in the **Source Edge** rollout, you can select the free edges to be stitched. The options in the **Target Geometry** rollout are used to select the edges/faces to which the source edge is to be stitched, refer to Figure 4-54. After selecting the edges and the target geometry, the edges will get stitched, as shown in Figure 4-55. The function of the **Tolerance** rollout has already been discussed.

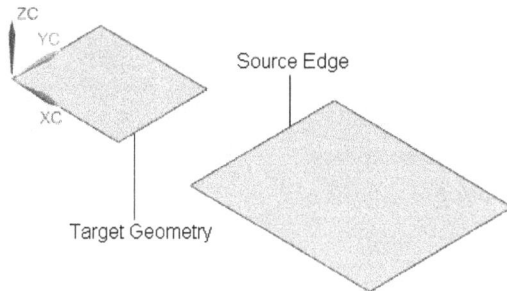

Figure 4-54 Selecting edge and target geometry

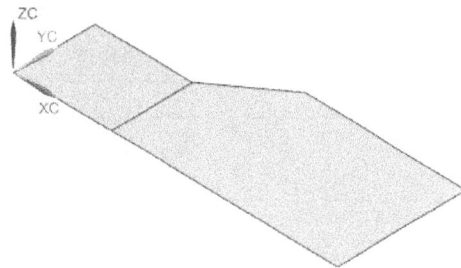

Figure 4-55 Edges after stitching

Unstitch Edge

Ribbon: Home > Polygon Geometry > More > Edge Operations > Unstitch Edge
Menu: Insert > Model Cleanup > Unstitch Edge

You can use the **Unstitch Edge** tool for unstitching an edge. Note that you can only use the **Unstitch Edge** tool to unstitch edges that you join with the **Stitch Edge** tool.

To unstitch edges, choose **Menu > Insert > Model Cleanup > Unstitch Edge** from the **Top Border Bar**; the **Unstitch Edge** dialog box will be displayed, as shown in Figure 4-56. Also, the edges that are stitched using the **Stitch Edge** tool will be highlighted in the model. Select the stitched edge from the model and choose the **Apply** button; the stitched edge will be unstitched.

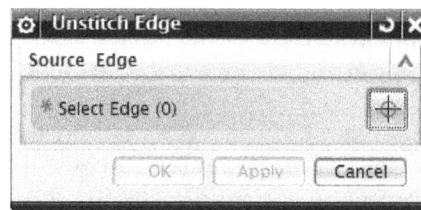

*Figure 4-56 The **Unstitch Edge** dialog box*

Collapse Edge

Ribbon: Home > Polygon Geometry > More > Edge Operations > Collapse Edge
Menu: Insert > Model Cleanup > Collapse Edge

You can use the **Collapse Edge** tool to collapse the edge at any selected point. This tool is also used to modify the mid-sheets or the meshing of the model.

To collapse edges, choose **Menu > Insert > Model Cleanup > Collapse Edge** from the **Top Border Bar**; the **Edge Collapse** dialog box will be displayed, as shown in Figure 4-57. Select the point along which you want to collapse the edge; the edge will be collapsed to the selected point. In Figure 4-58, the mid-sheet part is shown with the point selected for edge collapsing. Figure 4-59 shows the model after edge collapsing.

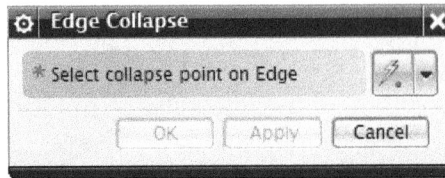

Figure 4-57 The **Edge Collapse** *dialog box*

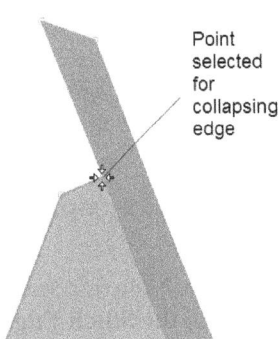

Figure 4-58 The mid-sheet part

Figure 4-59 The mid-sheet edge after collapsing

Face from Boundary

Ribbon:	Home > Polygon Geometry > More > Face Operations > Face from Boundary
Menu:	Insert > Model Cleanup > Face from Boundary

You can use the **Face from Boundary** tool to create a new polygon face from the boundary edges of the model. You can also create a new face to fill voids in the model.

To create a polygon face, choose **Menu > Insert > Model Cleanup > Face from Boundary** from the **Top Border Bar**; the **Face From Boundary Edges** dialog box will be displayed, as shown in Figure 4-60. The rollouts and options displayed in the dialog box are discussed next.

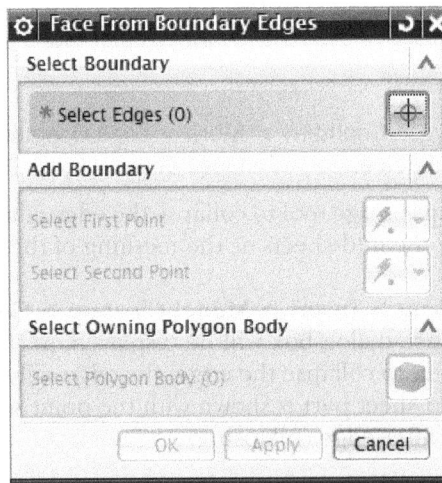

Figure 4-60 The **Face From Boundary Edges** *dialog box*

Select Boundary

In this rollout, the **Select Edges (0)** option is highlighted and you are prompted to select the boundary edges which are required to create the face.

Add Boundary

In this rollout, two options are available: **Select First Point** and **Select Second Point**. After selecting the edges, select the **Select First Point** option; you will be prompted to select the first point of the face. After selecting the first point, the **Select Second Point** option will be activated and you will be prompted to select the second point. Note that the first and second points can only be picked on the selected boundary edges.

Select Owning Polygon Body

In this rollout, the **Select Polygon Body (0)** option is available. This option will be activated on selecting the edges and start point. Select this option; you will be prompted to select the part that you want to attach with the created face.

In Figure 4-61, the mid-sheet part is shown with selected edges and point for creating face boundary. Figure 4-62 shows the model after creating face boundary.

Note

*If you select close boundary for face creation, then the options available in the **Add Boundary** rollout will not be activated.*

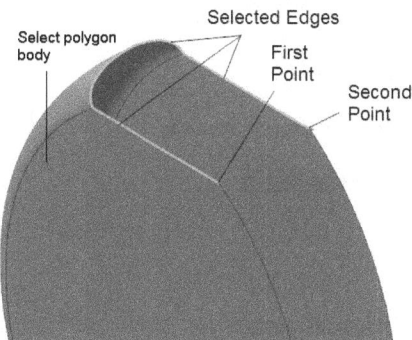

Figure 4-61 Edges and point of the model selected for creating boundary

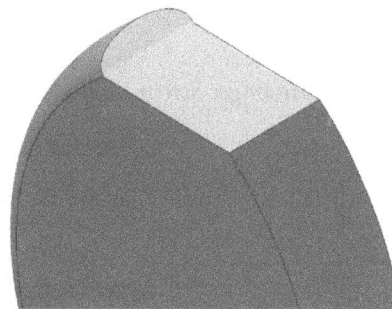

Figure 4-62 Model after creating boundary on the face

Delete Face

Ribbon:	Home > Polygon Geometry > Delete Face
Menu:	Insert > Model Cleanup > Delete Face

Sometimes while creating an FEM file, software automatically creates polygon faces in the model. You can use the **Delete Face** tool to delete such polygon faces. You can also delete faces while preparing a mid-sheet.

To delete faces, choose **Menu > Insert > Model Cleanup > Delete Face** from the **Top Border Bar**; the **Delete Polygon Face** dialog box will be displayed, as shown in Figure 4-63. Select the faces that you want to delete and choose the **Apply** button; the selected faces will be deleted from the model.

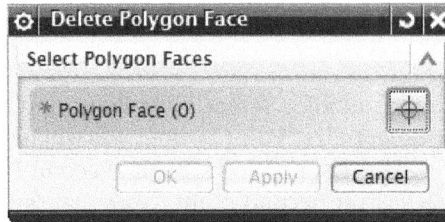

Figure 4-63 *The **Delete Polygon Face** dialog box*

Reset

Ribbon:	Home > Polygon Geometry > More > Abstraction > Reset
Menu:	Insert > Model Cleanup > Reset

You can use the **Reset** tool to restore abstracted polygon geometry to its original state. Using the **Reset** tool, you can remove changes in the polygon geometry that you made using the **Model Cleanup** tools.

To reset faces, choose **Menu > Insert > Model Cleanup > Reset** from the **Top Border Bar**; the **Reset** dialog box will be displayed, as shown in Figure 4-64. Select the faces that you want to reset or from which you want to remove the changes and then choose the **Apply** button; the changes in the respective faces will be removed.

Note
*The **Reset** tool can remove all changes in the polygon geometry except the changes made by the **Collapse Edge**, **Stitch Edge**, **Face from Boundary**, and **Delete Face** tools.*

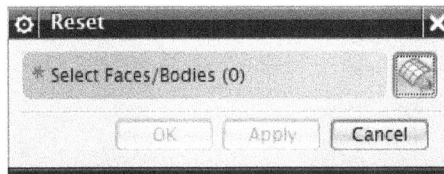

Figure 4-64 *The **Reset** dialog box*

Circular Imprint

Ribbon:	Home > Polygon Geometry > More > Holes > Circular Imprint
Menu:	Insert > Model Cleanup > Circular Imprint

Using the **Circular Imprint** tool, you can control the distribution of mesh/elements around holes in the model. The **Circular Imprint** tool is mainly helpful while using the **Bolt Connection** tool. By specifying the imprint, you can ensure that there are sufficient nodes in the mesh surrounding the bolt's head and nut.

Note
You will learn about creating the Bolt Connection in the later chapter in this textbook.

To imprint the circle around the holes, choose **Menu > Insert > Model Cleanup > Circular Imprint** from the **Top Border Bar**; the **Circular Imprint** dialog box will be displayed, as shown in Figure 4-65. The options and rollouts available in the dialog box are discussed next.

Surface to Imprint

Using the options available in the **Surface to Imprint** rollout, you can select the surface and point/edge, around which the circular imprinting needs to be done. Two options are available in this rollout: **Select Surface (0)** and **Select Point or Edge (0)**. Using these two options, you can select the surface and point/edge, respectively.

Diameter of Imprint

The **Diameter of Imprint** rollout is used to provide diameter around the selected point or edge. In this rollout, two areas are available: **Around Point** and **Around Edge**.

Figure 4-65 The Circular Imprint dialog box

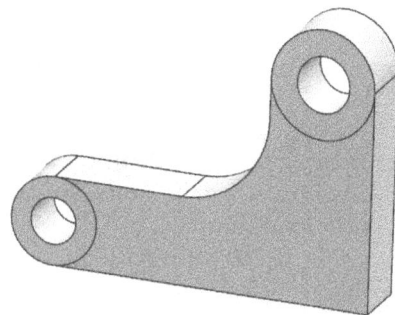

After you choose a point on the hole to create a circular imprint using the **Select Point or Edge (0)** option, you need to provide diameter value in the **Absolute Diameter** edit box available in the **Around Point** area.

If you choose edge, you will provide diameter value using the **Around Edge** area. You can specify the diameter value in the **Absolute Diameter** edit box if you choose the **Absolute** option from the **Type** drop-down list available in the area. If you choose the **Relative** option from the **Type** drop-down list, you have to provide scale value of the diameter of selected edge in the **Edge Diameter Scale Factor** edit box.

In Figure 4-66, the master part is shown with holes around which the circular imprinting needs to be done. Figure 4-67 shows the model after circular imprinting.

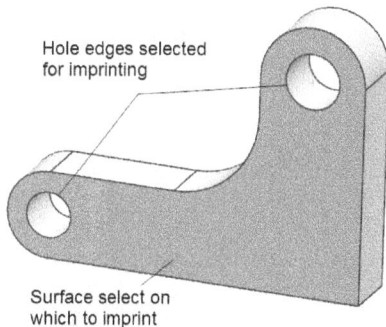

Figure 4-66 The master part

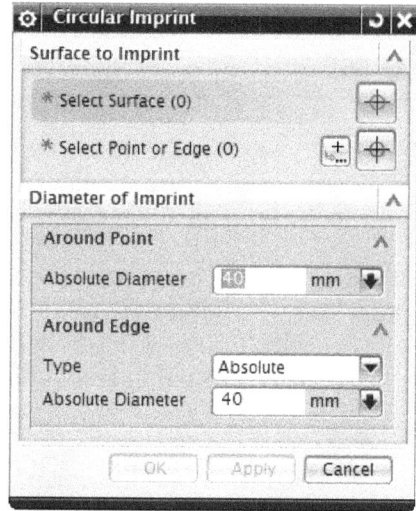

Figure 4-67 The master part after circular imprinting

Suppress Hole

| Ribbon: | Home > Polygon Geometry > More > Holes > Suppress Hole |
| Menu: | Insert > Model Cleanup > Suppress Hole |

After creating the midsurface, sometimes you need to remove small holes to eliminate unnecessary mesh density. In such cases, you can use the **Suppress Hole** tool to remove holes from the mid-sheet. The holes can be circular or non-circular.

To suppress holes, choose **Menu > Insert > Model Cleanup > Suppress Hole** from the **Top Border Bar**; the **Suppress Hole** dialog box will be displayed, as shown in Figure 4-68. For removing holes, you can select the **Automatic** or **Manual** option from the drop-down list available in the **Creation Method** rollout in the dialog box.

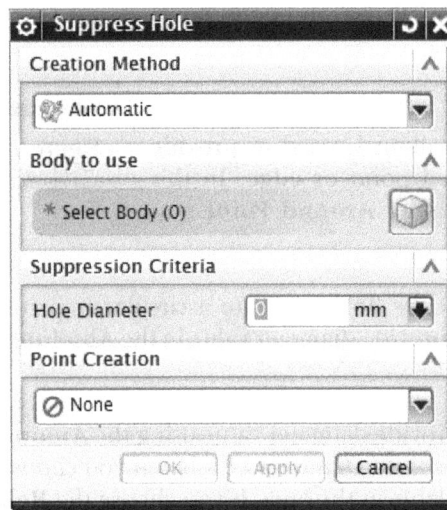

*Figure 4-68 The **Suppress Hole** dialog box*

If you select the **Automatic** option, three rollouts will be displayed: **Body to use**, **Suppression Criteria**, and **Point Creation**. Select the model using the **Body to use** rollout and specify the holes criteria in the **Hole Diameter** edit box available in the **Suppression Criteria** rollout. Choose the **Apply** button to remove the holes from the selected model.

If you choose the **Manual** option, two rollouts will be displayed: **Hole to Suppress** and **Point Creation**. Also, you will be prompted to select the edge of the hole to be removed. Select the edge of the hole to be removed and choose the **Apply** button to remove the holes.

In the drop-down list of the **Point Creation** rollout, three options are available: **None**, **Point**, and **Mesh Point**. If you select the **None** option, no point will be created at the center of the suppressed hole. If the **Point** option is selected, a point will be created at the center of the suppressed hole. If you select the **Mesh Point** option, a mesh point will be created at the center of the suppressed hole. During meshing, a node at each mesh point will be created.

TUTORIALS

To perform the tutorials you need to download the zipped file named as *c04_NX_Nastran_input* from the Input Files section of the CADCIM website. The complete path for downloading the file is:

Textbooks > CAE Simulation > NX Nastran > NX Nastran 9.0 > Input Files

After the file is downloaded, extract the folder to the location *C:\NX Nastran* and rename it as *c04*.

Tutorial 1

In this tutorial, you will import an IGES model shown in Figure 4-69 into NX Nastran. Next, you will idealize and then split it for swept meshing. Figure 4-70 shows the prepared model for swept meshing. **(Expected time: 30 min)**

The following steps are required to complete this tutorial:

a. Start NX and import the IGES file.
b. Start advanced simulation and create FEM and idealized part file.
c. Check the free edges of the model.
d. Enter in Ideal environment and promote part for idealizing.
e. Sew the model using the **Sew** tool
f. Remove the holes using the **Idealize Geometry** tool.
g. Remove the slot feature using the **Defeature Geometry** tool.
h. Split the body for swept meshing.
i. Save and close the model.

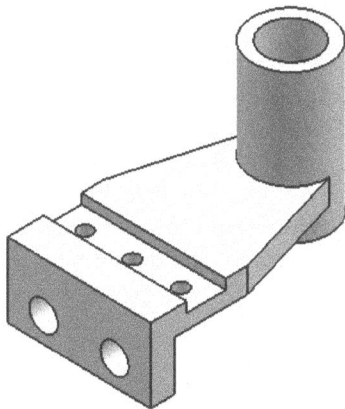

Figure 4-69 The imported IGES model   *Figure 4-70 The model prepared for meshing*

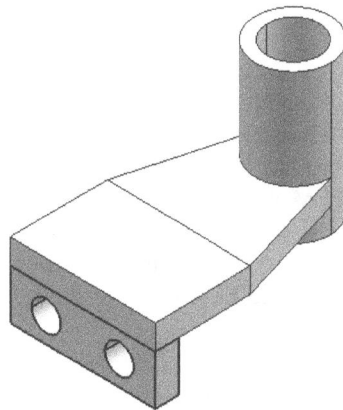

Importing the IGES File

1. Start NX 9.0 by double-clicking on the **NX 9.0** icon on the desktop of your computer.

2. Choose the **Open** button from the **Home** tab of the **Ribbon** or choose **Menu > File > Open** from the **Top Border Bar**; the **Open** dialog box is displayed.

3. Select the **IGES Files (*.igs)** option from the **Files of type** drop-down list. Next, select the **c04_tut01** file from *C:\NX Nastran\c04\Tut01* and then choose the **OK** button; the part opens in the Gateway environment of NX, as shown in Figure 4-71.

Creating FEM and the Idealized File

For creating FEM and the idealized file, you need to invoke the Advanced Simulation environment.

1. Choose **Advanced Simulation** from the **Applications** area of the **File** tab; the Advanced Simulation environment is invoked.

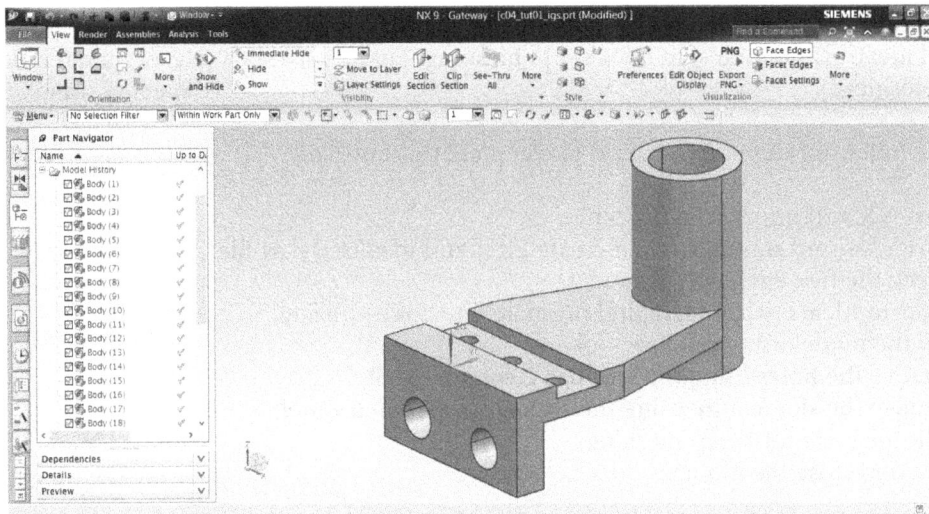

Figure 4-71 The part opened in the Gateway environment

2. Choose the **New FEM** tool from the **Change Window Drop-down** list in the **Context** group of the **Home** tab; the **New Part File** window is displayed, as shown in Figure 4-72. Select **NX Nastran** from the **Name** column of the **FEM** type and choose the **OK** button; the **New FEM** dialog box is displayed, as shown in Figure 4-73, along with the preview of the imported part in the **CAD Part** window. Select the **Create Idealized Part** check box, if it is not selected by default, and then choose the **OK** button from the dialog box; the part is opened in the FEM environment.

Checking the Free Edges

Now, you need to check the free edges for performing geometry idealization. Note that before performing geometry idealization, you must have a water tight solid part.

1. Choose **Menu > Preferences > Model Display** from the **Top Border Bar**; the **Model Display** dialog box is displayed. Choose the **Polygon Geometry** tab if it is not chosen by default and select the **Display Free Edges** check box available in the **Free Edges** rollout, refer to Figure 4-74. Choose the **OK** button; the free edges are displayed in the model.

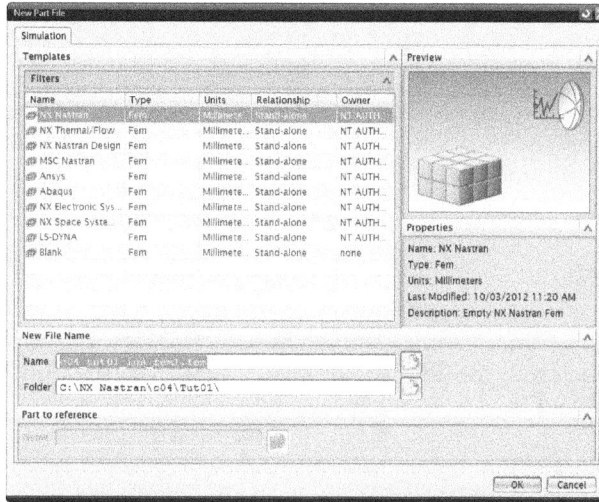

Figure 4-72 *The **New Part File** window*

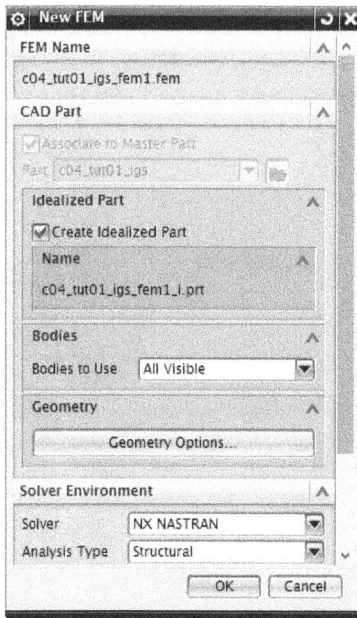

Figure 4-73 *The **New FEM** dialog box*

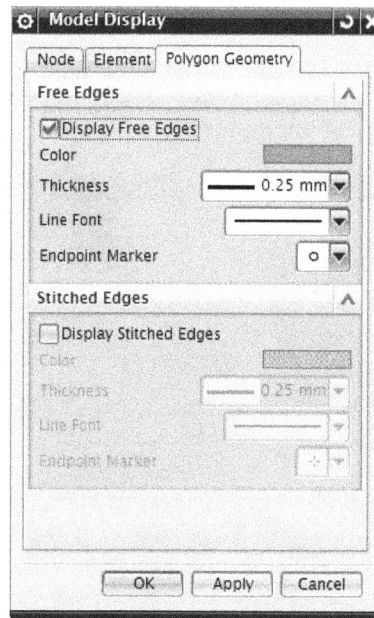

Figure 4-74 *The **Model Display** dialog box*

In this model, all the edges are free. So all edges are highlighted.

Promoting Part for Geometry Idealization

For geometry idealization, you need to invoke the Ideal environment using the created idealized part.

1. In the **Simulation File View** rollout of the **Simulation Navigator**, right-click on the *c04_tut01_igs_fem_i* file name; a shortcut menu is displayed. Choose the **Make Displayed**

Part option from the shortcut menu; the **Idealized Part Warning** message box is displayed. Choose the **OK** button from the window; the part is opened in the Ideal environment.

2. For promoting the part, choose **Menu > Insert > Associative Copy > Promote** from the **Top Border Bar**; the **Promote Body** dialog box is displayed. Select all the surfaces by dragging a window around the part and choose the **OK** button; an associative copy of the selected part is created.

Sewing the Model
Next, you need to sew the free edges of the model.

1. Choose **Menu > Insert > Model Preparation > Sew** from the **Top Border Bar**; the **Sew** dialog box is displayed.

2. In the dialog box, select the **Sheet** option from the drop-down list in the **Type** rollout; the corresponding options are displayed under the rollout, refer to Figure 4-75. Also, the **Select Sheet Body (0)** option is highlighted in the **Target** rollout and you are prompted to select the target face along which the remaining faces are to be sewed.

3. Choose the top face of the model, refer to Figure 4-76; the **Select Sheet Body (0)** option is highlighted in the **Tool** rollout and you are prompted to select the faces that you want to sew with the target face.

Figure 4-75 The **Sew** dialog box

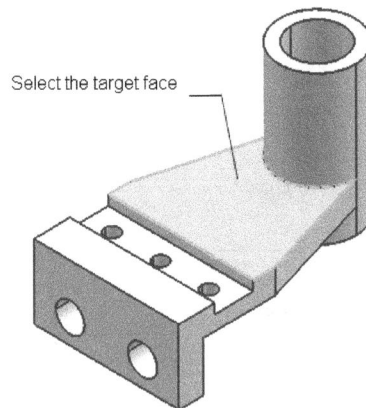

Figure 4-76 Selecting face for sewing

4. Drag a window around the model; remaining faces to be sewed with the target face of the model get selected. Next, set **0.0254** as the tolerance value in the **Tolerance** edit box available in the **Settings** rollout.

5. Choose the **OK** button from the dialog box; all the faces get sewed together.

Tip
*You can check whether the free edges are sewed or not using the **Model Display** dialog box available in the FEM Environment. Use the **Simulation File View** rollout in the **Simulation Navigator** to switch the environment.*

Removing Small Holes from the Model

Next, you need to remove small holes from the model.

1. Choose **Menu > Insert > Model Preparation > Idealize** from the **Top Border Bar**; the **Idealize Geometry** dialog box is displayed, as shown in Figure 4-77. Select the model from the window; other options in the dialog box get activated.

2. Choose the **Removed Faces (Optional)** button from the **Selection Steps** area and select the faces of the holes from the model, refer to Figure 4-78.

3. Choose the **OK** button from the dialog box; all the small holes will be removed from the model.

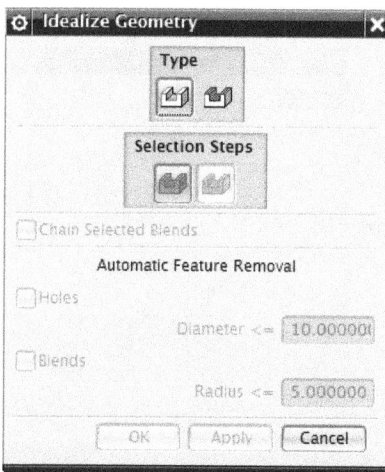

*Figure 4-77 The **Idealize Geometry** dialog box*

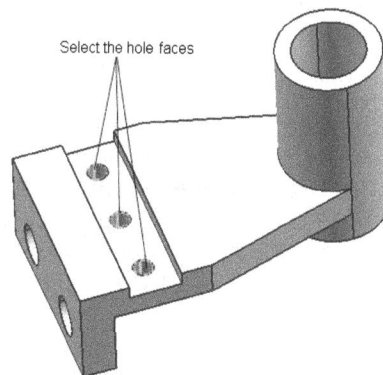

Figure 4-78 Selecting the faces of holes to idealize

Removing the Slot Feature

Next, you need to remove the slot feature of the model.

1. Choose **Menu > Insert > Model Preparation > Defeature** from the **Top Border Bar**; the **Defeature Geometry** dialog box is displayed and you are prompted to select the faces to be defeatured.

2. Select the faces of the slot from the model, refer to Figure 4-79 and choose the **OK** button from the dialog box; the slot is removed from the model. Choose the **Cancel** button to exit the dialog box.

    The model after idealizing and defeaturing is shown in Figure 4-80.

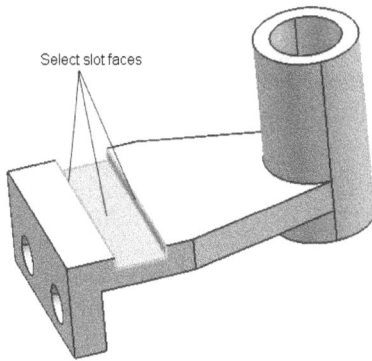

Figure 4-79 *Selecting slot faces to defeature*

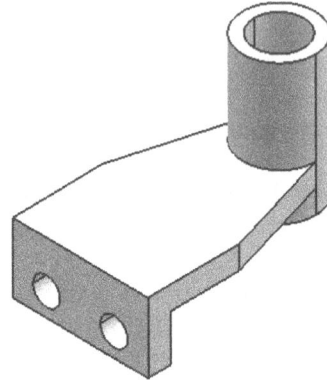

Figure 4-80 *Model after idealizing and defeaturing process*

Splitting Model for Swept Meshing

Now, you need to split the model for swept meshing.

1. Choose **Menu > Insert > Model Preparation > Trim > Split Body** from the **Top Border Bar**; the **Split Body** dialog box is displayed, refer to Figure 4-81.

2. Select the **Extrude** option from the **Tool Option** drop-down list in the **Tool** rollout, refer to Figure 4-81. Select the **Create Mesh Mating Conditions** and **Check for Sweepable Body** check boxes from the **Simulation Settings** rollout; the model turns red indicating that the model is not suitable for swept meshing.

3. Specify the tolerance value **0.0254** in the **Tolerance** edit box available in the **Settings** rollout.

4. Choose the **Select Body (0)** option from the **Target** rollout and select the model from the window. Then, choose the **Select Curve (0)** option from the **Section** sub-rollout displayed in the **Tool** rollout; you are prompted to select the edge to be extruded and the part to be split.

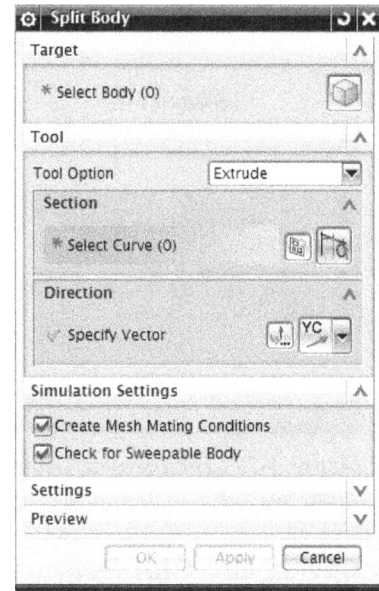

Figure 4-81 *The **Split Body** dialog box*

5. Select the bottom round edge of the model, refer to Figure 4-82; an arrow is displayed. Next, select the **Inferred Vector** option from the **Specify Vector** drop-down list displayed in the **Direction** sub-rollout of the **Tool** rollout; a 3D axis arrow is displayed. Select the Z axis from the 3D axis arrow to define the perpendicular direction of extrusion, if it is not selected by default.

6. Choose the **Apply** button from the dialog box; the model splits and the parts turn yellow and green in color.

    The green part indicates that the model is perfect for swept meshing while the yellow color indicates that the model is still not perfect for swept meshing. So, you need to further split the part.

7. Select the yellow part for further splitting, refer to Figure 4-83.

8. Select the **New Plane** option from the **Tool Option** drop-down list in the **Tool** rollout; you are prompted to select the face to specify the plane.

9. Select the bottom face of the model; a plane is displayed, refer to Figure 4-83.

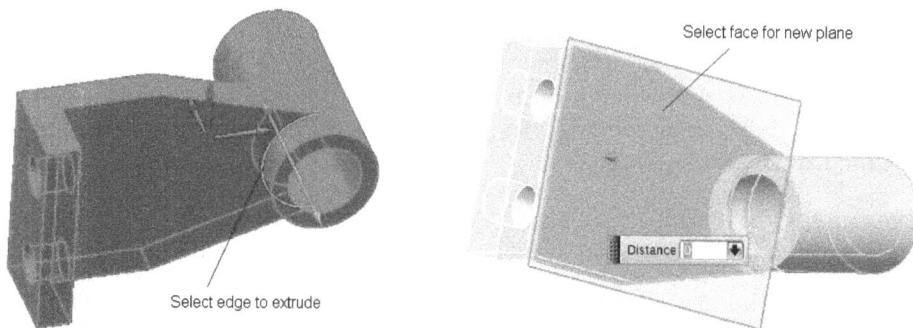

Figure 4-82 Selecting edge to extrude for splitting

Figure 4-83 Selecting face for creating a new plane to split

10. Choose the **Apply** button from the dialog box; the model splits along the plane. The part splits into green and yellow colored parts. Now, you need to split the remaining part.

11. Select the yellow part for further splitting, refer to Figure 4-84.

12. Select the **Extrude** option from the **Tool Option** drop-down list in the **Tool** rollout; you are prompted to select the edge to be extruded and the part to be split.

13. Select the side edge of the model; an arrow is displayed, refer to Figure 4-84. Select the **Inferred Vector** option from the **Specify Vector** drop-down list in the **Direction** sub-rollout of the **Tool** rollout; a 3D axis arrow is displayed, refer to Figure 4-84. Select the X axis arrow to define extrusion along the perpendicular direction.

14. Choose the **Apply** button from the dialog box; the model splits and turns green indicating that the model is ready for swept meshing. Choose the **Cancel** button to exit from the dialog box.

The final idealized model is shown in Figure 4-85.

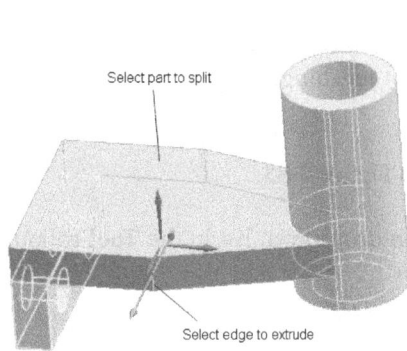

Figure 4-84 *Selecting edge for extruding*   Figure 4-85 *The final idealized model*

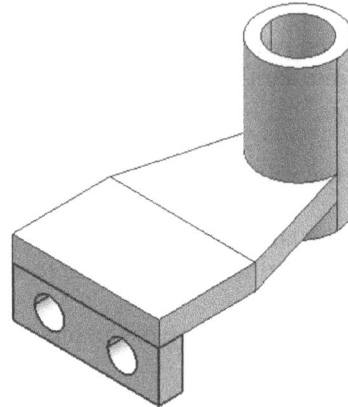

15. Switch to the FEM environment from the Ideal environment; the **Information** window is displayed, as shown in Figure 4-86, showing the description of the idealizing process. Now, close the **Information** window.

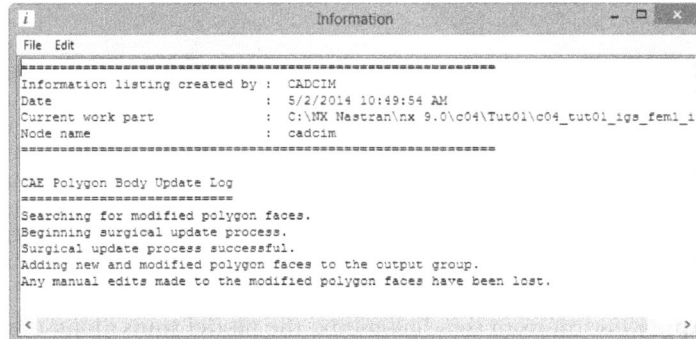

Figure 4-86 *The **Information** window*

Saving and Closing the Model

1. For saving the model, choose **File > Save > Save** from the **Ribbon**; the **Name Parts** window is displayed, as shown in Figure 4-87, showing the default file name of the idealized and FEM files. Choose the **OK** button to save all the files.

2. Choose **File > Close > All Parts** from the **Ribbon** to close all the parts.

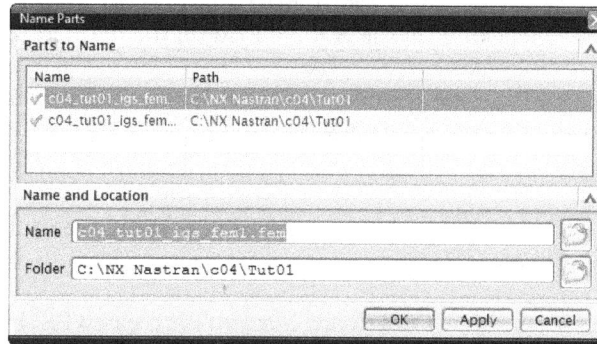

*Figure 4-87 The **Name Parts** dialog box*

Tutorial 2

In this tutorial, you will import an NX CAD file shown in Figure 4-88 into NX Nastran. After importing the part, you will create midsurface and use geometry abstraction tools to modify the midsurface. Figure 4-89 shows the prepared model. **(Expected time: 45 min)**

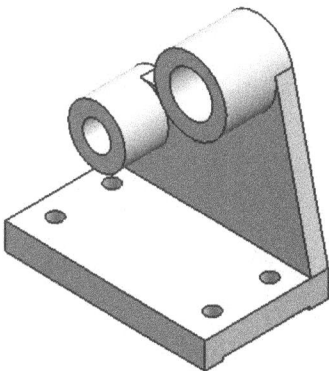

Figure 4-88 The NX CAD model to be imported

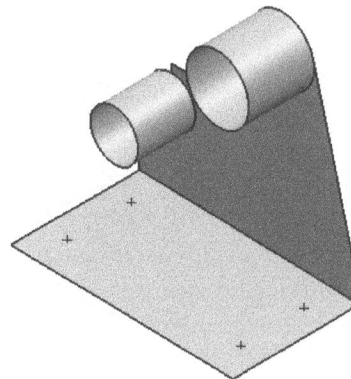

Figure 4-89 Final midsurface model

The following steps are required to complete this tutorial:

a. Start NX Nastran and import the NX CAD file. Next, create FEM and the idealized part files.
b. Create the midsurfaces.
c. Trim the midsurface.
d. Merge the horizontal faces.
e. Split and delete faces.
f. Suppress small holes.

Starting NX, and Creating New FEM and Idealized Files

1. Start NX 9.0 by double-clicking on the NX 9.0 icon on the desktop of your computer

2. Choose the **New** button from the **Home** tab of the **Ribbon**; the **New** dialog box is displayed. Choose the **Simulation** tab and select **NX Nastran** from the **Name** field corresponding to **Fem** in the **Type** field.

3. Enter **c04_tut02** in the **Name** edit box. Specify the location to save the file in the **Folder** edit box as *C:\NX Nastran\c04\Tut02*.

4. Choose the **OK** button; the **New FEM** dialog box with the created FEM file name is displayed on the screen.

5. Select the **Associate to Master Part** check box; other options are also displayed in the dialog box.

6. Choose the **Open Part** button, available next to the **Part** option; the **Open** window is displayed. Select the *c04_tut2* file from the location where you have saved it. Next, choose the **OK** button; the respective file name is displayed in the **Part** edit box, refer to Figure 4-90. Also, the preview of the part is displayed in a separate **CAD Part** window.

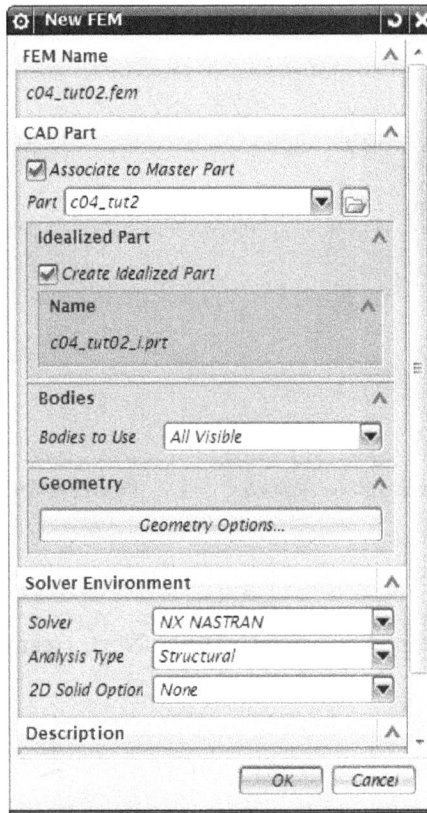

*Figure 4-90 The **New FEM** dialog box*

7. Select the **Create Idealized Part** check box available in the **Idealized Part** sub-rollout; a file with the name *c04_tut02_i* is created, refer to Figure 4-90. Choose the **OK** button from the dialog box; the CAD file is opened in the FEM environment of NX Nastran. Also, the created FEM file and idealized files are displayed in the **Simulation File View** rollout of the **Simulation Navigator**, refer to Figure 4-91.

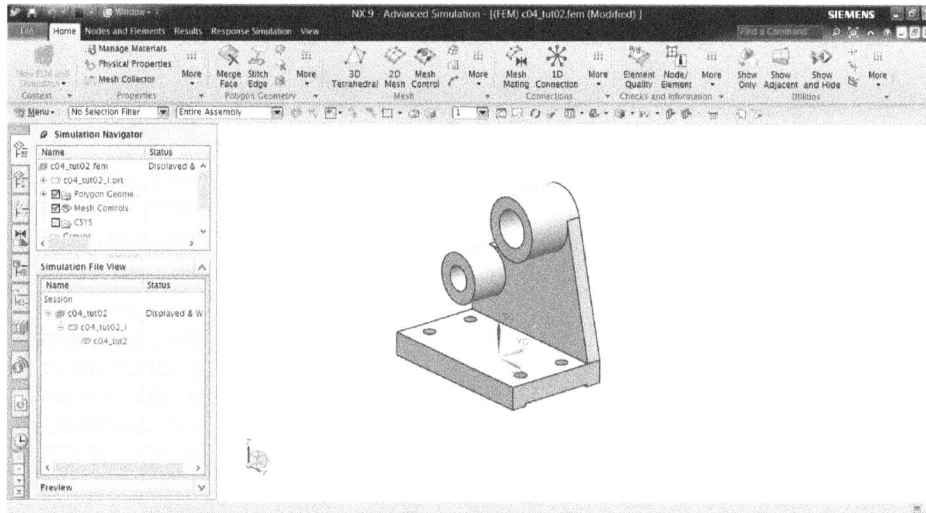

Figure 4-91 *The model opened in the FEM environment*

Promoting Part for Geometry Idealization

1. For geometry idealization, you need to invoke the Ideal environment. In the **Simulation File View** rollout of the **Simulation Navigator**, right-click on the *c04_tut02_i* file name; a shortcut menu is displayed. Choose the **Make Displayed Part** option from the menu; the **Idealized Part Warning** message box is displayed. Choose the **OK** button from the window; the part is opened in the Ideal environment.

2. For promoting the part, choose **Menu > Insert > Associative Copy > Promote** from the **Top Border Bar**; the **Promote Body** dialog box is displayed. Select the part from the Ideal environment and choose the **OK** button; an associative copy of the selected part is created.

Creating Midsurface

1. To create the midsurface, choose **Menu > Insert > Model Preparation > Midsurface > Face Pairs** from the **Top Border Bar**; the **Midsurface by Face Pairs** dialog box is displayed, refer to Figure 4-92. Also, you are prompted to select the model.

2. Select the CAD part from the window; other options and rollouts become available in the dialog box, refer to Figure 4-92.

3. Select the **Manual** option from the **Strategy** drop-down list available in the **Face Pairing** rollout.

4. Clear the **Hide Solid Body Upon Apply** check box from the **Display Options** rollout.

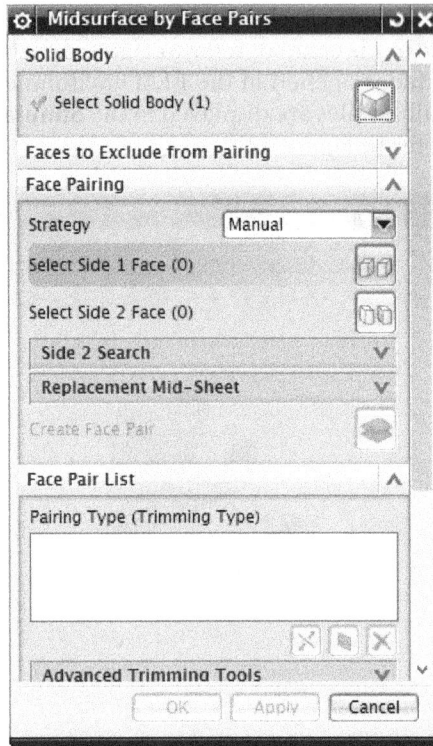

*Figure 4-92 The **Midsurface by Face Pairs** dialog box*

5. Choose the **Select Side 1 Face (0)** option from the **Face Pairing** rollout and select the inner face of the large cylinder. Now, choose the **Select Side 2 Face (0)** option and select outer face of the large cylinder, refer to Figure 4-93.

6. Choose the **Create Face Pair** button from the **Face Pairing** rollout; the preview of the midsurface is displayed. Choose the **Apply** button to create the midsurface.

7. Similarly, select the model and faces of the smaller cylinder to create the midsurface. Figure 4-94 shows the midsurfaces created on both the cylindrical faces.

Note
*For the single surface selection, you may have to choose the **Single Face** option from the **Face Rule** drop-down of the **Selection Group** of the **Top Border Bar**.*

Now, you need to create other midsurface.

8. Select the model. Next, choose the **Select Side 1 Face (0)** option and select the inner face of the vertical wall. Now, choose the **Select Side 2 Face (0)** option and select the outer face of the vertical wall, refer to Figure 4-95.

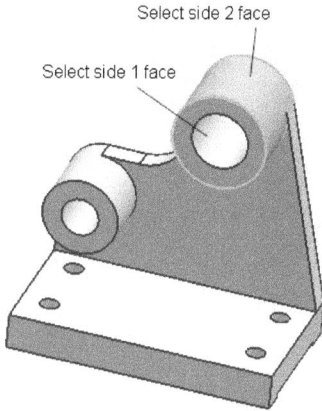

Figure 4-93 *Selecting surfaces for creating midsurface on cylinder*

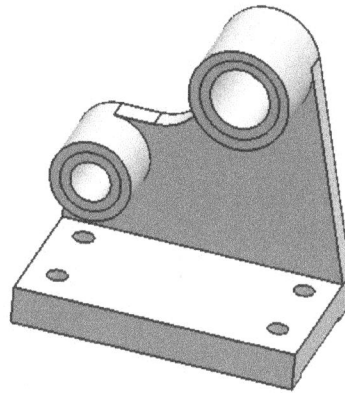

Figure 4-94 *Midsurface created on the cylindrical faces*

9. Choose the **Create Face Pair** button and then the **Apply** button to create the midsurface.

10. Similarly, select the surface of horizontal wall and choose the **Create Face Pair** button, refer to Figure 4-96.

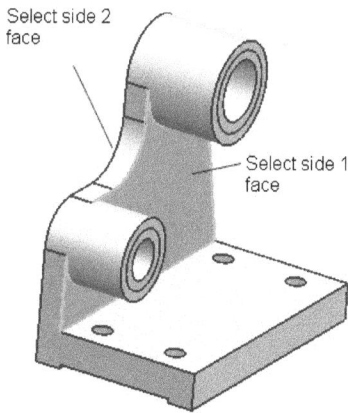

Figure 4-95 *Selection of surfaces for creating midsurface on a wall*

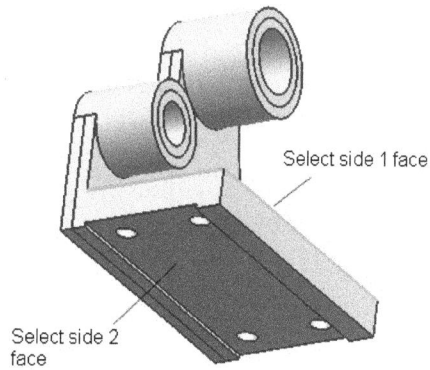

Figure 4-96 *Selection of surfaces for creating midsurface on horizontal wall*

11. Select the **Hide Solid Body Upon Apply** check box and choose the **OK** button; the solid part of the model gets hidden and the created midsurface is displayed, as shown in Figure 4-97.

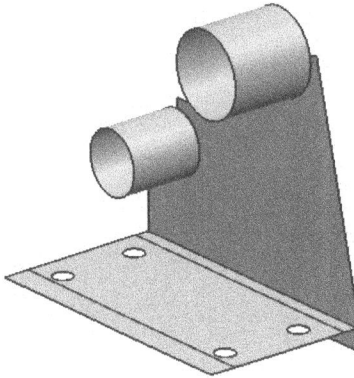

Figure 4-97 *The final midsurface model*

Trimming the Midsurface

Now, you will trim the extended surface which is at the bottom of the midsurface model.

1. Choose **Menu > Insert > Model Preparation > Trim > Trim and Extend** from the **Top Border Bar**; the **Trim and Extend** dialog box is displayed, refer to Figure 4-98.

2. Select the **Until Selected** option from the drop-down list available in the **Type** rollout, refer to Figure 4-98; you are prompted to select the target face.

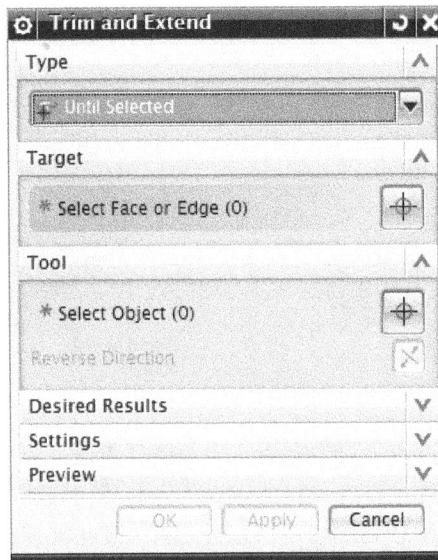

Figure 4-98 *The **Trim and Extend** dialog box*

3. Select the **Tangent Faces** option from the **Face Rule** drop-down list available in the **Selection Group** of the **Top Border Bar**.

4. Select the vertical face as the target face, refer to Figure 4-99. The vertical face is selected in the **Target** rollout.

5. Select the **Select Object (0)** option from the **Tool** rollout. Next, select the horizontal face as the tool face to trim, refer to Figure 4-99; an arrow is displayed to show the direction of trim. Choose the **Reverse Direction** button to flip the direction upward, if required.

6. Choose the **OK** button to trim the midsurface. The final trimmed and extended model is shown in Figure 4-100.

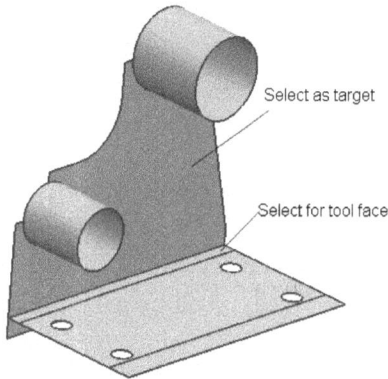

Figure 4-99 *Selection of surfaces for trimming*   *Figure 4-100* *Model after trimming*

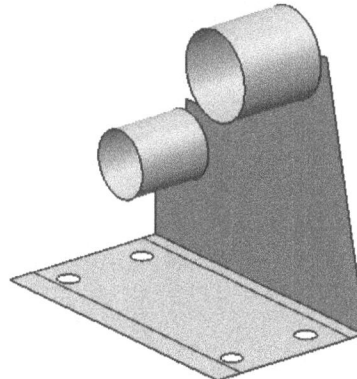

7. Switch to the FEM environment from the Ideal environment using the **Simulation File View** rollout of the **Simulation Navigator**.

8. In the FEM environment, solid body is also displayed along with the midsurface. To hide the solid body, clear the **Polygon Body (1)** check box available in the **Simulation Navigator**, refer to Figure 4-101. Only the midsurface remains in the window.

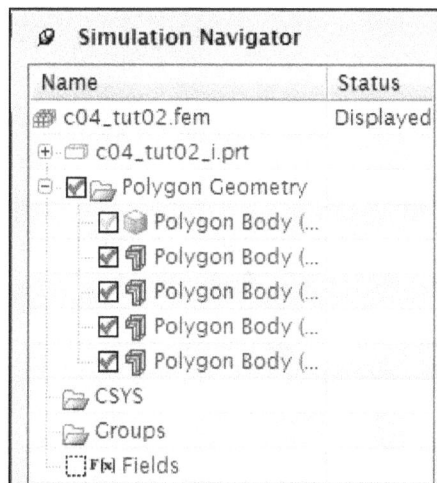

Figure 4-101 *The Simulation Navigator*

Merging Faces

Next, you will merge the horizontal faces.

1. Choose **Menu > Insert > Model Cleanup > Merge Face** from the **Top Border Bar**; the **Face Merge** dialog box is displayed, refer to Figure 4-102. Choose the **Merge** button available in the dialog box; you are prompted to select the edges to merge the face.

Figure 4-102 *The Face Merge dialog box*

2. Select the two edges of the horizontal face, refer to Figure 4-103. Choose the **OK** button to merge the faces. The final merged model is shown in Figure 4-104.

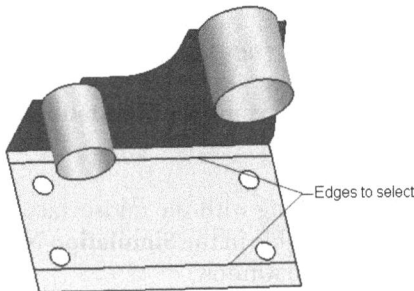

Figure 4-103 *Edges to be selected for merging face*

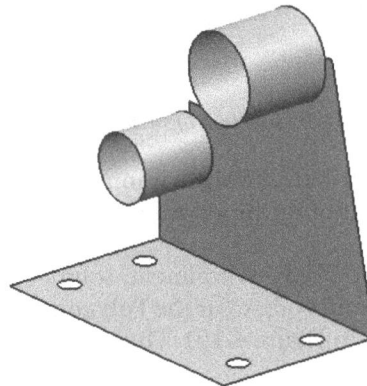

Figure 4-104 *Model after merging faces*

Splitting Faces

Now, you will split the right side face.

1. Choose **Menu > Insert > Model Cleanup > Split Face** from the **Top Border Bar**; the **Split Face** dialog box is displayed, refer to Figure 4-105. Choose the **Split face by points** option from the drop-down list available in the **Type** rollout.

2. Select the two right corner points of the model, refer to Figure 4-106, as the start and end points of the merging line. Choose the **OK** button to split the face. The final split model is shown in Figure 4-107.

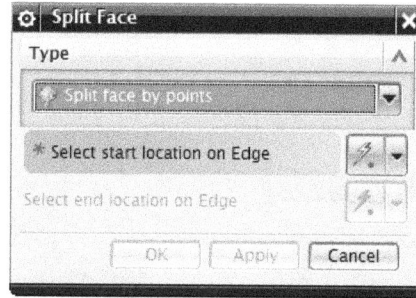

Figure 4-105 *The **Split Face** dialog box*

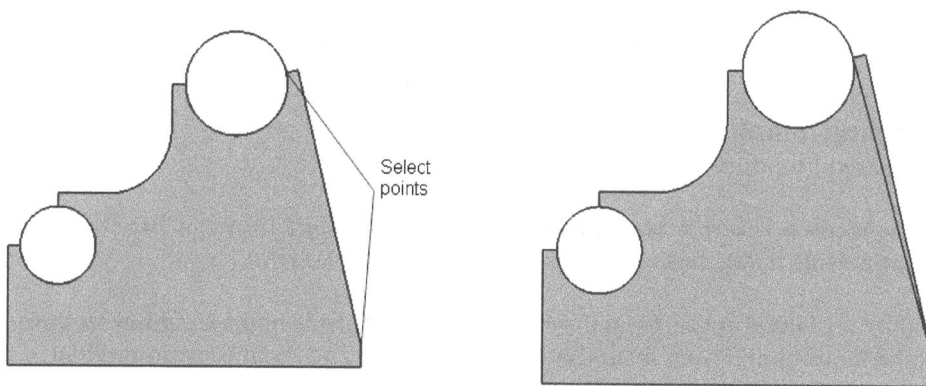

Figure 4-106 *Points to be selected for splitting*

Figure 4-107 *Model face after splitting*

Deleting the Split Face

Next, you will delete the face that was splitted previously.

1. Choose **Menu > Insert > Model Cleanup > Delete Face** from the **Top Border Bar**; the **Delete Polygon Face** dialog box is displayed, refer to Figure 4-108. You are prompted to select the split face.

2. Select the split face and choose the **OK** button; the face is deleted. The final model after deleting the face is shown in Figure 4-109.

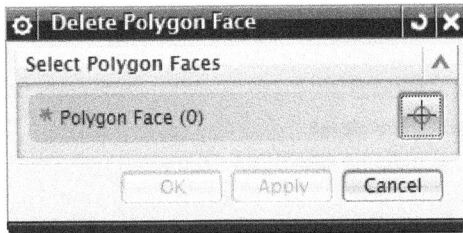

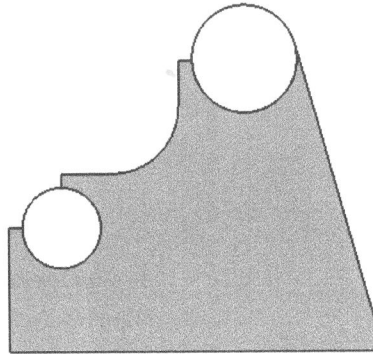

Figure 4-108 The *Delete Ploygon Face*   *Figure 4-109* *Model after deleting the splitted face*
dialog box

Suppressing Small Holes

Now, you will suppress small holes shown in the model.

1. Choose **Menu > Insert > Model Cleanup > Suppress Hole** from the **Top Border Bar**; the **Suppress Hole** dialog box is displayed, refer to Figure 4-110.

2. Choose the **Manual** option from the drop-down list available in the **Creation Method** rollout and the **Point** option from drop-down list available in the **Point Creation** rollout.

3. Select edges of all small holes, refer to Figure 4-111, and choose the **OK** button from the dialog box; all selected holes are suppressed.

The final model after completing all the operation is shown in Figure 4-112.

Figure 4-110 The *Suppress Hole* dialog box

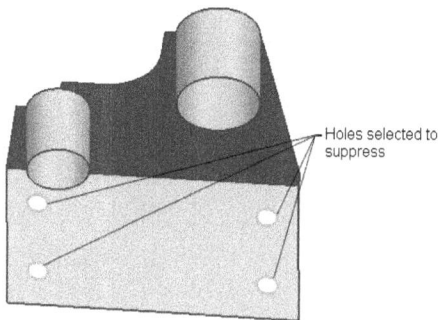

Figure 4-111 *Holes selected to suppress*

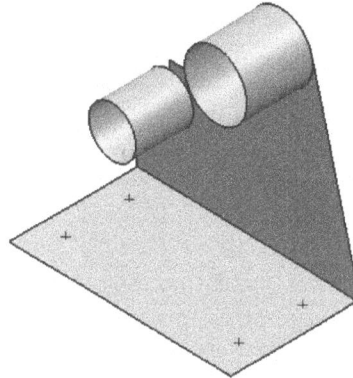

Figure 4-112 *Final midsurface model*

Saving and Closing the Model

1. Choose **File > Save > Save** from the **Ribbon** to save the model.

2. Choose **File > Close > All Parts** from the **Ribbon** to close the file.

Self-Evaluation Test

Answer the following questions and then compare them to those given at the end of this chapter:

1. Before geometry idealization, you need to _____ the model for creating an associative copy of the model.

2. Using the _____ tool, you can split an edge.

3. While splitting a model, the _____ color of the model indicates that it is not ready for swept meshing.

4. The _____ tool is used to split a face into two parts.

5. You can use the _____ tool to delete a face.

6. You can merge two edges into a single entity by using the _____ tool.

7. Geometry idealization is a part of model preparation process. (T/F)

8. You can use the **Defeature** tool to remove a feature from a geometry. (T/F)

9. The **Split Body** tool is used to split a body for swept meshing. (T/F)

10. The **Create Mesh Mating Conditions** check box is used to make a connection between two parts while doing swept meshing. (T/F)

Review Questions

Answer the following questions:

1. Which of the following colors is not temporarily present while splitting a body?

    (a) Red   (b) Yellow
    (c) Green   (d) Blue

2. Which of the following tools is used to merge a face?

    (a) **Merge Face**   (b) **Reset**
    (c) **Face from Boundary**   (d) None of these

3. Which of the following tools is used to control the distribution of mesh/elements around holes in the model?

    (a) **Face from Boundary**   (b) **Delete Face**
    (c) **Circular Imprint**   (d) None of these

4. Which of the following tools is used to remove holes from a model?

    (a) **Suppress Hole**   (b) **Face Pair**
    (c) **Split Body**   (d) None of these

5. Which of the following tools is used to collapse an edge?

    (a) **Collapse Edge**   (b) **Suppress Hole**
    (c) **Circular Imprint**   (d) None of these

6. The **Check for Sweepable Body** check box is used to check whether the parts are split correctly for swept meshing or not. (T/F)

7. Using the **Sew** tool, you can sew the free edges of any model. (T/F)

8. The **Reset** tool is used to restore abstracted polygon geometry to its original state. (T/F)

9. You cannot use the **Stitch Edge** tool to join two separate edges into a single edge. (T/F)

10. You can use the **Face Pairs** tool to create a midsurface on a part by selecting the faces/surfaces. (T/F)

EXERCISES

To perform the exercises, you need to extract the *c04_NX_Nastran_input .zip* file downloaded at the beginning of the tutorials.

Exercise 1

In this exercise, you will import the *c04_exr01.prt* file into NX Nastran from the *c04_Nastran_input* folder that you have extracted. After importing the part, you will idealize the part and create midsurface. Figure 4-113 shows the imported model and Figure 4-114 shows the created midsurface. **(Expected time: 45 min)**

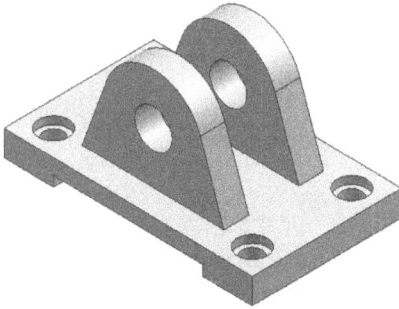

Figure 4-113 Model for Exercise 1

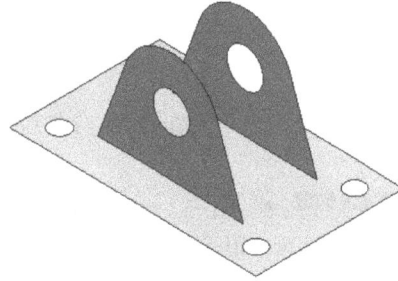

Figure 4-114 Final midsurface model

Hint

1. Import the NX CAD model. Create the FEM and idealized part files.
2. Invoke the Ideal environment and promote the part for idealizing.
3. Remove counterbore hole feature using the **Defeature Geometry** tool. Select the faces, refer to Figure 4-115.
4. Create a midsurface of the part. Choose the **Manual** strategy. Select the faces of the horizontal part, refer to Figure 4-116.
5. Similarly, create the midsurface of the vertical part of the model.
6. Switch to FEM environment and choose the **Merge Face** tool to merge horizontal midsurface. Select the edges shown in Figure 4-117 to merge face.
7. Save and close the model.

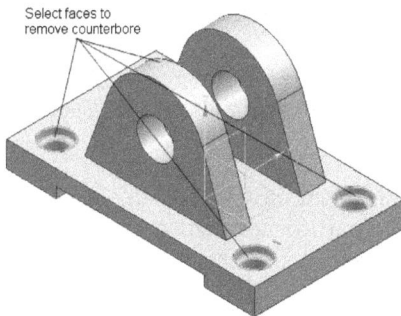

Figure 4-115 Faces selected for removing counterbore

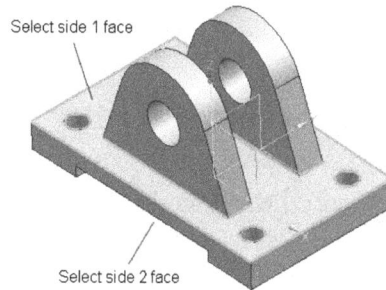

Figure 4-116 Faces selected for creating midsurface

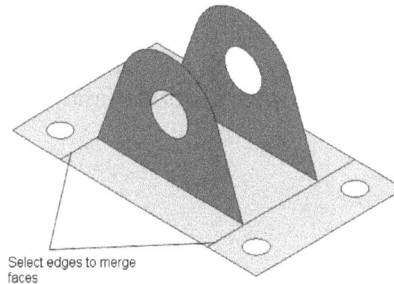

Figure 4-117 *Edges to be selected to merge faces*

Exercise 2

In this exercise, you will import the *c04_exr02.igs* file into NX Nastran from the *c04_Nastran_input* folder that you have extracted. After importing the part, you will idealize the part and split it for swept meshing. Figure 4-118 shows the imported model and Figure 4-119 shows the prepared model for swept meshing.   **(Expected time: 45 min)**

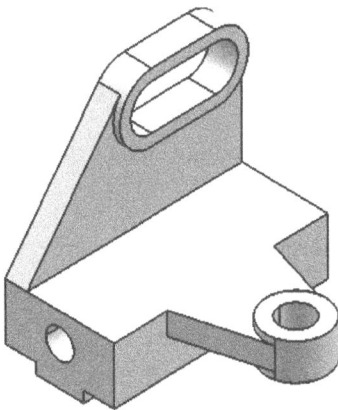

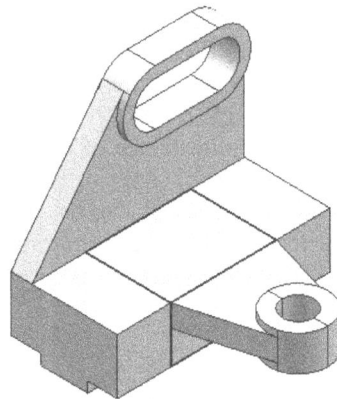

Figure 4-118 *Model for Exercise 2*   *Figure 4-119* *Final split model*

Hint
1. Import the IGS model. Create the FEM and idealized part files.
2. Invoke the Ideal environment and promote the part for idealizing.
3. Remove hole feature using the **Defeature Geometry** tool. Select the hole face, refer to Figure 4-120.
4. Split the body for creating swept meshing. Choose the **Extrude** tool from the **Split Body** dialog box. Select the round edges and set the arrow direction, refer to Figure 4-121.

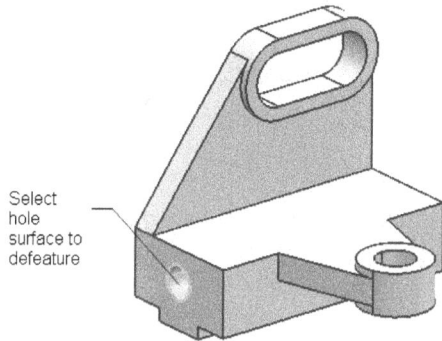

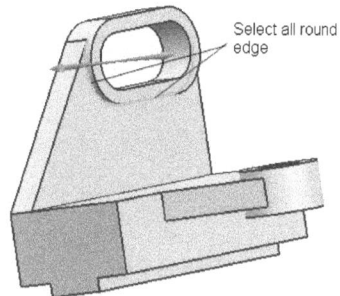

Figure 4-120 *Hole face to be selected for defeaturing*

Figure 4-121 *Round edges selected for splitting*

5. Choose the **Extrude** tool. Select the round edge and set the arrow direction, refer to Figure 4-122.
6. Choose the **New Plane** tool. Select the horizontal face to split, refer to Figure 4-123.

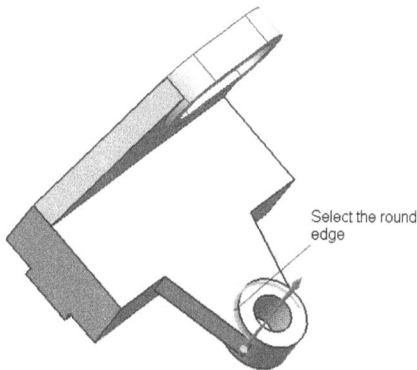

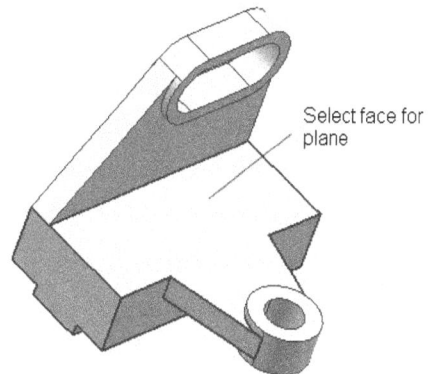

Figure 4-122 *Round edge to be selected for splitting*

Figure 4-123 *Face to be selected for splitting*

7. Choose the **New Plane** tool. Select the front face to split, refer to Figure 4-124.
8. Choose the **Extrude** tool. Select the front face, refer to Figure 4-125 and create the sketch, refer to Figure 4-126.

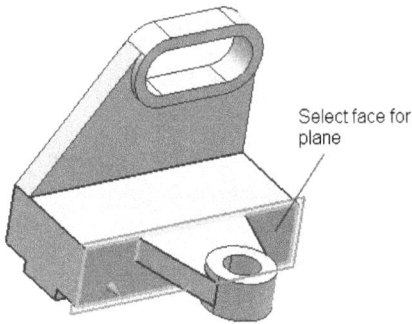

Figure 4-124 *Face to be selected for splitting*

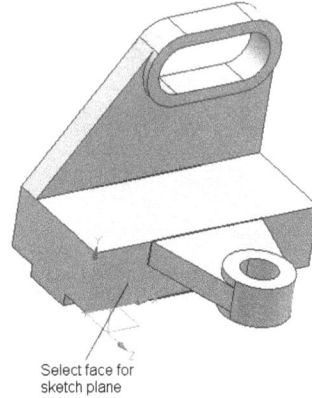

Figure 4-125 *Faceto be selected for creating a sketch plane*

9. Choose the **Extrude** tool. Select the inner edge of the hole and set the direction of arrow, refer to Figure 4-127.

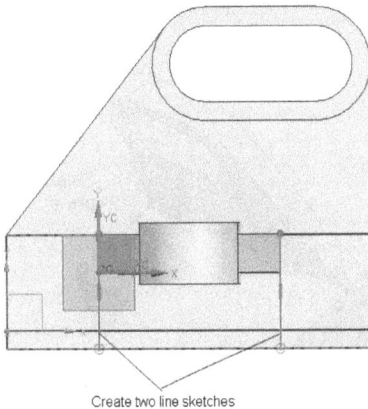

Figure 4-126 *Sketch selected for splitting*

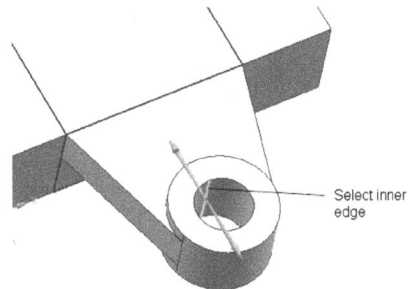

Figure 4-127 *Inner edge selected for splitting*

10. Switch to the FEM environment. Save and close the model.

Answers to Self-Evaluation Test

1. promote, **2. Split Edge**, **3.** red, **4. Split Face**, **5. Delete Face**, **6. Merge Edge**, **7.** T, **8.** T, **9.** T, **10.** T

Chapter 5

Meshing-I

Learning Objectives
After completing this chapter, you will be able to:
- *Create nodes*
- *Create 0D, 1D, 2D, 3D elements*
- *Create elements*
- *Create Mesh Collector for 0D, 1D, 2D, 3D elements*
- *Create 0D mesh*
- *Create 1D mesh*
- *Create 2D mesh and its types*
- *Create 3D mesh and its types*

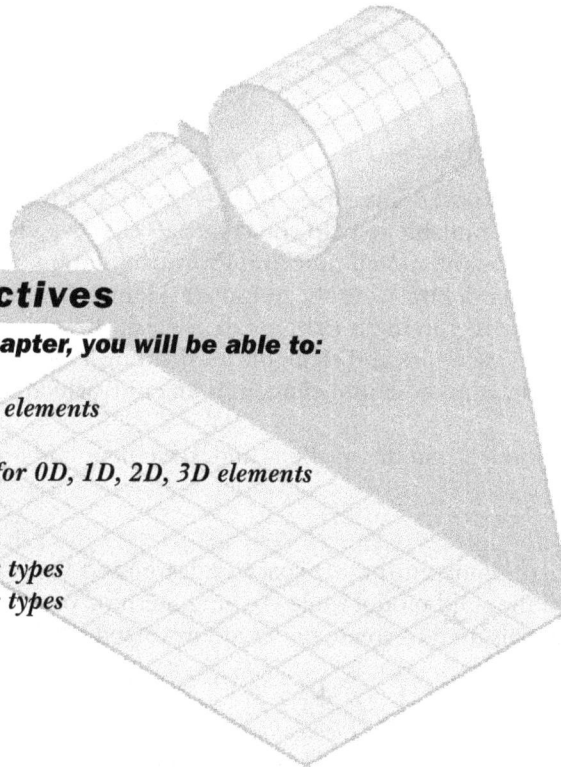

INTRODUCTION

Meshing is the process of discretization of a component into a number of small elements of defined size and the component thus resulted is called a mesh. The elements in a mesh are connected to each other at points called nodes. Each node may have two or more elements connected to it. Figure 5-1 shows a solid model and Figure 5-2 shows mesh created on the solid model.

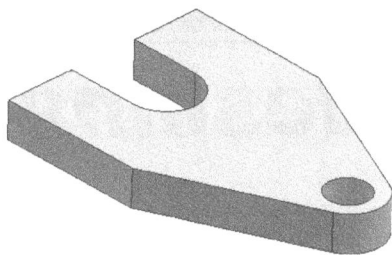

Figure 5-1 A solid model

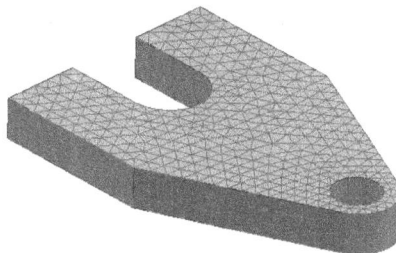

Figure 5-2 Mesh created on the solid model

Meshing is an important part of pre-processing in any FEA software. In NX Nastran, there are many tools and options available to help you create an effective mesh. An effective mesh is the one that requires less computational time and gives maximum accuracy. In NX Nastran, you can create mesh on a model by using the automatic element size calculation. You can also set parameters as per your requirement to generate a mesh. Besides directly creating mesh on a model, you can also create nodes and elements for dividing a model. Nodes define and connect elements in a finite element mesh and element is an entity which comprises nodes.

In this chapter you will learn in detail about nodes, element and meshing.

NODES

Nodes are coordinate points in the 3D space that are used to create elements. The software generates nodes on the solid model while creating mesh or you can also create nodes directly on the model. The tools used to create nodes are discussed next.

Note
Nodes are created in the Finite Element Model. So to create them in the model, you must be in the FEM environment as discussed in the previous chapters.

Create

Ribbon:	Nodes and Elements > Nodes > Node Create
Menu:	Insert > Node > Create

You can use this tool to create nodes. To do so, choose **Menu > Insert > Node > Create** from the **Top Border Bar**; the **Node Create** dialog box will be displayed, as shown in Figure 5-3. The options and rollouts available in the dialog box are discussed next.

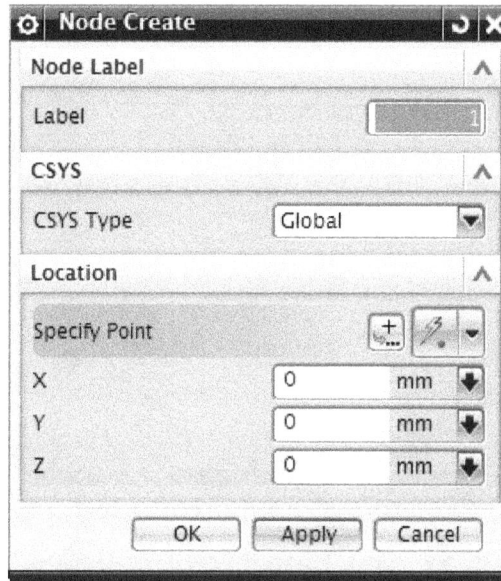

Figure 5-3 *The **Node Create** dialog box*

Node Label

In this rollout, you can assign a label to the node. If you assign a label that already exist then a warning message box will be displayed and software will provide another label to that node.

CSYS

In this rollout, the **CSYS Type** drop-down list is available. There are four options available in this drop-down list: **Global**, **Cartesian**, **Cylindrical**, and **Spherical**. You can create nodes using these options.

Location

In this rollout, options for specifying the position of nodes are available. The options in this rollout change depending upon the type of CSYS chosen from the **CSYS Type** drop-down list in the **CSYS** rollout. You can create the node by selecting any point, edge, or free point in the graphics window. You can also create nodes by specifying coordinate values in their respective edit boxes in this rollout.

Between Nodes

Ribbon:	Nodes and Elements > Nodes > Between Nodes
Menu:	Insert > Node > Between Nodes

Using this tool, you can create a node between two nodes. To create a node between two nodes, choose **Menu > Insert > Node > Between Nodes** from the **Top Border Bar**; the **Node Between Nodes** dialog box will be displayed, as shown in Figure 5-4. For creating nodes, there are two options available in the drop-down list of the **Type** rollout: **Between Two Nodes** and **At Center of Three Nodes**. The methods to create a node using these options are discussed next.

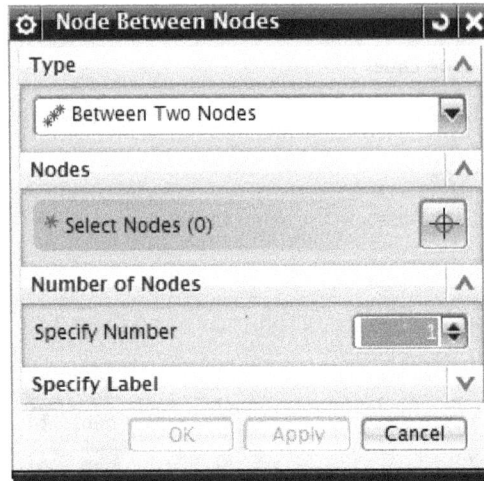

*Figure 5-4 The **Node Between Nodes** dialog box*

Between Two Nodes

When you select this option from the drop-down list in the **Type** rollout, three rollouts will be displayed: **Nodes**, **Number of Nodes**, and **Specify Label**. Also, the **Select Nodes (0)** option in the **Nodes** rollout will be highlighted and you will be required to select two nodes between which you want to create node. You can specify the number of nodes to be created in the **Specify Number** edit box in the **Number of Nodes** rollout. You can also specify the labels to be assigned to each node in the **Label** edit box and the increment value between the node label in the **Increment** edit box of the **Specify Label** rollout. After specifying all options choose the **OK** button to create nodes.

At Center of Three Nodes

When you select this option from the drop-down list in the **Type** rollout, two rollouts will be displayed: **Nodes** and **Specify Label**. Also, the **Select Nodes (0)** option in the **Nodes** rollout will be highlighted and you are prompted to select three nodes between which you want to create node. You can also assign labels to the nodes in the **Label** edit box of the **Specify Label** rollout. Choose the **OK** button to create nodes after specifying all the options.

On Point/Curve/Edge

Ribbon: Nodes and Elements > Nodes > More > Create > On Curve/Edge
Menu: Insert > Node > On Point/Curve/Edge

Using this tool, you can create a node along the point, curve, or edge. To create a node along any geometry, choose **Menu > Insert > Node > On Point/Curve/Edge** from the **Top Border Bar**; the **Node on Point/Curve/Edge** dialog box will be displayed, as shown in Figure 5-5. The options and rollouts available in this dialog box are discussed next.

Point/Edge/Curve

In this rollout, the **Select Point/Edge/Curve (0)** option is highlighted and you are prompted to select the geometry along which you want to create nodes. When you select any curve or edge, the length of the selected geometry is also be displayed in this rollout.

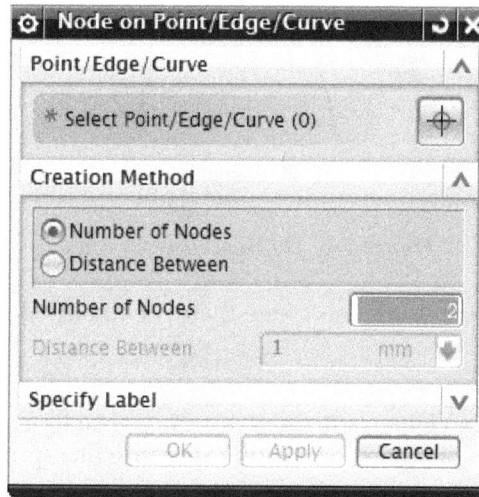

*Figure 5-5 The **Node on Point/Edge/Curve** dialog box*

Creation Method

In this rollout, two radio buttons are available: **Number of Nodes** and **Distance Between**. When you select the **Number of Nodes** radio button, the **Number of Nodes** edit box will be enabled in this rollout. In this edit box, you can specify the number of nodes to be created on the selected geometry. The nodes are created at equal distance according to the length of the selected geometry.

When you select the **Distance Between** radio button, the **Distance Between** edit box appears. Using this edit box, you can specify the distance between two nodes to be created on the selected geometry.

Specify Label

In this rollout, you can specify the labels to each node by using the **Label** edit box. Also, you can specify the increment between the node label using the **Increment** edit box in this rollout.

Choose the **OK** button to create nodes after specifying all the options.

ELEMENTS

An element is created using nodes. Element is an entity into which the part under study is divided. The shape (length, area, and volume) of an element depends upon the nodes used in making the element. Figure 5-6 shows the three different element types.

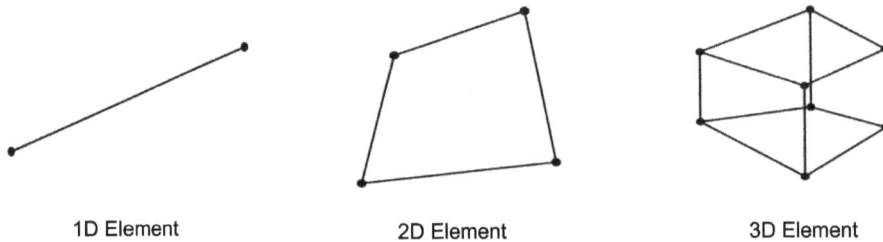

1D Element   2D Element   3D Element

Figure 5-6 *Different elements types*

The elements types and the tools used to create them are discussed next.

0D Elements

0D elements also called scalar elements are used to create concentrated masses on a point. These element connect two points of a part for creating structure like springs and dampers. Various 0D elements and their uses in analysis are as follows:

Element	Analysis Type	Use in Analysis
CBUSH	Structural Analysis	Used to define spring connect from a node to ground in linear or non-linear analysis
CBUSH1D	Structural Analysis	Used to define spring and viscous damper connection
CDAMP1, CDAMP2	Structural Analysis	Used to define damper connect from a node to ground
CELAS1, CELAS2	Structural Analysis	Used to define a scalar spring element
CMASS1, CMASS2	Structural Analysis	Used to define a scalar mass element
CONM1, CONM2	Structural Analysis	Used to define concentrated mass element

1D Elements

1D elements are also called line elements and are used to create elements for rod or beam type behavior. The different type of 1D elements and their uses in analysis are as follows:

Element	Analysis Type	Use in Analysis
CBAR	Linear Structural Analysis Thermal Analysis	Used for beam structure
CBEAM	Linear and Nonlinear Structural Analysis Thermal Analysis	Used for beam structure

Element	Analysis Type	Use in Analysis
CROD, CONROD	Linear and Nonlinear Structural Analysis Thermal Analysis	Used for rod structure in tension, compression or torsion analysis
RBE2	Linear and Nonlinear Structural Analysis Thermal Analysis	Used for rigid body link
CBUSH, CBUSH1D	Linear and Nonlinear Structural Analysis Thermal Analysis	Used for spring and damper type connection
CELAS1, CELAS2	Linear Structural Analysis Thermal Analysis	Used for defining scalar spring element
CMASS1, CMASS2	Linear Structural Analysis Thermal Analysis	Used for defining scalar mass element
CDAMP1, CDAMP2	Linear Structural Analysis Thermal Analysis	Used for defining a scalar damper element
CTUBE	Linear and Nonlinear Structural Analysis Thermal Analysis	Used for tube structure in tension, compression or torsion problem
PLOTEL	Structural Analysis	Used for defining one-dimensional dummy element
RBAR	Linear and Nonlinear Structural Analysis Thermal Analysis	Used for defining rigid bar element
RROD	Linear and Nonlinear Structural Analysis Thermal Analysis	Used for defining rigid pin-ended element
CGAP	Linear, Nonlinear Structural and Axisymmetric Structural Analysis	Used for defining gap or friction element
CVISC	Linear Structural Analysis	Used for defining a viscous damper element

2D Elements

2D elements are also called surface elements and are used to create elements for structure whose thickness is smaller such as sheets or shells. Various 2D elements and their uses in analysis are as follows:

Elements	Analysis Type	Use in Analysis
CQUAD4, CQUAD8	Linear and Nonlinear Structural Analysis Thermal Analysis	Used to create quadrilateral element in thin shell structure
CQUADR	Linear Structural Analysis	Used to create quadrilateral element in thin shell structure
CTRIA3, CTRIA6	Linear and Nonlinear Structural Analysis Thermal Analysis	Used to create triangular element in thin shell structure
CTRIAR	Linear Structural Analysis	Used to create triangular element in thin shell structure
CSHEAR	Linear Structural Analysis	Used to create non-transition quadrilateral element in shear panel structure

3D Elements

3D elements are also called solid elements and are used to create elements for solid structure. Various 3D elements and their uses in analysis are as follows.

Element	Analysis Type	Use in Analysis
CTETRA(4), CTETRA(10)	Linear and Nonlinear Structural Analysis Thermal Analysis	Used to create four-sided elements in solid structure with four to ten grid points
CHEXA(8), CHEXA(10)	Linear and Nonlinear Structural Analysis Thermal Analysis	Used to create six-sided elements in solid structure with eight to twenty grid points
CPENTA(6), CPENTA(15)	Linear and Nonlinear Structural Analysis Thermal Analysis	Used to create five-sided elements in solid structure with six to fifteen grid points

Create

Ribbon: Nodes and Elements > Elements > Element Create
Menu: Insert > Element > Create

You can create an element by using existing nodes or points in the environment. To do so, choose **Menu > Insert > Element > Create** from the **Top Border Bar**; the **Element Create** dialog box will be displayed, as shown in Figure 5-7. The options and rollouts available in the dialog box are discussed next.

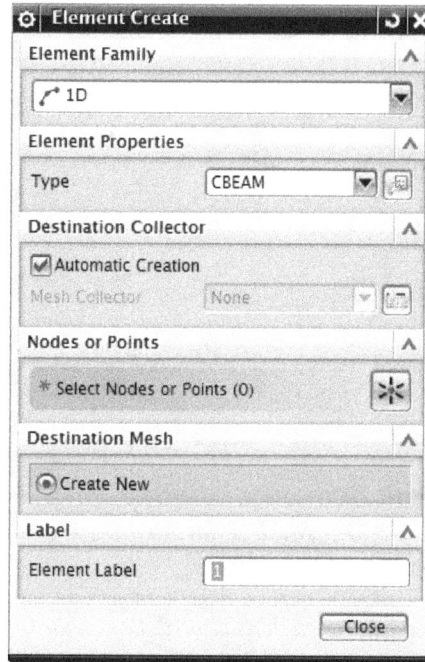

*Figure 5-7 The **Element Create** dialog box*

Element Family

In this rollout, you can specify the element family. Four types of elements are available in the drop-down list available in this rollout: **0D**, **1D**, **2D**, and **3D**. You can choose the desired element type from this drop-down list to create that type of elements. The options related to the selected element type will be displayed in other rollouts.

Element Properties

After choosing the element type from the drop-down list available in the **Element Family** rollout, the related elements will be displayed in the **Type** drop-down list of the **Element Properties** rollout. In this rollout, the **Edit Mesh Associated Data** button is also available. When you choose this button, the **Mesh Associated Data** dialog box will be displayed, refer to Figure 5-8. Using this dialog box, you can provide additional information related to the elements. After defining data, choose the **OK** button to return to the **Element Create** dialog box.

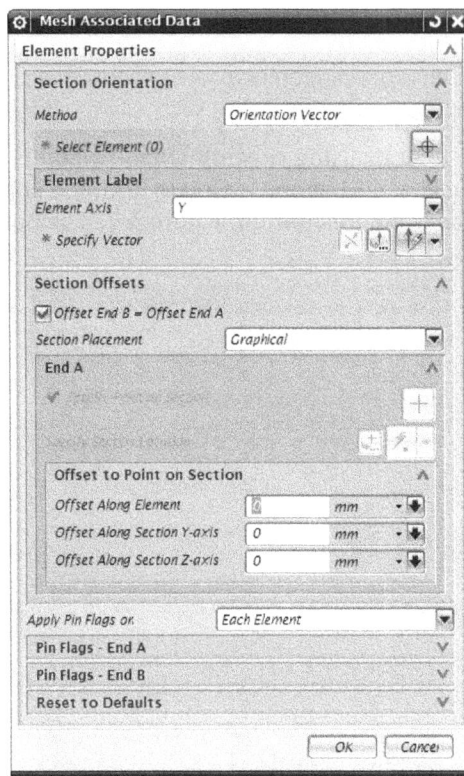

*Figure 5-8 The **Mesh Associated Data** dialog box*

Note
You will learn more about editing mesh associated data in the later chapters of this book.

Destination Collector

The **Automatic Creation** check box is available in the **Destination Collector** rollout. When you select this check box, the element is created at the default location. In this rollout, the **New Collector** button is also displayed. When you choose this button, the **Mesh Collector** dialog box will be displayed, refer to Figure 5-9. Using this dialog box, you can define physical properties of the elements. After defining properties, choose the **OK** button from the dialog box; mesh collector name will be displayed in the **Mesh Collector** drop-down list of the **Destination Collector** rollout of the **Element Create** dialog box.

Note
*You will learn more about **Mesh Collector** later in this chapter.*

Nodes or Points

In this rollout, the **Select Nodes or Points (0)** option is highlighted and you are prompted to select the node or point to create element. Select the node or point, the element will be created at defined destination. Note that, for creating an element, you have to select node as per the element type chosen in the drop-down list. For example, CQUAD4 element needs four nodes and CQUAD8 element needs eight nodes to create the element.

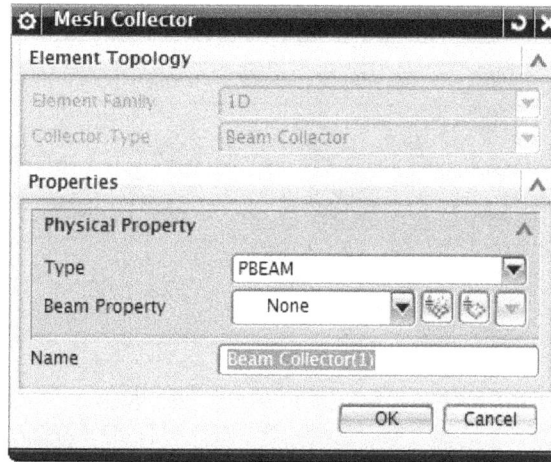

Figure 5-9 *The **Mesh Collector** dialog box*

Destination Mesh

This rollout is used to provide destination of mesh in the **Simulation Navigator**. By default, the **Create New** radio button is available in this rollout. Select this radio button to place the created element in the new node in the **Simulation Navigator**. If you clear the **Automatic Creation** check box from the **Destination Collector** rollout, another radio button named **Add to Existing** is also displayed. When you select this radio button, the **Mesh Name** drop-down will be displayed. You can select the previous destination from this box.

Note
*You will be able to clear the **Automatic Creation** check box only if you have created any mesh collector or any element.*

Label

In this rollout, the **Element Label** edit box is available. It is used to provide number label to the element. If two nodes have similar label, a warning will be displayed and you will be provided another label for the element.

After specifying all the options, choose the nodes; the relevant element will be created.

Extrude

Ribbon:	Nodes and Elements > Elements > Extrude
Menu:	Insert > Element > Extrude

Using this tool, you can extrude an element upto a defined distance. To extrude an element, choose **Menu > Insert > Element > Extrude** from the **Top Border Bar**; the **Element Extrude** dialog box will be displayed, as shown in Figure 5-10. The options and rollouts available in the dialog box are discussed next.

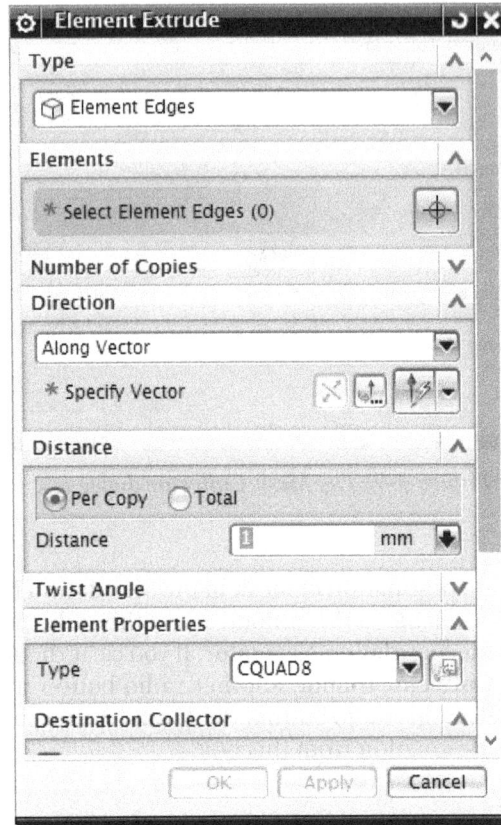

Figure 5-10 *The **Element Extrude** dialog box*

Type

In the drop-down list of the **Type** rollout, two options are available: **Element Edges** and **Element Faces**. The **Element Edges** option is used to extrude the edges to create new 2D elements. The **Element Faces** option is used to extrude the faces to create new 3D elements.

Elements

The options in this rollout are used to select the edges or faces to be extruded. If you select the **Element Edges** option from the drop-down list of the **Type** rollout, the **Select Element Edges (0)** option will be highlighted and you will be prompted to select the edges to extrude.

Number of Copies

In this rollout, the **Number of Copies** edit box is available. In this edit box, you can specify the numbers to be copied while extruding elements.

Direction

In this rollout, you can specify the direction of the element extrusion. Four options are displayed in the drop-down list available in this rollout: **Along Vector**, **Along Path**, **Project to Surface**, and **Element Normal**.

Along Vector

Using this option, you can extrude the elements along a vector. When you choose this option, the **Distance** and **Twist Angle** rollouts are displayed. The **Distance** rollout is used to specify the options of the distance of extrusion. Using the radio buttons available in this rollout, you can specify the extrusion distance as **Per Copy** or **Total**. The options in the **Twist Angle** are used to twist the element along specified vector.

Along Path

Using this option, you can extrude the elements along a specified path. The path may be any geometric entity of the model.

Project to Surface

Using this option, you can project the elements along a specified vector on a target surface. Also, you can specify the distance percentage for projection using the **% Offset to Surface** edit box. For example, if you specify 50% value, the projection will happen between the surface and the element.

Element Normal

Using this option, you can extrude the elements along their normals. This option will be available only if the **Element Faces** option is chosen from the drop-down list in the **Type** rollout.

Label

In this rollout, two edit boxes are available: **Element Label** and **Increment**. Using the **Element Label** edit box, you can provide label numbers to the element. You can also specify the increment in label while creating or copying the elements by using the **Increment** edit box.

Preview

The **Show Result** button in this rollout is used to preview the extruded element. If you select the **Preview** check box, you can view the extruding element.

The options in the rest of the rollouts are same as as discussed in the **Element Create** dialog box. After specifying all the settings, choose the **OK** button to create extruded element from the selected element.

Revolve

Ribbon:	Nodes and Elements > Elements > Revolve
Menu:	Insert > Element > Revolve

By using this tool, you can revolve the created element around a specified vector. To revolve an element, choose **Menu > Insert > Element > Revolve** from the **Top Border Bar**; the **Element Revolve** dialog box will be displayed, as shown in Figure 5-11. Most of the rollouts in this dialog box have options same as those in the **Element Extrude** dialog box and have been discussed earlier. Remaining options and rollouts available in the dialog box are discussed next.

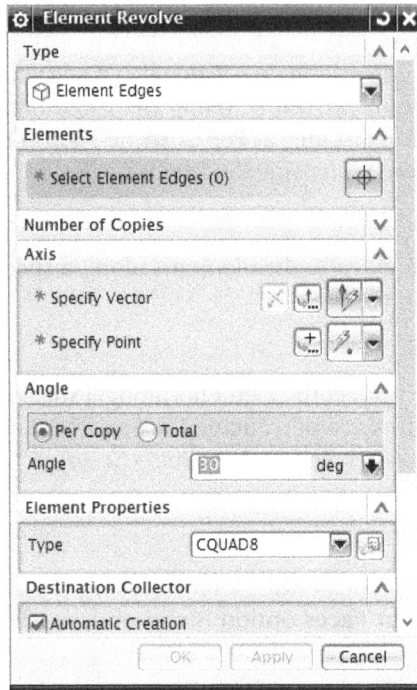

Figure 5-11 The **Element Revolve** *dialog box*

Axis

Using the **Specify Vector** option from the **Axis** rollout, you can specify a vector axis around which the elements need to be revolved. The **Specify Point** option is used to select the revolving point.

Angle

In this rollout, you can specify the angle value in the **Angle** edit box. This is the angle by which the element will be revolved. The **Per Copy** radio button is use to specify the angle between each copy while the **Total** radio button is use to specify the total angle.

After specifying all the options, choose the **OK** button to create revolved element from the selected elements.

3D Sweep Between

Ribbon:	Nodes and Elements > Elements > More > Create > 3D Sweep
Menu:	Insert > Element > 3D Sweep Between

By using this tool, you can create elements by sweeping the existing elements upto the surface. To create elements by sweeping, choose **3D Sweep Between** tool from **Menu > Insert > Element** in the **Top Border Bar**; the **3D Sweep Between** dialog box will be displayed, as shown in Figure 5-12. In the drop-down list of the **Type** rollout, two options are available for creating sweep elements: **Automatic** and **Manual**. These options are discussed next.

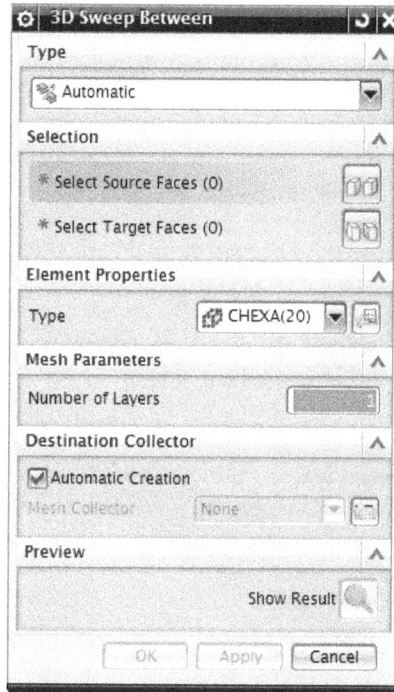

Figure 5-12 The **3D Sweep Between** *dialog box*

Automatic

Use this option to automatically select the edges/loop between the source and target faces. This is applied only for parallel and closed positioned source and target faces. When you select this option, the **Select Source Faces (0)** option gets highlighted in the **Selection** rollout and you are prompted to select the source face of the element. After selecting the source face, choose the **Select Target Faces (0)** option and select the target face upto which the elements need to be swept. Note that the source face must be meshed while the target face can be either meshed or unmeshed.

Manual

Use this option to manually select the edges/loop between the source and target faces. When you select this option, the **Select Source Face (0)** option gets highlighted in the **Source and Target Selection** rollout and you are prompted to select the source face of element. After selecting the source face, choose the **Select Target Face (0)** option and select the target face upto which the elements need to be swept. Note that the source face must be meshed while the target face can be either meshed or unmeshed.

After selecting the source and target faces, select the source edge and target edge for creating loop and direction for sweeping elements by using the options available in the **Match Loops** sub-rollout of the **Source and Target Selection** rollout. You can also flip the direction of sweep corresponding to the relevant edge. The created loop with its name will be displayed in the **List** rollout area. Note that for parallel faces, the loop will be selected automatically.

After selecting the faces and edges, you need to select the **Select Wall Face (0)** option from the **Wall Selection** rollout to create sweep between faces and edges. If you select the **Dynamic Wall Selection** check box available in the **Automatic Wall Selection** sub-rollout, the software will automatically select the wall faces that are attached to the source and target faces.

Using the **Number of Layers** edit box available in the **Mesh Parameters** rollout, you can specify the number of element layers. The layers will be created between the source and target faces.

After specifying all options, choose the **OK** button to create sweep elements. The options in the other rollouts of this dialog box are same as those discussed earlier.

Figure 5-13 shows a model with source element face and the target face and Figure 5-14 shows 3D sweep elements created upto the target face.

Note
For creating 3D sweep elements between faces, you need to choose the source and target face elements in sequential order. For example, first you need to select source element face then the target face.

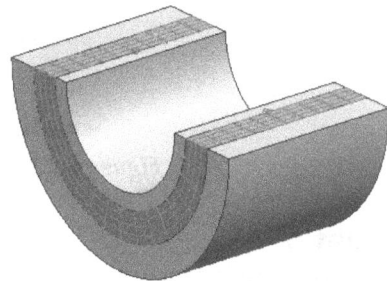

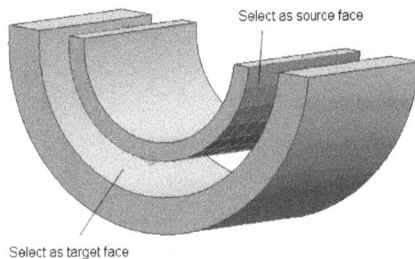

Figure 5-13 *A model with source element face and target face*

Figure 5-14 *Model after creating 3D sweep element*

MESHING

Meshing is the stage of the finite element modeling process in which you divide a continuous structure into a finite number of regions. These regions are known as elements and are connected together by nodes.

The NX Nastran provides four types of meshing in the model: **0D Meshing**, **1D Meshing**, **2D Meshing**, and **3D Meshing**. For creating mesh you need to assign some physical property to the model depending on your requirement. These properties can be assigned using the mesh collector. The mesh collector and different types of meshing are discussed next.

MESH COLLECTOR

Ribbon: Home > Properties > Mesh Collector
Menu: Insert > Mesh Collector

The mesh collector is a collection of nodes in the **Simulation Navigator** that contains meshes and elements with their properties such as materials and physical properties. You can create a mesh collector by using the **Mesh Collector** tool. You can also create a mesh collector during the creation of element or mesh by using the **New Collector** button in the **Destination Collector** rollout

To create a mesh collector, choose **Menu > Insert > Mesh Collector** from the **Top Border Bar**; the **Mesh Collector** dialog box will be displayed, as shown in Figure 5-15. You can select the element type from the **Element Family** drop-down list available in the **Element Topology** rollout. The appearance of the **Mesh Collector** dialog box changes depending on the element selection. The methods to create a mesh collector for different elements are discussed next.

*Figure 5-15 The **Mesh Collector** dialog box*

Creating a Mesh Collector for 0D Elements

As discussed earlier, 0D elements are used to create concentrated masses on a point. To create a mesh collector for 0D element, choose the **0D** option from the **Element Family** drop-down list; the **Mesh Collector** dialog box will be modified. The **Collector Type** drop-down list contains the option of relevant element type collector. Figure 5-16 shows the **Mesh Collector** dialog box on choosing the **0D** element type and the **Cmass1 Collector (Grounded)** collector type from the **Element Family** and **Collector Type** drop-down lists, respectively.

You can define the properties of the elements chosen from the **Collector Type** drop-down list by using the options available in the **Properties** rollout. Figure 5-16 shows the PMASS option in the **Type** drop-down list under the **Physical Property** sub-rollout. You can edit the relevant physical properties using the **Create Physical** 🖾 button available in the sub-rollout. When you choose this button, the relevant element property dialog box will be displayed. Figure 5-17 shows the **PMASS** dialog box. You can specify the name, label and other properties using this dialog box.

Figure 5-16 The Mesh Collector dialog box for the 0D element

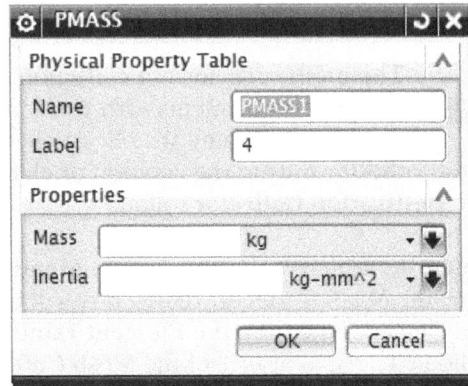

Figure 5-17 The PMASS dialog box

After choosing the **OK** button from the dialog box, the created property name will be displayed in the **Cmass Property** drop-down list under the **Physical Property** sub-rollout of the **Mesh Collector** dialog box. You can also edit the properties using the **Edit** button that will be available when you choose any created property from this drop-down list. You can specify the collector name in the **Name** edit box of the dialog box. Choose the **OK** button from the dialog box; the mesh collector with the specified name will be created in the **Simulation Navigator**.

To assign this mesh collector to any 0D element/mesh while creating the element/mesh from the relevant dialog box, clear the **Automatic Creation** check box available in the **Destination Collector** rollout; the created mesh collector will be displayed in the **Mesh Collector** drop-down list in the rollout. Choose the collector name; the created mesh collector settings will be applied to the element/mesh.

Creating a Mesh Collector for 1D Elements

As discussed earlier, 1D elements are used for rod or beam behavior. To create a mesh collector for 1D element, choose the **1D** option from the **Element Family** drop-down list of the **Mesh Collector** dialog box. On doing so, the dialog box will be modified for the element type. The **Collector Type** drop-down list contains the option for the relevant element type collector. Figure 5-18 shows the **Mesh Collector** dialog box that is displayed on choosing the **1D** element type and the **Beam Collector** collector type from the **Element Family** and **Collector Type** drop-down lists, respectively.

You can also define the properties of the elements chosen from the **Collector Type** drop-down list for 1D element using the options available in the **Properties** rollout. Figure 5-18 shows the PBEAM option selected in the **Type** drop-down list of the **Physical Property** sub-rollout. You can edit the physical property using the **Create Physical** button available in the sub-rollout. When you choose this button, the corresponding element property dialog box will open. Figure 5-19 shows the **PBEAM** dialog box. You can specify the name, label and other properties using this dialog box.

Figure 5-18 *The Mesh Collector dialog box for 1D element*

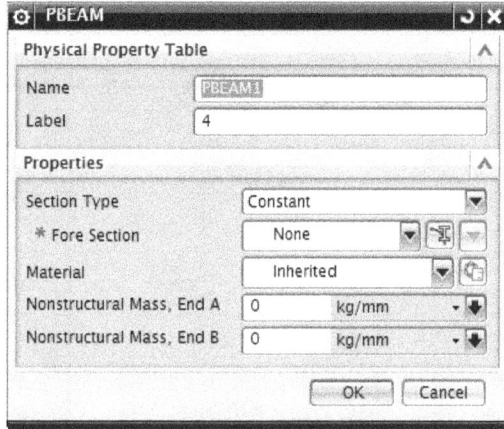

Figure 5-19 *The PBEAM dialog box*

In case of some 1D mesh/element, you may need to define the section for the analysis. The steps required to create section for one dimensional beam element are as follows.

1. Choose the **Show Section Manager** button available in the **Properties** rollout of the **PBEAM** dialog box; the **Beam Section Manager** dialog box will be displayed, as shown in Figure 5-20.

2. Choose the **Create Section** button from the dialog box; the **Beam Section** dialog box will be displayed, as shown in Figure 5-21.

3. Different beam sections are available in the drop-down list available in the **Type** rollout. Select the desired section from the list and specify the dimension using edit boxes available in the **Dimension** sub-rollout of the **Properties** rollout.

4. Choose the **OK** button; the name of the created section will be displayed under the **Cross Section List** rollout of the **Beam Section Manager** dialog box.

5. Choose the **Close** button from the dialog box; the created beam section name will be displayed in the **Fore Section** drop-down list of the **Properties** rollout of the **PBEAM** dialog box.

6. Choose the **OK** button from the dialog box; the created beam name will be displayed in the **Beam Property** drop-down list of the **Physical Properties** sub-rollout in the **Properties** rollout of the **Mesh Collector** dialog box. Choose the **OK** button from the dialog box; the mesh collector with the specified name will be created in the **Simulation Navigator**.

7. While creating the 1D element/mesh from the dialog box, clear the **Automatic Creation** check box in the **Destination Collector** rollout. On doing so, the created beam section name will be displayed in the **Mesh Collector** drop-down list in the rollout. Now, choose the collector name from the drop-down list that will create the element/mesh on defined node/edge.

The Figure 5-22 shows a beam element with tubular cross-section.

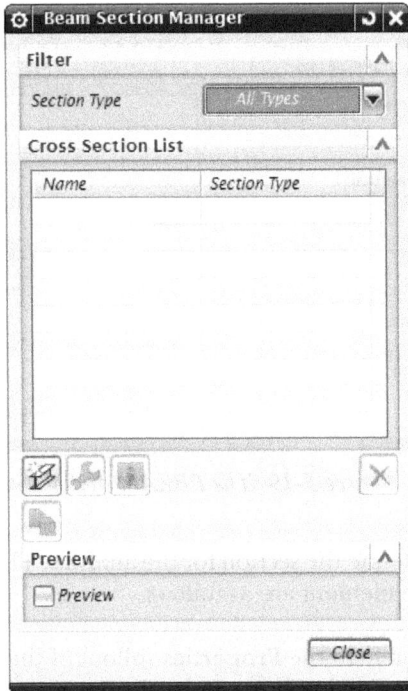

Figure 5-20 The Beam Section Manager
dialog box

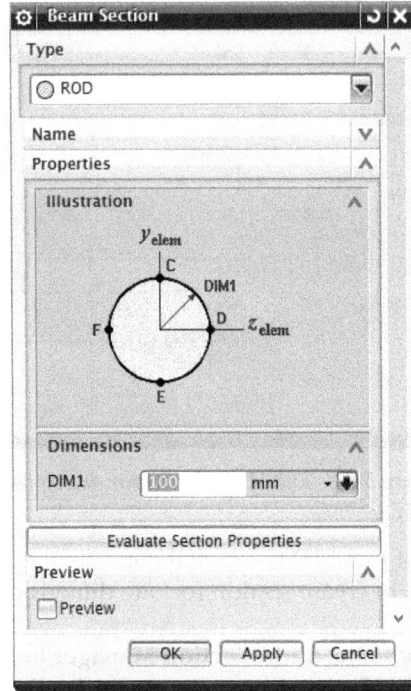

Figure 5-21 The Beam Section dialog box

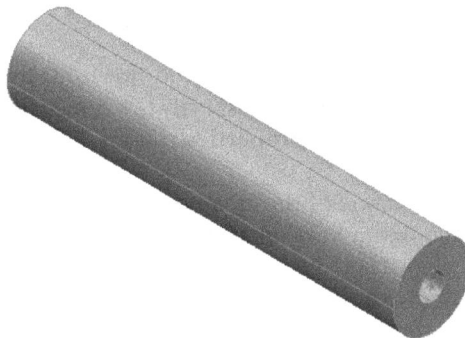

Figure 5-22 *The 1D element with tubular section*

You can also assign material to the elements. For specifying the material to the PBEAM physical property of the element, choose the **Choose Material** 🔲 button available on the right of the **Material** drop-down list in the **PBEAM** dialog box, refer to Figure 5-19; the **Material List** dialog box will be displayed, as shown in Figure 5-23. You can choose the required material from this dialog box.

Note
You will learn more about specifying material to the element, mesh, or model in the later chapters of this book.

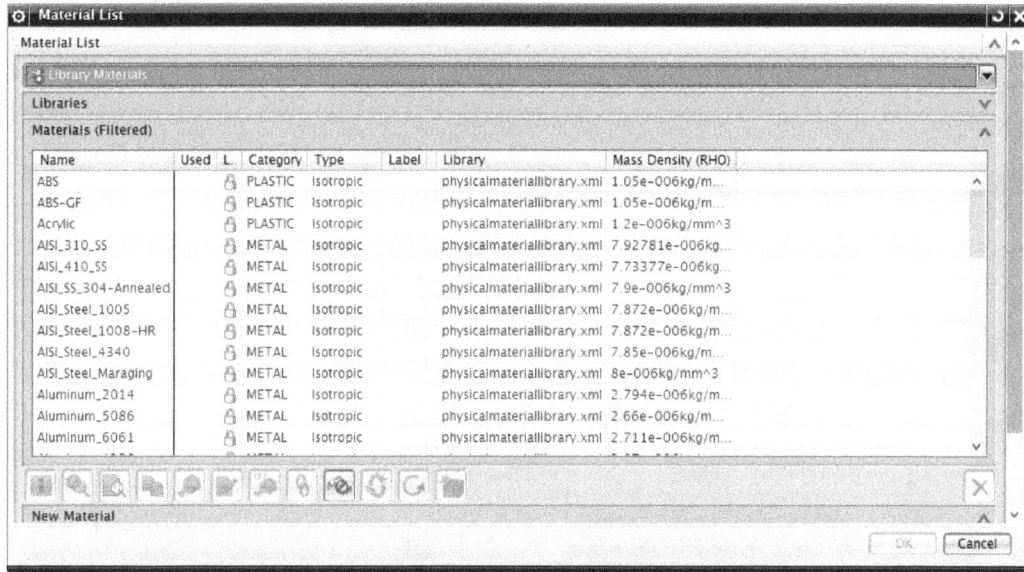

Figure 5-23 The **Material List** *dialog box*

Creating a Mesh Collector for 2D Elements

As discussed earlier, 2D elements are used for creating thin shell structure or surface structure. To create a mesh collector for 2D elements, choose the **2D** option from the **Element Family** drop-down list of the **Mesh Collector** dialog box; the dialog box will be modified according to the element type chosen for 2D element. The **Collector Type** drop-down list contains the options for the relevant element type collector. Figure 5-24 shows the **Mesh Collector** dialog box on choosing **2D** element type and the **Thin Shell** collector type from the **Element Family** and **Collector Type** drop-down lists, respectively.

Similar to defining elements, you can also define the properties of the 2D element chosen from the **Collector Type** drop-down list. Figure 5-24 shows the PSHELL in the **Type** drop-down list of the **Physical Property** sub-rollout. You can also create the relevant physical property using the **Create Physical** button available in the **Physical Property** sub-rollout. When you choose this button, the corresponding element property dialog box will be displayed. Figure 5-25 shows the **PSHELL** dialog box. You can specify the name, label, and other properties using this dialog box. As in shell geometry, you can also provide the thickness value in the **Default Thickness** edit box of the dialog box.

Figure 5-24 *The **Mesh Collector** dialog box for 2D element*

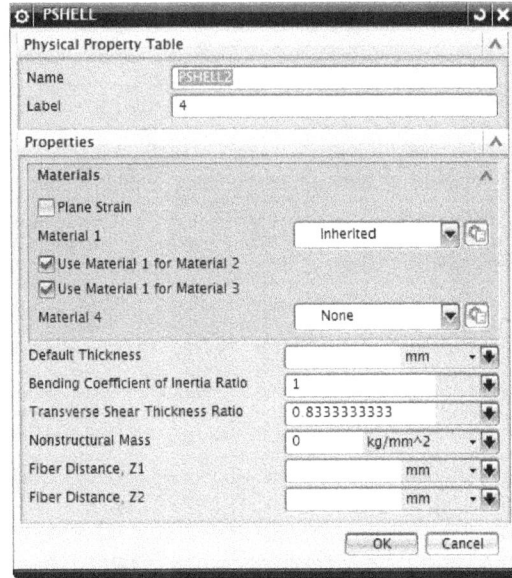

Figure 5-25 *The **PSHELL** dialog box*

After specifying all settings, choose the **OK** button from the **PSHELL** dialog box; the created property name will be displayed in the **Shell Property** drop-down list in the **Physical Property** sub-rollout of the **Mesh Collector** dialog box. You can edit the properties by using the **Edit** button available next to the **Shell Property** drop-down list. This button will be available only when you select the created property from this drop-down list. You can specify the collector name in the **Name** edit box of the dialog box. Next, choose the **OK** button from the dialog box; the mesh collector with the specified name will be created in the **Simulation Navigator**.

While creating the 2D element/mesh, clear the **Automatic Creation** check box in the **Destination Collector** rollout from the respective dialog box. The created mesh collector will be displayed in the **Mesh Collector** drop-down list in the rollout. Choose the collector while creating element/mesh; the created mesh collector settings will be applied to the element/mesh.

Figure 5-26 shows a meshed shell geometry after applying thickness to it.

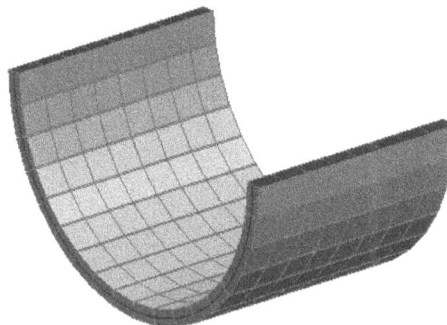

Figure 5-26 *The 2D element with defined thickness*

Tip

After creating the mesh/element, you can also change the display setting. To change the display settings of the mesh/element, choose **Menu > Preferences > Mesh Display** *from the* **Top Border Bar**; *the* **Mesh Display** *dialog box will be displayed, as shown in Figure 5-27. This dialog box will display the tab settings for the created mesh/element. Figure 5-27, shows the 2D mesh/element settings in the* **2D** *tab chosen in the dialog box. For displaying the thickness, select the* **Element Thickness and Offset** *check box and choose the* **Apply** *button; the geometry will be displayed with the defined thickness in the mesh collector. If the geometry has none of the mesh/element created, then the dialog box displayed will have no tab, as shown in Figure 5-28.*

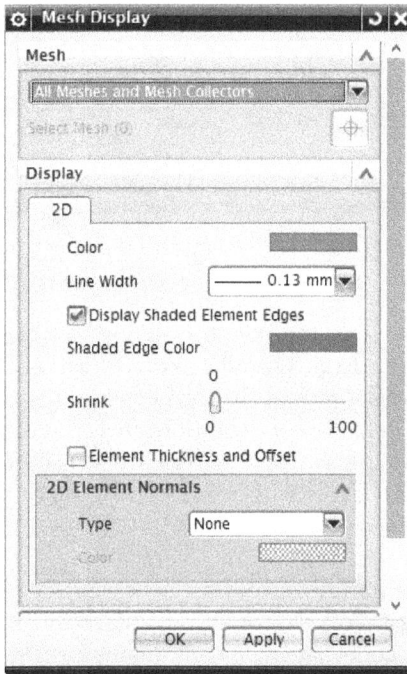

Figure 5-27 The **Mesh Display** *dialog box with* **2D** *tab*

Figure 5-28 The **Mesh Display** *dialog box*

Creating a Mesh Collector for 3D Element

3D elements are used for solid structure. To create a mesh collector for 3D element, choose the **3D** option from the **Element Family** drop-down list; the modified dialog box will be displayed for the element type. The **Collector Type** drop-down list contains the option for the relevant element type collector. Figure 5-29 shows the **Mesh Collector** dialog box is displayed on choosing the **3D** element type and the **Solid** collector type from the **Element Family** and the **Collector Type** drop-down lists, respectively.

Similar to defining 0D or 1D elements, you can also define the properties of the 3D elements available in the **Collector Type** drop-down list. Figure 5-29 shows the PSOLID in the **Type** drop-down list of the **Physical Property** sub-rollout. You can edit the relevant physical property using the **Create Physical** button available in the **Physical Property** sub-rollout. When you choose this button, the corresponding element property dialog box will open. Figure 5-30 shows the **PSOLID** dialog box. You can specify the name, label and other properties using this dialog box.

Figure 5-29 The **Mesh Collector** dialog box for 3D element

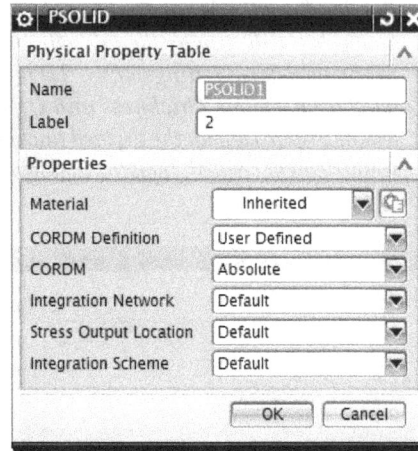

Figure 5-30 The **PSOLID** dialog box

After specifying all the settings, choose the **OK** button from the **PSOLID** dialog box; the created property name will be displayed in the **Solid Property** drop-down list in the **Physical Property** sub-rollout of the **Mesh Collector** dialog box. You can also edit the property using the **Edit** button that will be available on selecting the property from this drop-down list. You can specify the collector name in the **Name** edit box of the dialog box. Next, choose the **OK** button from the dialog box; the mesh collector with the specified name will be created in the **Simulation Navigator**.

While creating the 3D element/mesh, clear the **Automatic Creation** check box available in the **Destination Collector** rollout. The created mesh collector will be displayed in the **Mesh Collector** drop-down list in the rollout. Choose the collector while creating element/mesh; the created mesh collector settings will be applied to the element/mesh.

0D MESHING

0D meshing is used to create point based elements at specific nodes. It is also called scalar elements meshing. The 0D mesh is used for meshing the structure that has two different degrees of freedom. The 0D mesh is used in springs, masses, and viscous dampers type structure. The method to create 0D meshing is as follows.

0D Mesh

Ribbon: Home > Mesh > More > 1D and 2D > 0D Mesh
Menu: Insert > Mesh > 0D Mesh

You can create a 0D mesh by choosing **Menu > Insert > Mesh > 0D Mesh** from the **Top Border Bar**; the **0D Mesh** dialog box will be displayed, as shown in Figure 5-31. You can use the options available in the dialog box for creating mesh. These options are discussed next.

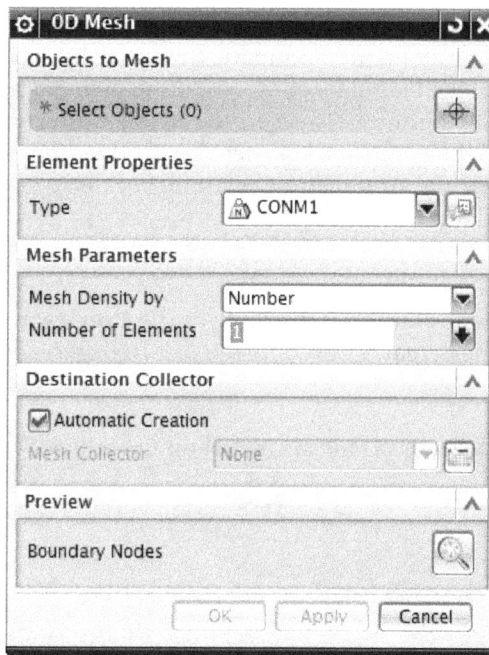

*Figure 5-31 The **0D Mesh** dialog box*

Objects to Mesh

In this rollout, the **Select Objects (0)** option is highlighted by default and you are prompted to select the geometry to create 0D mesh. The geometry may be line, point, or edge of the model. If you select surface of the model, the software will find the edges of surface for creating the 0D elements.

Element Properties

You can select the element type from the **Type** drop-down list of this rollout. The **Edit Mesh Associated Data** button available next to the **Type** drop-down list is used to provide additional information related to the element.

> **Note**
> *You will learn about editing mesh associated data in the later chapters of this book.*

Mesh Parameters

You can control mesh creation by using the options available in this rollout. There are two options available in the **Mesh Density by** drop-down list of the **Mesh Parameters** rollout: **Size** and **Number**. If you select the **Size** option, the software will create the elements at approximate distance between each other. The **Element Size** edit box is used to specify the size of element. On selecting the **Number** option from the drop-down list, you can specify the total number of elements created on the selected geometry. The **Number of Elements** edit box is used to specify the number of elements to be created.

Destination Collector

In the **Destination Collector** rollout, the **Automatic Creation** check box is available. If this check box is selected, the software will automatically create the element at the default location. On choosing the **New Collector** button ▣ next to the **Mesh Collector** drop-down list, you can define physical property of the elements. You can also set the predefined 0D mesh collector to the meshing. To do so, clear the **Automatic Creation** check box and select the collector from the **Mesh Collector** drop-down list.

Preview

The **Preview Boundary Nodes** button is available in the **Preview** rollout. You can choose this button to preview the nodes creation around the selected geometry to define size/number elements.

Figure 5-32 shows the model with 0D mesh created on the edges.

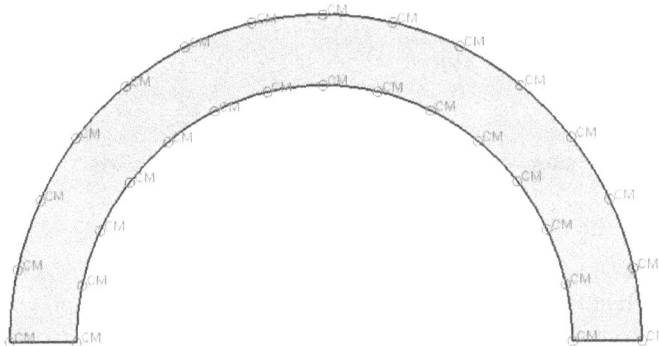

Figure 5-32 *The 0D mesh created*

1D MESHING

1D meshing is used to create one dimensional (line) element along curves or polygon edges. Generally, the line elements are used to represent rod and beam behavior. The method to create 1D meshing is as follows.

Creating 1D Mesh

Ribbon: Home > Mesh > More > 1D and 0D > 1D Mesh
Menu: Insert > Mesh > 1D Mesh

You can create 1D meshing by choosing **Menu > Insert > Mesh > 1D Mesh** from the **Top Border Bar**; the **1D Mesh** dialog box will be displayed, as shown in Figure 5-33. You can use the option and rollouts available in the dialog box for creating the mesh. These options are discussed next.

Objects to Mesh

In the **Objects to Mesh** rollout, the **Select Objects (0)** option is highlighted by default. You can select curves, polygon edges, or polygon faces to create 1D mesh. For a polygon face, 1D mesh will be created along each edge of the face. If you select the **Auto Chain Selection** check box, the edges of the face will be selected automatically in sequential order for mesh creation. Note that the **Auto Chain Selection** check box does not work for polygon face selection. The **Reverse Direction** button is used to reverse the direction of edge or curve selection.

Figure 5-33 The 1D Mesh dialog box

Element Properties

You can select the 1D element type from the **Type** drop-down list of this rollout. The **Edit Mesh Associated Data** button available next to the **Type** drop-down list is used to provide additional information related to the element.

Mesh Parameters

You can control the mesh creation by using the options available in this rollout. In the **Mesh Density by** drop-down list of the **Mesh Parameter** rollout, two options are available: **Size** and **Number**. Using the **Size** option, you can create the 1D element at approximate distance between each other. The **Element Size** edit box is used to specify the element size. If you select the **Number** option from the drop-down list, the software will create the total specified 1D elements on the selected geometry. The **Number of Elements** edit box is used to specify the number of elements to be created.

Destination Collector

The **Automatic Creation** check box is available in the **Destination Collector** rollout. When you select this check box, the software will create 1D element at default location. In this rollout, the **New Collector** button is used to define physical properties of the 1D elements. You can also set the predefined 1D mesh collector to the meshing. To do so, clear the **Automatic Creation** check box and select the collector from the **Mesh Collector** drop-down list.

Preview

The **Preview Boundary Nodes** button is available in the **Preview** rollout. You can choose this button, to preview the nodes creation around the selected geometry to define size/number elements.

2D MESHING

2D meshing is used to create two dimensional (surface) elements on the selected surfaces. You can create 2D elements on plates, flat sheets, or shell models. The tools to create different 2D mesh are discussed next.

2D Mesh Tool

Ribbon:	Home > Mesh > 2D Mesh
Menu:	Insert > Mesh > 2D Mesh

You can create a 2D mesh by using the **2D Mesh** tool. To do so, choose **Menu > Insert > Mesh > 2D Mesh** from the **Top Border Bar**; the **2D Mesh** dialog box will be displayed, as shown in Figure 5-34. You can use the options available in the dialog box for creating a mesh. The options available in the dialog box are discussed next.

Figure 5-34 The **2D Mesh** dialog box

Objects to Mesh

In this rollout, the **Select Objects (0)** option is highlighted. You can select polygon faces to create 2D mesh.

Element Properties

You can select the 2D element type from the **Type** drop-down list of this rollout. The **Edit Mesh Associated Data** button available next to the drop-down list is used to provide additional information related to the element. If you select CSHEAR from the **Type** drop-down list, the **Edit Mesh Associated Data** button will not be available. This element does not contain any transition element so you can not edit it.

Mesh Parameters

In this rollout, two options are available in the **Meshing Method** drop-down list to create meshing: **Subdivision** and **Paver**. These options are discussed next.

Subdivision

When you select the **Subdivision** option, the software will create meshes by using repeated subdivision method. In this procedure, first software creates mesh and then performs cleaning and smoothing operation to create quality mesh.

Paver

When you select the **Paver** option, the software will create meshes by combining them. In this procedure, first the mesh is created using the paving technique and then through repeated subdivisions, a quality mesh is created.

Figure 5-35 shows a model meshed using the **Subdivision** option and Figure 5-36 shows mesh created using the **Paver** option.

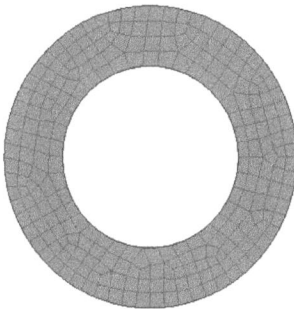

Figure 5-35 *2D mesh created using the* **Subdivision** *method*

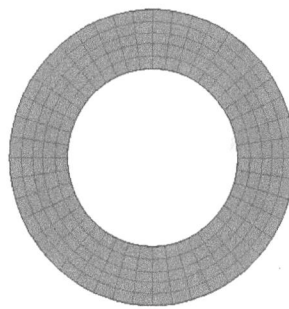

Figure 5-36 *2D mesh created using the* **Paver** *method*

You can specify the size of the element in the **Element Size** edit box. The **Automatic Element Size** button available next to the edit box is used to examine the selected face and calculate proper element size according to the face size. The **Attempt Free Mapped Meshing** check box is used to create mapped meshing with default settings. The mapped mesh contains either quadrilateral elements or triangular elements. If you select any quadrilateral element from the **Type** drop-down list of the **Element Properties** rollout, the **Attempt Quad Only** will be activated in the rollout. In this rollout, three options are available: **Off - Allow Triangles**, **On - Zero Triangles**, and **On - Single Triangle**. These options are discussed next.

Off - Allow Triangles

When you choose this option, the software will create a quadrilateral mesh that also contains some triangular elements.

On - Zero Triangles

When you choose this option, the software will create a mesh that does not contain any triangular element.

On - Single Triangle

When you choose this option, the software will create a mesh with maximum one triangular element on each selected face.

Mesh Quality Options

In this rollout, the options are available for defining the quality of mesh. The options in this rollout will be available only when you choose midnode element (like CTRIA6) or quadrilateral element from the **Type** drop-down list of the **Element Properties** rollout. The methods of defining quality by using the options available in the rollout are discussed next.

Midnode Method

This drop-down list will be available only for midnode elements (like CTRIA6) and is used to define the midnode creation of elements. Three options are available in this drop-down list to specify midnode creation: **Mixed**, **Curved**, and **Linear**. These options are discussed next.

Mixed: When you select this option, the **Max Jacobian** edit box is displayed in the rollout. You can specify the max jacobian value for element creation. Using this option, the elements will be randomly generated on the geometry within the specified jacobian value. Note that the specified jacobian value must be greater than zero.

Curved: Using this option, the midnodes will be created on the geometry irrespective of the element quality.

Linear: Using this option, the midnodes will be created on the geometry linearly.

Split Quads

This check box will be available only for quadrilateral elements. When you select this check box, the excess quadrilateral elements around the geometry are split into triangular elements within the specified range. Below this check box, two edit boxes are available for defining quality elements: **Max Warp Threshold** and **Max Jacobian**. The **Max Warp Threshold** edit box is used for specifying element angle deviation along a planar surface. While the **Max Jacobian** edit box is used for specifying jacobian for an element within the specified warp threshold value.

Mesh Settings

Using the options available in this rollout, you can define further settings for meshing. The options available in this rollout for settings are discussed next.

Export Mesh to Solver

If you select this check box, the created mesh will be exported to specific solver. But clear this check box if you want to use 2D mesh for creating 3D mesh.

Transition Element Size

When you select this check box, the software will create transition layers between the elements for smooth creation of the elements.

Curvature Based Size Variation

This slider will be available only for triangular elements. It controls the element refinement and position around the curved surfaces. Moving the slider toward the minimum settings (0%) results in less refinement and also the elements size does not get varied from their original size. While moving the slider toward the maximum settings (100%) results in for more refined elements and the element size will vary from their original size according to the curvature surface.

Model Cleanup Options

Using the options available in this rollout, you can define further settings for meshing. The options available in this rollout are discussed next.

Match Edges

If you select this check box, the software will create continuous mesh with the previously created mesh. Also the **Match Edge Tolerance** edit box will be displayed. You can specify the value in this edit box to match the edges of the mesh.

Small Feature Tolerance (% of Element Size)

Using this slider, you can eliminate the small feature of the geometry by specifying its percentage value. For example, your slider is at 10% value. You have a feature of 3mm in the geometry and you have specified element size 30mm. Now, while meshing, the software will eliminate this feature.

Merge Edges

If you select this check box, the **Vertex Angle** edit box will be displayed. In this edit box, you can specify the angle value between two edges to merge the edges while meshing.

Destination Collector

The **Automatic Creation** check box is available in the **Destination Collector** rollout. When you select this check box, the software will create 2D element at a default location. You can use the **New Collector** button 🔳 to define physical properties of the 2D elements. You can also set the predefined 2D mesh collector to the meshing. To do so, clear the **Automatic Creation** check box and select the collector from the **Mesh Collector** drop-down list.

Preview

In this rollout, the **Show Result** button is used to preview the mesh creation on the selected faces.

2D Mapped Mesh Tool

Ribbon: Home > Mesh > 2D Mapped
Menu: Insert > Mesh > 2D Mapped Mesh

The 2D mapped meshing is a process of dividing elements into triangular or quadrilateral elements. To create a 2D mesh, choose **Menu > Insert > Mesh > 2D Mapped Mesh** from the **Top Border Bar**; the **2D Mapped Mesh** dialog box will be displayed, as shown in Figure 5-37. The options and rollouts available in the dialog box are discussed next.

Figure 5-37 *The* **2D Mapped Mesh** *dialog box*

Element Properties

In this rollout, you can select the 2D element type from the **Type** drop-down list. The **Edit Mesh Associated Data** button available next to the **Type** drop-down list is used to provide additional information related to the element. If you select CSHEAR from the **Type** drop-down list, the **Edit Mesh Associated Data** button will not be displayed. This element does not contain any transition element so you can not edit it.

Destination Collector

In the **Destination Collector** rollout, the **Automatic Creation** check box is available. This check box is selected by default. As a result, the 2D element will be created at default location. In this rollout the **New Collector** button is also available. Using this, you can define physical property of the 2D elements. You can also set the preassign 2D mesh collector to the meshing. To do so, clear the **Automatic Creation** check box and select the collector from the **Mesh Collector** drop-down list.

Selection

In this rollout, the **Select Objects (0)** option is highlighted. You can select polygon faces to create the 2D mapped mesh.

Meshing Parameters

Using the options available in this rollout, you can specify mesh parameters. The **Element Size** edit box is used to specify the size of element. Beside the edit box, the **Automatic Element Size** button is available. When you choose this button after selecting the face, the software examines the face and calculates proper element size according to the face size. Below the edit box, the **Export Mesh to Solver** check box is available. If you select this check box, the software will export the created mesh to specific solver. But clear this check box if you want to use the 2D mesh for creating 3D meshing.

Preview and Modify Mesh Constraints

Using the options available in this rollout, you can preview and edit the created mesh. The **Show Result** button in this rollout is used to preview the mapped mesh. Two other sub-rollouts are also available in this rollout: **Define Corners** and **Edge Density**. When you choose the **Show Result** button to view the mesh, the **Select Face Corner 1 of 4** option will be available in the **Define Corners** sub-rollout. You can select the four corners of the geometry to define the elements. The **Edge Density** edit box will be available when you select the diamond symbol on the edge of meshed surface. After selecting the diamond symbol the created elements number will be shown in this edit box. You can decrease or increase the number of elements using the edit box. Choose the **Show Result** button to view the updated mesh.

Mesh Options

In this rollout, the options are used for defining mesh creation. The check boxes and options available in this rollout are discussed next.

Merge Edges

When you select this check box, the **Vertex Angle** edit box will be available. While meshing, you can merge the edges of the selected face by specifying the angle value between the faces in this edit box.

Match Edges

When you select this check box, the **Match Edge Tolerance** edit box will be available. For creating continuous mesh, you can match the nodes of edges by specifying the tolerance value in this edit box.

Quad Only on 3 Sided Faces

If you select this check box, the software will create only quad element on the three sided selected faces.

Store Free Mesh If Mapped Mesh Fails

If you select this check box, the software will create free meshes on the selected faces in case the mapped mesh fails.

Edge Mapping

If you select this check box, the software will project vertices along the opposite edges of the selected face. By using these vertices, the software will create more uniform mesh.

Flip Diagonals

If you select this check box, the software will reverse the direction of the element creation around their diagonals.

Figure 5-38 shows a 2D mapped meshed model with triangular elements and Figure 5-39 shows 2D mapped meshed model with quadrilateral elements.

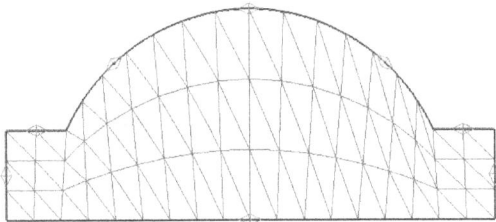

Figure 5-38 *2D mapped mesh with triangular elements*

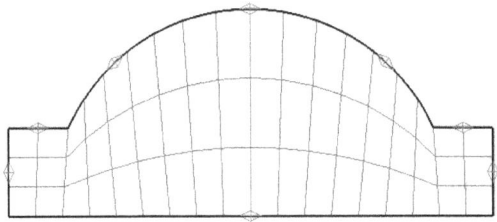

Figure 5-39 *2D mapped mesh with quadrilateral elements*

2D Dependent Mesh Tool

Ribbon: Home > Mesh > More > 2D > 2D Dependent
Menu: Insert > Mesh > 2D Dependent Mesh

The **2D Dependent Mesh** tool is used to create similar free or mapped meshes on selected faces of a model. In this method, the faces selected need not have common edges. To create a 2D dependent mesh, choose **Menu > Insert > Mesh > 2D Dependent Mesh** from the **Top Border Bar**; the **2D Dependent Mesh** dialog box will be displayed, as shown in Figure 5-40. You can use the options available in this dialog box for creating dependent mesh. The options and rollouts available in the dialog box are discussed next.

Type

In this rollout, two options are available in the **Type** drop-down list: **General** and **Symmetric**. The **General** option is used to create dependent mesh on the selected surface which is equally scaled from the master surface. The **Symmetric** option is used to create dependent mesh on the selected surface whose size is equal size to the master surface.

Select Master and Target Face

In this rollout, choose the **Select Master Face (0)** option; you will be prompted to select master face that will act as an independent face of meshing. After selecting the master face, choose the

Select Target Face (0) option; you are prompted to select target face that will act as a dependent face and meshing will be done on it similar to master face.

Figure 5-40 *The* **2D Dependent Mesh** *dialog box*

Note
You can also select the meshed face as master face and then create frozen mesh (default mesh) on the selected target face. This method is discussed further. But, you can not select the mesh face as target face.

Match Loops

After selecting the master and target face for dependent mesh, two other options will be displayed in this rollout: **Select Master Edge (0)** and **Select Target Edge (0)**. The **Select Master Edge (0)** option is used to select an edge from the master face for defining the meshing direction. While the **Select Target Edge (0)** option is used to select an edge from the target face. Using the **Flip Direction** check box, you can change the direction of the target edge. In the **List** sub-rollout area, the selected face will be displayed in a loop.

Direction

This rollout will be available only when you select the **Symmetric** option from the **Type** drop-down list of the **Type** rollout. Using the options available in this rollout, you can specify the CSYS along which the dependent mesh will be created.

Mesh Type

For specifying dependent mesh type, four options are available in the **Mesh Type** drop-down list of this rollout: **Free Mesh**, **Mapped Mesh**, **Frozen Free Mesh**, and **Frozen Mapped Mesh**. These options are discussed next.

Free Mesh

This option is used to create free mesh on the target face. After choosing the **OK** button from the **2D Dependent Mesh** dialog box, the **2D Mesh** dialog box will be displayed. Specify the settings in the dialog box and choose the **OK** button to confirm. The 2D free meshing will be created on the master and target faces.

Mapped Mesh

This option is used to create mapped mesh on the target face. After choosing the **OK** button from the **2D Dependent Mesh** dialog box, the **2D Mapped Mesh** dialog box will be displayed. Specify the settings in the dialog box and choose the **OK** button to confirm. The 2D mapped meshing will be created on the master and target faces.

Frozen Free Mesh

This option will be available only when the master face has free meshing. It will create a free mesh with default settings on the selected target face.

Frozen Mapped Mesh

This option will be available if the master face has mapped mesh. It will create the mapped mesh with default settings on the selected target face.

Figure 5-41 shows a model for master and target face selection and Figure 5-42 shows the model after creating 2D dependent mapped meshing.

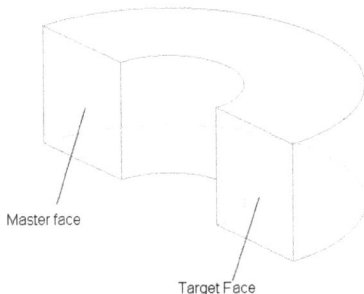

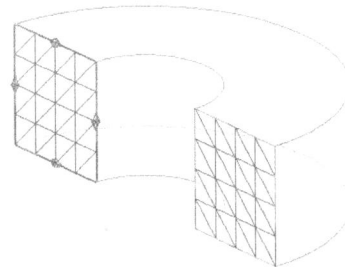

Figure 5-41 *Model faces for 2D dependent meshing*   *Figure 5-42* *2D dependent mapped meshing*

> **Tip**
> *After creating the dependent mesh, a diamond symbol will be created on the edges of the master face, refer to Figure 5-42. When you double-click on this symbol, the **Mesh Control** dialog box will be displayed. You can change the element density on the face using this dialog box. You will learn more about changing mesh density in later chapters.*

Surface Coat Tool

Ribbon:	Home > Mesh > More > 2D > Surface Coat
Menu:	Insert > Mesh > Surface Coat

Using the **Surface Coat** tool, you can create surface coating of 2D elements on solid elements. The coating of 2D elements may be used either on the free faces of the selected solid elements or on the entire solid mesh. To create surface coating of 2D elements, choose **Menu > Insert > Mesh > Surface Coat** from the **Top Border Bar**; the **Surface Coat** dialog box will be displayed, as shown in Figure 5-43. You can use the options available in the dialog box for creating coating of 2D elements. The options and rollouts available in the dialog box are discussed next.

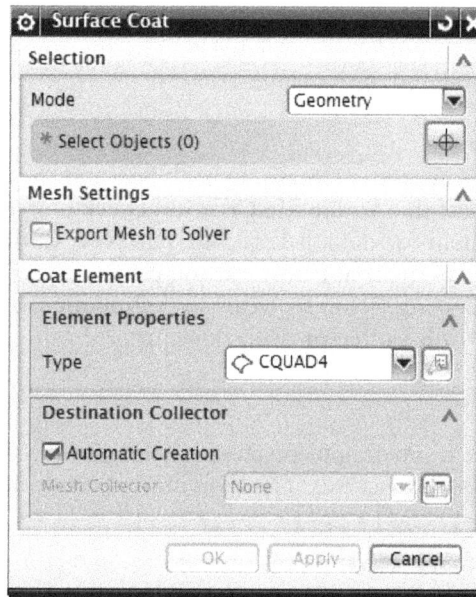

Figure 5-43 *The **Surface Coat** dialog box*

Selection

The options in this rollout are used for selecting the faces of elements or the entire solid elements. The options for selecting elements are available in the **Mode** drop-down list of this rollout. These options are discussed next.

3D Mesh

This option is used to select the entire solid mesh on which you want to create surface coat.

Element Face

This option is used to select individual 3D element face on which you want to create surface coat.

Geometry

This option is used to select the entire solid part or individual faces of the part on which you want to create surface coat.

Mesh Settings

In this rollout, the **Export Mesh to Solver** check box is available. If you select this check box, the software will export the created surface coat to specific solver. But clear this check box if you want to use these surface coatings for further use.

Coat Element

In this rollout, two sub-rollouts are available: **Element Properties** and **Destination Collector**. Usage of these sub-rollouts is discussed next.

Element Properties

You can select the required element from the **Type** drop-down list of this sub-rollout. The **Edit Mesh Associated Data** button available next to this drop-down list is used to provide additional information related to the element.

Destination Collector

In this sub-rollout, the **Automatic Creation** check box and the **New Collector** button are available. When you select the **Automatic Creation** check box, the software will create the surface coat of 2D elements at default location. The **New Collector** button is used to define physical property of the coat elements. You can also assign the predefined 2D mesh collector to the surface coating. To do so, clear the **Automatic Creation** check box and select the collector from the **Mesh Collector** drop-down list.

3D MESHING

3D meshing is used to create three dimensional (solid) elements on the selected solid body. You can create 3D elements on solid structures. The tools used to create 3D mesh are discussed next.

3D Tetrahedral Mesh Tool

Ribbon: Home > Mesh > 3D Tetrahedral
Menu: Insert > Mesh > 3D Tetrahedral Mesh

This tool is used to create 3D tetrahedral mesh. To do so, choose **Menu > Insert > Mesh > 3D Tetrahedral Mesh** from the **Top Border Bar**; the **3D Tetrahedral Mesh** dialog box will be displayed, as shown in Figure 5-44. You can use the options available in the dialog box for creating mesh. The options and rollouts available in the dialog box are discussed next.

Objects to Mesh

In this rollout, the **Select Bodies (0)** option is highlighted. You need to select a solid body to create 3D tetrahedral mesh.

Element Properties

You can select the required tetrahedral element from the **Type** drop-down list of this rollout. The **Edit Mesh Associated Data** button available next to the drop-down list is used to provide additional information related to the element.

Mesh Parameters

You can specify mesh parameters by using the options available in this rollout. The **Element Size** edit box is used to specify the size of element. Beside this edit box, the **Auto Element Size** button is available. When you choose this after selecting the solid body, the software examines the solid body and calculates proper element size according to the solid body. Below the edit box, the **Attempt Free Mapped Meshing** check box is available. When you select this check box, the software will create free mapped meshing with default settings.

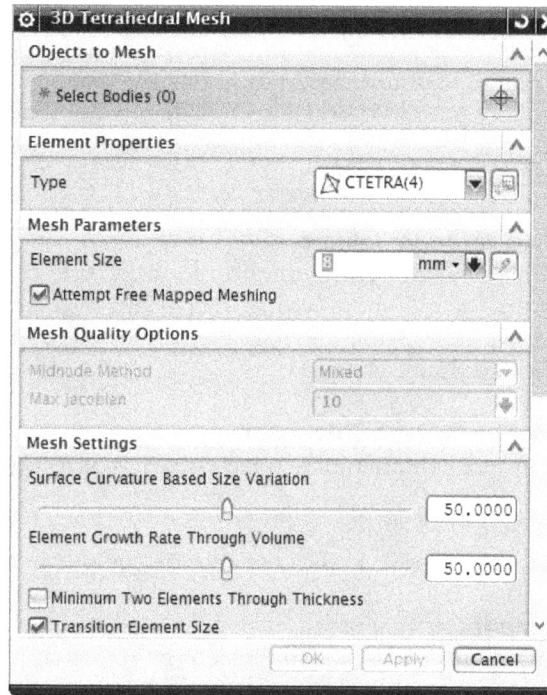

Figure 5-44 The **3D Tetrahedral Mesh** *dialog box*

Mesh Quality Options

The options in this rollout will be available only when you choose CTETRA(10) (midnode) element from the **Type** drop-down list of the **Element Properties** rollout. In this rollout, the options are available for creating midnodes of the element on the geometry. In the **Midnode Method** drop-down list, three options are available: **Mixed**, **Curved**, and **Linear**. These options are discussed next.

Mixed

When you choose this option, the **Max Jacobian** edit box will also be available below this drop-down list. You can specify the max jacobian value for element creation using this option. The elements will be generated randomly on the geometry within the specified jacobian value. Note that the specified jacobian value must be greater than zero.

Curved

This option is used to create midnode on the geometry irrespective of the element quality.

Linear

This option is used to create midnodes on the geometry in a linear form.

Mesh Settings

The check boxes and sliders available in this rollout are used for the refinement of the mesh to be created. All these options are discussed next.

Surface Curvature Based Size Variation

This slider controls the element refinement and position around the curved surfaces. Moving the slider toward the minimum setting (0%) results in less element refinement and the size of the elements does not vary from its original size. While moving slider toward the maximum setting (100%) results in more refined element and the size of the elements may vary from its original size toward the curvature surface.

Element Growth Rate Through Volume

This slider controls the element density toward the inner side of the geometry surfaces. Moving the slider toward the minimum setting (0%) means the element density will remain constant toward inner surfaces. While the maximum setting (100%) is used for increasing the element density toward inner surfaces.

Minimum Two Elements Through Thickness

Select this check box to create minimum two elements even on a very thin area of the model.

Transition Element Size

Select this check box to create layers in the geometry for smooth generation of the elements.

Auto Fix Failed Elements

Select this check box to auto fix the failed element by reducing its size if quality issue of elements are detected while meshing.

Model Cleanup Options

In this rollout, the **Small Feature Tolerance (% of Element Size)** slider is available. Using this slider, you can eliminate the small feature of the geometry by specifying the percentage value. For example, your slider is at 10% value. You have a feature of 3mm in the geometry and you have specified element size as 30mm. Now, while meshing, the software will eliminate this feature.

Destination Collector

The **Automatic Creation** check box in this rollout is used to create a 3D tetrahedral element at default location. The **New Collector** button 🔲 is used to define physical property of 3D elements. You can also set the predefined 3D mesh collector for the tetrahedral meshing. To do so, clear the **Automatic Creation** check box and select the collector from the **Mesh Collector** drop-down list.

Preview

In this rollout, the **Preview Boundary Nodes** button is available which is used to preview the nodes around the edges of the model.

Figure 5-45 shows a model with 3D tetrahedral meshing.

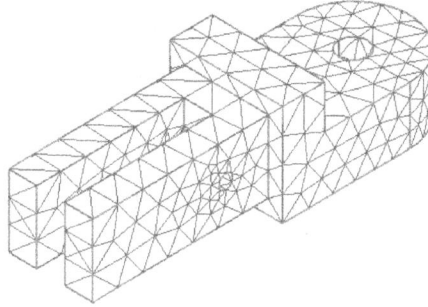

Figure 5-45 3D tetrahedral meshed model

Solid from Shell Mesh Tool

Ribbon: Home > Mesh > More > 3D > Solid from Shell Mesh
Menu: Insert > Mesh > Solid from Shell Mesh

The **Solid from Shell Mesh** tool is used to create 3D tetrahedral mesh using triangular shell elements. Note that for creating 3D tetrahedral mesh, the 2D triangular shell elements should be present in a closed loop. To create solid element using shell meshing, choose **Menu > Insert > Mesh > Solid from Shell Mesh** from the **Top Border Bar**; the **Solid From Shell** dialog box will be displayed, as shown in Figure 5-46. You can use the options available in this dialog box for creating mesh. The options and rollouts available in this dialog box are discussed next.

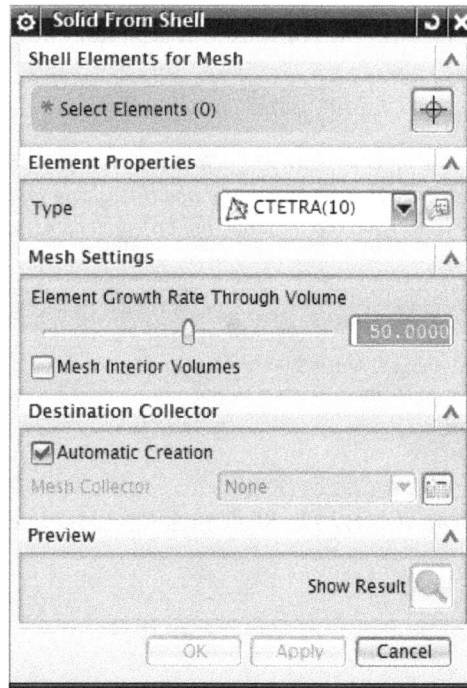

*Figure 5-46 The **Solid From Shell** dialog box*

Shell Elements for Mesh

In this rollout, the **Select Elements (0)** option is highlighted. You need to select 2D triangular shell elements to create 3D tetrahedral mesh. You must select all the 2D shell elements that form a closed loop.

Element Properties

In this rollout, you can select the required tetrahedral element from the **Type** drop-down list. The **Edit Mesh Associated Data** button 🖾 is used to provide additional information related to the element.

> **Note**
> *While creating the element, you can not create mixed element type. For example, if you have created CTRIA3 type shell element, then you can not create CTETRA(10) (midnode) element.*

Mesh Settings

The check box and slider available in this rollout are used for refinement of the mesh to be created. These options are discussed next.

Element Growth Rate Through Volume

This slider controls the element density toward inner side of the geometry surfaces. If you move the slider toward 0%, the density will remain constant and moving it toward 100%, causes the element density to increase toward the inner surfaces.

Mesh Interior Volumes

When you select this check box, the software will generate meshing in the interior volume of the model.

> **Note**
> *If the shell elements do not form a closed loop, then while creating the solid elements, the **Shell To Solid** window will be displayed with a warning about meshing failure, as shown in Figure 5-47. If you choose the **Yes** button from the window, the failed elements will create a group node in the **Simulation Navigator**.*

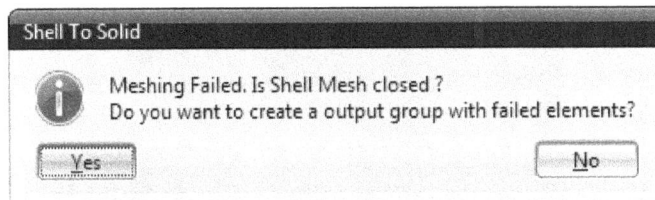

*Figure 5-47 The **Shell To Solid** dialog box*

3D Swept Mesh Tool

Ribbon: Home > Mesh > 3D Swept Mesh
Menu: Insert > Mesh > 3D Swept Mesh

The **3D Swept Mesh** tool is used to create mapped hexahedral mesh in the model. For creating 3D sweep mesh, the model must be sweepable. The method to create sweepable body by using the **Split Body** tool has already been discussed in the previous chapter. To create 3D swept meshing, choose **Menu > Insert > Mesh > 3D Swept Mesh** from the **Top Border Bar**; the **3D Swept Mesh** dialog box will be displayed, as shown in Figure 5-48. You can use the options available in this dialog box for creating the mesh. The options and rollout available in this dialog box are discussed next.

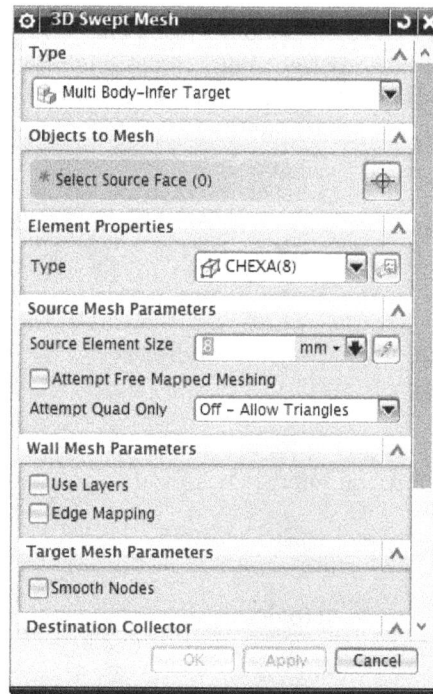

Figure 5-48 The **3D Swept Mesh** dialog box

Type
In this rollout, two options are available in the drop-down list: **Multi Body-Infer Target** and **Until Target**. Usage of these options are discussed next.

Multi Body-Infer Target
When you choose this option, the **Select Source Face (0)** option is highlighted in the **Objects to Mesh** rollout and you need to select the face for creating swept mesh. The software finds the feasible face upto which the swept meshing have to be done.

Until Target
When you choose this option, the **Select Source Face (0)** and **Select Target Face (0)** options will be displayed in the **Objects to Mesh** rollout. Using the **Select Source Face (0)** option,

you can select the face for creating swept mesh. While the **Select Target Face (0)** option is used to select the face upto which the swept meshing is to be done.

Element Properties

For creating swept mesh, you can select the required hexahedral element from the **Type** drop-down list of this rollout. The **Edit Mesh Associated Data** button available next to this drop-down list is used to provide additional information related to the element.

Source Mesh Parameters

You can specify the size of element in the **Source Element Size** edit box. Beside the edit box, the **Automatic Element Size** button is available. When you choose this button after selecting the face, the software examines the face and calculates proper element size according to the face size. Below the edit box, the **Attempt Free Mapped Meshing** check box is available. When you select this check box, the software will attempt free mapped meshing with default settings. The **Attempt Quad Only** drop-down list contains three options: **Off - Allow Triangles**, **On - Zero Triangles**, and **On - Single Triangle**. These options are discussed next.

Off - Allow Triangles

When you choose this option, the software will create a quadrilateral mesh that also contains some triangular element.

On - Zero Triangles

When you choose this option, the software will create a mesh that does not contain any triangular elements.

On - Single Triangle

When you choose this option, the software will create a mesh with maximum one triangular element per selected face.

Wall Mesh Parameters

The check boxes in this rollout are used to define mesh parameters around the walls of the part. These check boxes are discussed next.

Use Layers

This check box is used to divide the wall into layers between a source face and a target face. When you select this check box, the **Number of Layers** edit box will be displayed. Using this edit box, you can specify the value of layers to be created between the source and target faces.

Edge Mapping

Select this check box to project the mesh node vertically from the source face to the target face.

Target Mesh Parameters

The **Smooth Nodes** check box available in this rollout is used to control the mesh generation on the target face. Select this check box to smoothen the nodes of the mesh on the target face by doing some modifications. As a result, the generated mesh on the target face will be different from that on the source face.

Destination Collector

In the **Destination Collector** rollout, the **Automatic Creation** check box is available. When you select this check box, the software will create 3D element at default location. In this rollout, the **New Collector** button 📇 is also available. Using this, you can define physical properties of the 3D elements. You can also assign the predefined 3D mesh collector to the meshing. To do so, clear the **Automatic Creation** check box and select the pre-defined collector from the **Mesh Collector** drop-down list.

Preview

Using the options available in this rollout, you can preview the mesh creation. The **Show Result** button available in the rollout is used to preview the final meshed model.

Figure 5-49 shows a model for source face selection for 3D swept meshing and Figure 5-50 shows the model after creating 3D swept mesh.

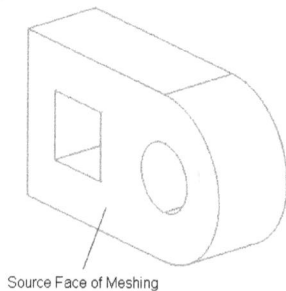

Figure 5-49 Source face of the model for 3D swept meshing

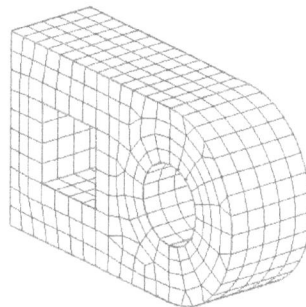

Figure 5-50 Model with 3D swept mesh

TUTORIALS

Tutorial 1

In this tutorial, you will create nodes at (0,0,0), (100,0,0), (200,0,0), (200,100,0), (200,200,0), (300,200,0), and (400,200,0) coordinates and then define mesh collector for 1D elements. Next, you will create 1D elements and assign mesh collector to them. The final created FEM model is shown in Figure 5-51.   **(Expected time: 30 min)**

The following steps are required to complete this tutorial:

a. Invoke FEM Environment.
b. Create nodes at coordinate points (0,0,0), (100,0,0), (200,0,0), (200,100,0), (200,200,0), (300,200,0), and (400,200,0).
c. Define mesh collector for 1D elements in PBAR section.
d. Create 1D elements through created nodes and assign mesh collector to them.
e. Save and close the model.

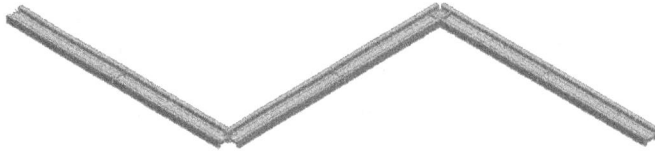

Figure 5-51 Final FEM model

Invoking the FEM Environment

1. Start NX 9.0 by double-clicking on its shortcut icon at the desktop of your computer.

2. Choose **File > New** from the **Ribbon**; the **New** dialog box is displayed. Choose the **Simulation** tab and select **NX Nastran** from the **Name** column and **Fem** from the **Type** column, refer to Figure 5-52.

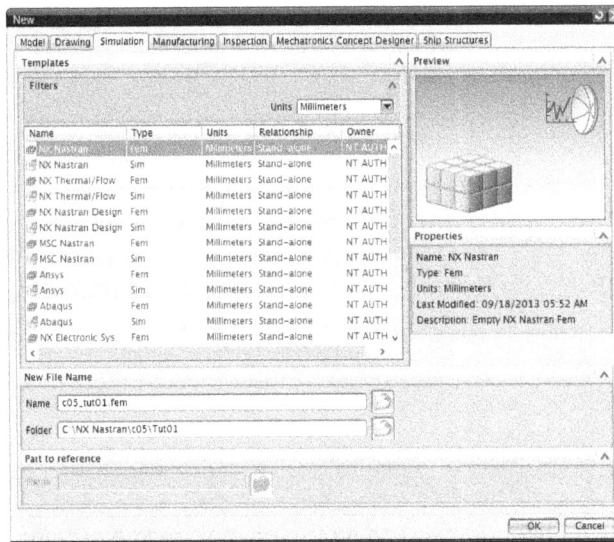

*Figure 5-52 Options selected in the **New** dialog box*

3. Enter **c05_tut01** in the **Name** edit box and specify the location *C:\NX Nastran\c05\Tut01* in the **Folder** edit box. Choose the **OK** button; the **New FEM** dialog box is displayed. Retain the default settings and choose the **OK** button; the FEM environment is invoked.

Note
You must create the Tut01 folder in c05 folder at the location C:\NX Nastran\ as created in the previous chapter.

Creating Nodes

1. Choose **Menu > Insert > Node > Create** from the **Top Border Bar**; the **Node Create** dialog box is displayed, as shown in Figure 5-53.

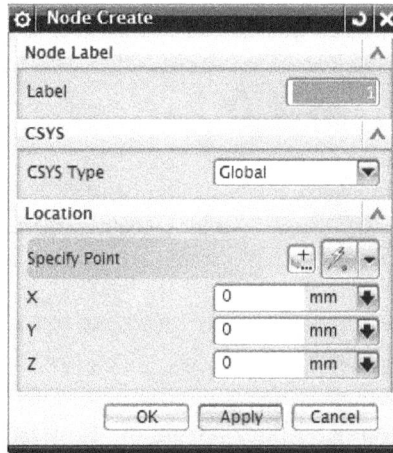

Figure 5-53 The Node Create dialog box

2. Select **Global** from the **CSYS Type** drop-down list in the **CSYS** rollout, if it is not selected by default.

3. Specify the coordinates as (0,0,0) in the **X, Y, Z** edit boxes, respectively, of the **Location** rollout and choose the **Apply** button; the node is created at the specified location.

4. Similarly, create nodes at (100,0,0), (200,0,0), (200,100,0), (200,200,0), (300,200,0), and (400,200,0) locations. After creating all the nodes, choose the **Cancel** button to exit the dialog box.

5. Choose the **TOP** view from the **Orient View Drop-down** list; the created nodes are displayed parallel to the screen.

Creating Mesh Collector for 1D Element

Before creating 1D elements on the created nodes, you need to create the mesh collector and define section for them.

1. Choose **Menu > Insert > Mesh Collector** from the **Top Border Bar**; the **Mesh Collector** dialog box is displayed.

2. Choose the **1D** option from the **Element Family** drop-down list and **Bar Collector** from the **Collector Type** drop-down list; the dialog box gets modified, refer to Figure 5-54.

3. Choose the **Create Physical** button available in the **Physical Property** sub-rollout of the **Properties** rollout; the **PBAR** dialog box is displayed, as shown in Figure 5-55.

Figure 5-54 *The **Mesh Collector** dialog box*

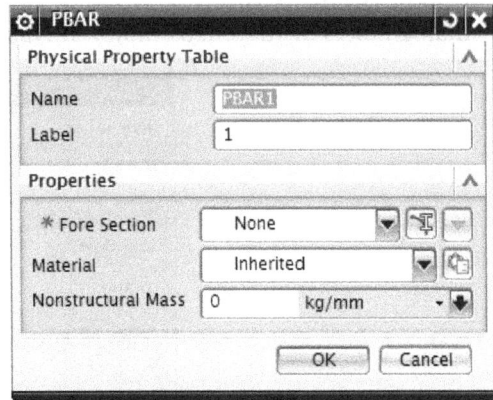

Figure 5-55 *The **PBAR** dialog box*

4. Choose the **Show Section Manager** button from the **Properties** rollout of the **PBAR** dialog box; the **Beam Section Manager** dialog box is displayed, as shown in Figure 5-56.

5. Choose the **Create Section** button from the dialog box; the **Beam Section** dialog box is displayed. Choose the **I** section type from the drop-down list in the **Type** rollout; the dialog box gets modified, refer to Figure 5-57.

Figure 5-56 *The **Beam Section** Manager dialog box*

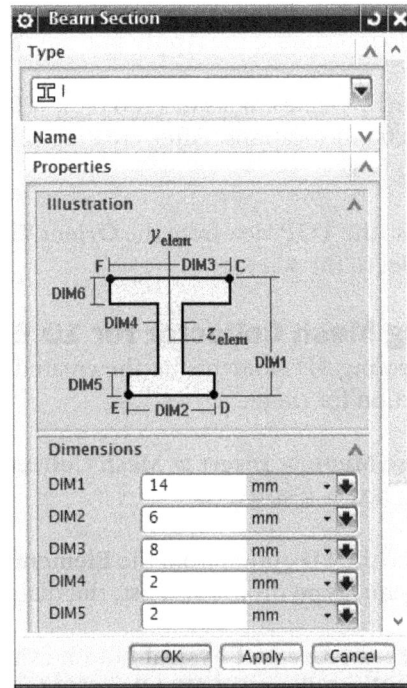

Figure 5-57 *The **Beam Section** dialog box*

6. Specify the dimensions in the edit boxes available in the **Dimensions** sub-rollout of the **Properties** rollout as given below.
   DIM1= 14, **DIM2**= 6, **DIM3**= 8, **DIM4**= 2, **DIM5**= 2, **DIM6**= 3

Tip
*You can evaluate the defined section properties such as Inertia, Centroid, and Area. To evaluate, choose the **Evaluate Section Properties** button from the **Beam Section** dialog box; the **Information** window will be displayed with all the physical properties of the section.*

7. Choose the **OK** button from the dialog box; the **Beam Section Manager** dialog box is displayed again and the created section name is displayed in the area available in the **Cross Section List** rollout.

8. Choose the **Close** button from dialog box; the **PBAR** dialog box is displayed again. Also, the created section name is displayed in the **Fore Section** drop-down list.

9. Choose the **OK** button from the dialog box; the **Mesh Collector** dialog box is displayed again. Now to create this mesh collector, choose the **OK** button from the dialog box; the mesh collector node with the default name is created in the **Simulation Navigator**.

Creating 1D Elements

1. For creating 1D elements, choose **Menu > Insert > Element > Create** from the **Top Border Bar**; the **Element Create** dialog box is displayed, refer to Figure 5-58.

2. Choose the **CBAR** from the **Type** drop-down list of the **Element Properties** rollout and clear the **Automatic Creation** check box from the **Destination Collector** rollout; the created mesh collector is shown in the **Mesh Collector** drop-down list.

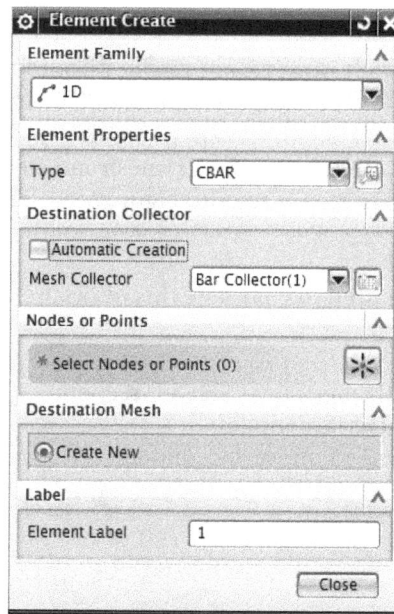

*Figure 5-58 The **Element Create** dialog box*

The **Select Nodes or Points (0)** option is highlighted in the **Nodes or Points** rollout and therefore, you need to select nodes or points to create elements.

3. Select the nodes in a sequential order like (0,0,0) to (100,0,0), then (100,0,0) to (200,0,0), then (200,0,0) to (200,100,0), and so on.

4. Choose the **Close** button from the dialog box; the FEM model after creating elements is shown in Figure 5-59.

Figure 5-59 *Final FEM model*

Saving and Closing the Model

1. Choose **File > Save > Save** from the **Ribbon** to save the model.

2. Choose **File > Close > All Parts** from the **Ribbon**; all displayed parts are closed.

Tutorial 2

In this tutorial, you will define mesh collector for 2D mesh and create 2D mesh on the midsurface model created in Tutorial 2 of Chapter 4, refer to Figure 5-60. The final meshed FEM model is shown in Figure 5-61.   **(Expected time: 30 min)**

The following steps are required to complete this tutorial:

a. Open the FEM model.
b. Define mesh collector for 2D mesh for PSHELL section.
c. Create 2D mesh on the midsurface model.
d. Display thickness of the meshed midsurface model.
e. Save and close the model.

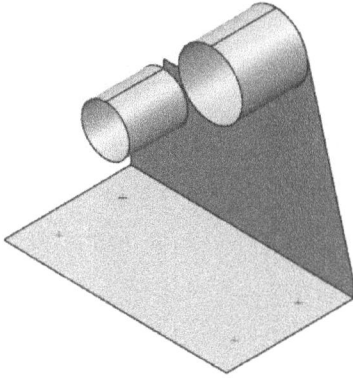

Figure 5-60 Tutorial 2 of chapter 4

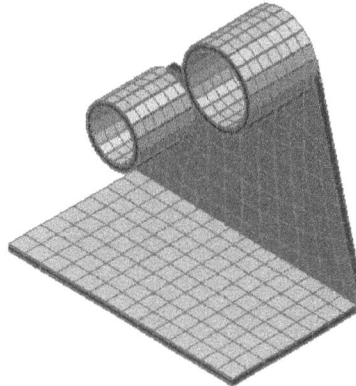

Figure 5-61 Final meshed midsurface model

Opening the FEM File

1. Start NX 9.0 by double-clicking on its shortcut icon located on the desktop of your computer.

2. Copy the **c04_tut02.fem** file from the location *C:\NX Nastran\c04\Tut02* and paste it to *C:\NX Nastran\c05\Tut02*. Change the file name to **c05_tut02.fem**. You can also download this file from *www.cadcim.com*.

> **Note**
> *You must create the Tut02 folder in c05 folder at the location C:\NX Nastran\ as created in the previous chapter.*

3. Choose **File > Open** from the **Ribbon**; the **Open** dialog box is displayed. Select the **c05_tut02.fem** file from the location and choose the **OK** button; the file is open in the FEM environment.

Creating Mesh Collector for 2D Mesh

Before creating 2D mesh on the midsurface model, you need to create the mesh collector and define thickness for the final meshed model.

1. Choose **Menu > Insert > Mesh Collector** from the **Top Border Bar**; the **Mesh Collector** dialog box is displayed, refer to Figure 5-62.

2. Choose the **2D** option from the **Element Family** drop-down list and the **ThinShell** from the **Collector Type** drop-down list; other options are updated in the dialog box, refer to Figure 5-62.

3. Choose the **Create Physical** button from the **Physical Property** sub-rollout of the **Properties** rollout; the **PSHELL** dialog box is displayed, as shown in Figure 5-63.

*Figure 5-62 The **Mesh Collector** dialog box*

4. Specify the thickness value of the meshed midsurface model as **4** mm in the **Default Thickness** edit box of the **Properties** rollout.

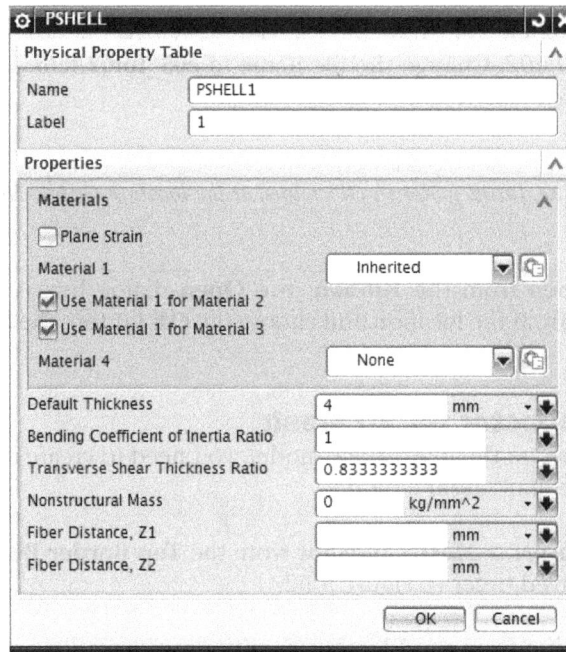

*Figure 5-63 The **PSHELL** dialog box*

5. Choose the **OK** button to exit the **PSHELL** dialog box; the **Mesh Collector** dialog box is displayed again and the created property name is displayed in the **Shell Property** drop-down list of the **Physical Property** sub-rollout of the **Properties** rollout.

6. Choose the **OK** button from the dialog box; the mesh collector node with the default name is created in the **Simulation Navigator**.

Creating 2D Mesh

1. Choose **Menu > Insert > Mesh > 2D Mesh** from the **Top Border Bar**; the **2D Mesh** dialog box is displayed, as shown in Figure 5-64. Also, you are prompted to select the model.

2. Select all the surfaces of the model from the drawing area.

3. Select the CQUAD4 element type from the **Type** drop-down list of the **Element Properties** rollout, if not selected by default, and specify the element size as **10** in the **Element Size** edit box of the **Mesh Parameters** rollout.

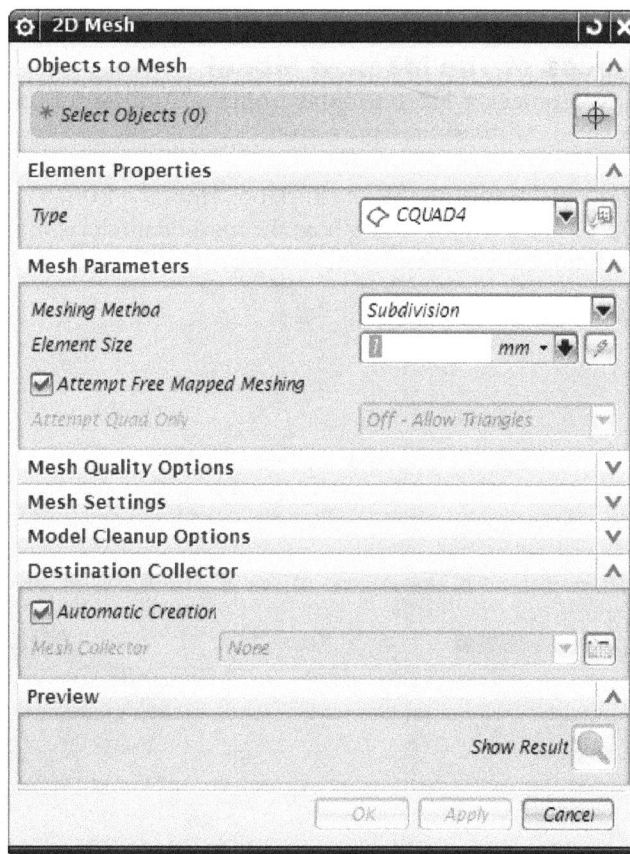

*Figure 5-64 The **2D Mesh** dialog box*

4. Clear the **Automatic Creation** check box of the **Destination Collector** rollout; the created mesh collector is displayed in the **Mesh Collector** drop-down list of this rollout.

5. Retain other default values and choose the **OK** button; the mesh is created on the midsurface model, as shown in Figure 5-65.

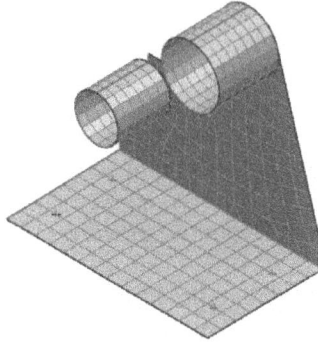

Figure 5-65 *The meshed midsurface model*

Displaying Thickness of The Meshed Model

1. Choose **Menu > Preferences > Mesh Display** from the **Top Border Bar**; the **Mesh Display** dialog box is displayed, as shown in Figure 5-66.

2. Select the **Element Thickness and Offset** check box from the **2D** tab of the **Display** rollout and choose the **OK** button from the dialog box; the meshed model with the defined thickness is created, as shown in Figure 5-67.

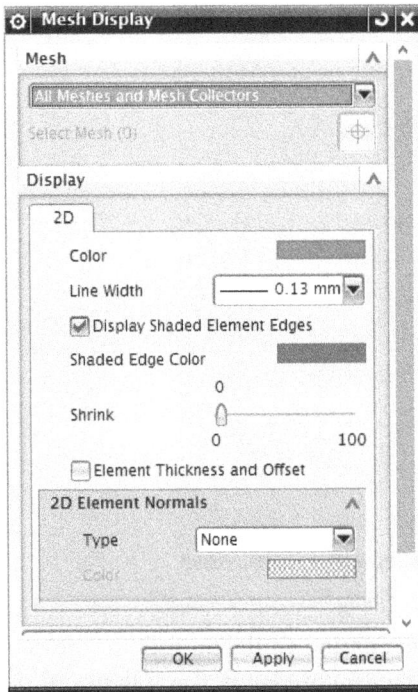

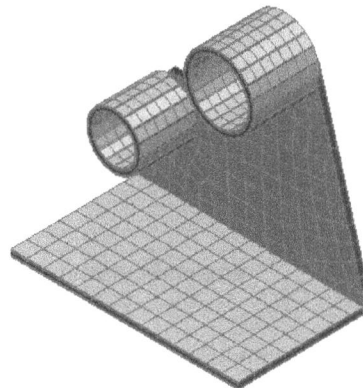

Figure 5-66 *The **Mesh Display** dialog box*

Figure 5-67 *The meshed midsurface model with the defined thickness*

Saving and Closing the Model

1. Choose **File > Save > Save** from the **Ribbon** to save the model.

2. Choose **File > Close > All Parts** from the **Ribbon**; all displayed parts are closed.

Tutorial 3

In this tutorial, you will create 3D swept mesh on the model splitted in Tutorial 1 of Chapter 4, refer to Figure 5-68. The final meshed FEM model is shown in Figure 5-69.

(Expected time: 30 min)

The following steps are required to complete this tutorial:

a. Open the FEM model.
b. Create the 3D swept mesh on the model.
c. Save and close the model.

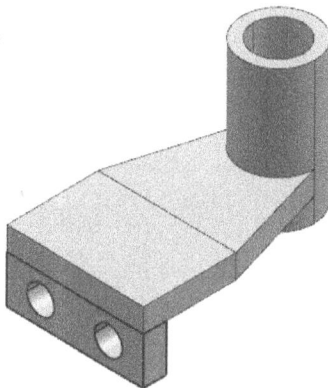

Figure 5-68 Final meshed model   *Figure 5-69 Final meshed model*

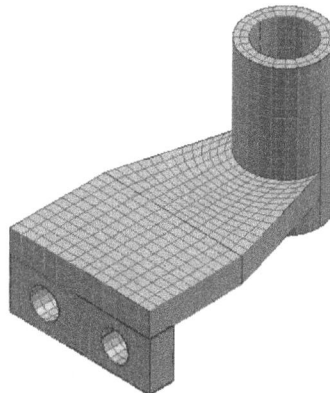

Opening the FEM File

1. Start NX 9.0 by double-clicking on its shortcut icon on the desktop of your computer.

2. Copy the *c04_tut01_igs_fem1.fem* file from *C:\NX Nastran\c04\Tut01* location and paste it to *C:\NX Nastran\c05\Tut03*. Change the file name to **c05_tut03.fem**. You can also download this file from *www.cadcim.com*.

> **Note**
> *You must create the Tut03 folder in c05 folder at the location C:\NX Nastran\ as created in the previous chapter.*

3. To open FEM file, choose **File > Open** from the **Ribbon**; the **Open** dialog box is displayed. Select the **c05_tut03.fem** file from the location and choose the **OK** button; the file is opened in the FEM environment.

Creating 3D Swept Mesh

1. Choose **Menu > Insert > Mesh > 3D Swept Mesh** from the **Top Border Bar**; the **3D Swept Mesh** dialog box is displayed, as shown in Figure 5-70. Also, you are prompted to select the face for swept mesh.

Figure 5-70 The 3D Swept Mesh dialog box

2. Select the top surface of the cylindrical feature of the model, as shown in Figure 5-71.

3. Select the CHEXA(8) element type from the **Type** drop-down list of the **Element Properties** rollout. Next, specify the element size **4** in the **Source Element Size** edit box and choose the **On-Zero Triangles** option from the **Attempt Quad Only** drop-down list of the **Source Mesh Parameters** rollout. Also, select the **Attempt Free Mapped Meshing** check box available in this rollout.

4. Retain all other default values and choose the **Apply** button; the mesh is created on the model, as shown in Figure 5-72.

5.   Select the top face of the model, as shown in Figure 5-73 and choose the **Apply** button; the mesh is created on the half of the model.

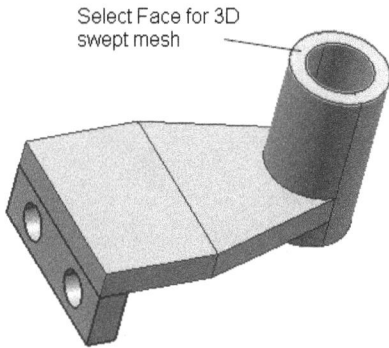

Figure 5-71   Faces to be selected to create swept mesh

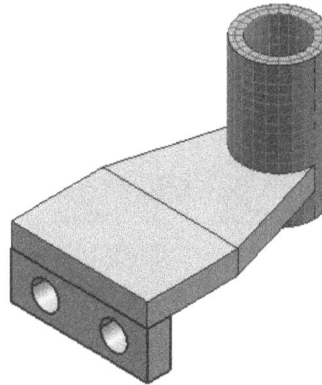

Figure 5-72   Swept mesh created on the cylindrical feature

6.   Similarly, create swept mesh on the side face of the top feature and side face of the bottom feature of the model, as shown in Figures 5-74 and 5-75. Choose the **Cancel** button to exit from the dialog box. The final swept mesh model is shown in Figure 5-76.

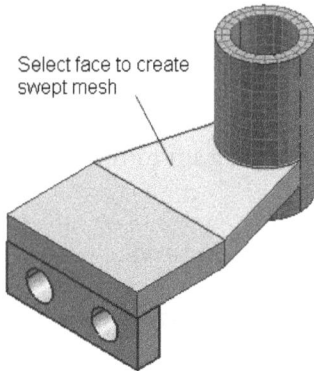

Figure 5-73   Top face to be selected to create swept mesh

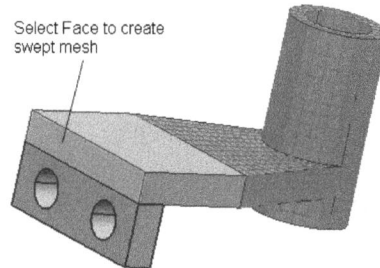

Figure 5-74   Side face to be selected of top face to create swept mesh

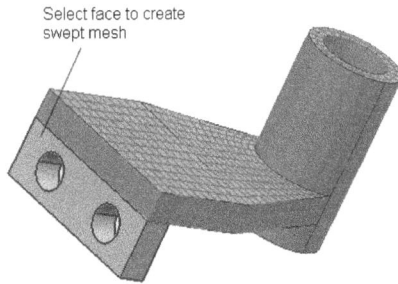

Figure 5-75 *Side face of the bottom feature to create swept mesh*

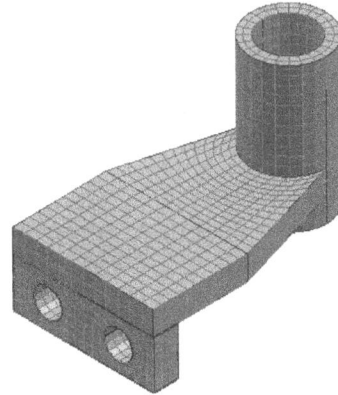

Figure 5-76 *Final meshed model*

Saving and Closing the Model

1. Choose **File > Save > Save** from the **Ribbon** to save the model.

2. Choose **File > Close > All Parts** from the **Ribbon**; all displayed parts are closed.

Self-Evaluation Test

Answer the following questions and then compare them to those given at the end of this chapter:

1. The _____ tool is used to create elements by sweeping up to the specified surface using the existing elements.

2. The _____ are used to create elements for rod or beam behavior.

3. _____ and _____ are two options to define 2D meshing.

4. In the _____, you can define physical properties of the elements.

5. The _____ tool is used to create 3D tetrahedral mesh using available triangular shell elements.

6. In a finite element model, you need to create nodes for performing analysis. (T/F)

7. The **Between Nodes** option is used to create node between two existing nodes. (T/F)

8. You can provide similar number label to two different elements. (T/F)

9. The 0D elements are used to create concentrated masses on a point. (T/F)

10. The 3D elements are used to create elements for solid structure. (T/F)

Review Questions

Answer the following questions:

1. Which of the following tools is used to create an element by selecting existing nodes or points in the environment?

    (a) **Element Revolve**   (b) **Copy And Project**
    (c) **Copy And Translate**   (d) **Element Create**

2. Which of the following tools is used to create elements by sweeping elements up to a surface?

    (a) **Copy And Translate**   (b) **Copy and Reflect**
    (c) **3D Sweep Between**   (d) **Element Extrude**

3. Which of the following tools is used create mapped hexahedral meshing on the model?

    (a) **Solid from Shell Mesh**   (b) **3D Tetrahedral Mesh**
    (c) **3D Swept Mesh**   (d) None of these

4. Which of the following tools is used to create a division of continuous triangular or quadrilateral elements?

    (a) **2D Mapped Mesh**   (b) **2D Mesh**
    (c) **Solid from Shell Mesh**   (d) **Surface Coat**

5. Which of the following tools is used to create coating of 2D elements on the solid elements?

    (a) **Element Revolve**   (b) **Surface Coat**
    (c) **3D Tetrahedral Mesh**   (d) None of these

6. Using the **Auto Element Size** button, you can examines the solid body and calculate proper element size. (T/F)

7. By using the **Mesh Collector**, you can define the thickness of the 2D meshed surface. (T/F)

8. For creating 3D sweep elements between faces, you need to choose the source and target face elements. (T/F)

9. Nodes are used to create elements. (T/F)

10. The **Element Extrude** tool is used to extrude an element upto a defined distance. (T/F)

EXERCISES

Exercise 1

In this exercise, you will define mesh collector for 2D mesh and create 2D mesh on the midsurface model created in Exercise 1 of Chapter 4, refer to Figure 5-77. The final meshed FEM model shown in Figure 5-78. **(Expected time: 45 min)**

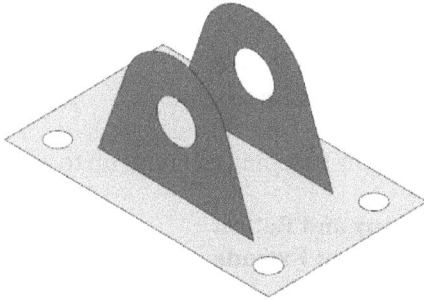

Figure 5-77 Final midsurface model

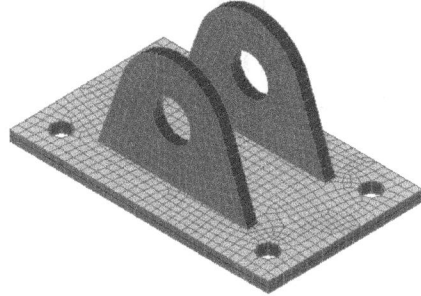

Figure 5-78 Final meshed midsurface model

Hint

1. Copy the **c04_exr01.fem** file from the location *C:\NX Nastran\c04\Exr01* and paste it to *C:\NX Nastran\c05\Exr01*. Change the file name to **c05_exr01.fem**.
2. Open the file in NX.
3. Open the **Mesh Collector** dialog box and specify the thickness of the meshed midsurface model **4** mm in the **Default Thickness** edit box for the **2D** option and the **ThinShell** collector type.
4. Create 2D mesh with CQUAD4 element type and 3 element size. Also, choose the **On-Zero Triangles** option from the **Attempt Quad Only** drop-down list.
5. Display the thickness of the meshed model using the **Mesh Display** dialog box.
6. Save and close the model.

Exercise 2

In this exercise, you will create 3D swept mesh on the model splitted in Exercise 2 of Chapter 4, refer to Figure 5-79. The final meshed FEM model shown in Figure 5-80. **(Expected time: 45 min)**

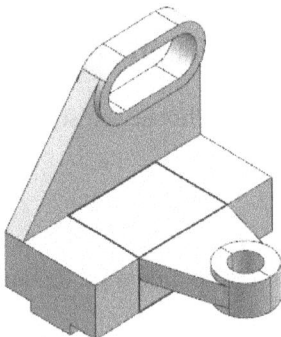

Figure 5-79 Final midsurface model

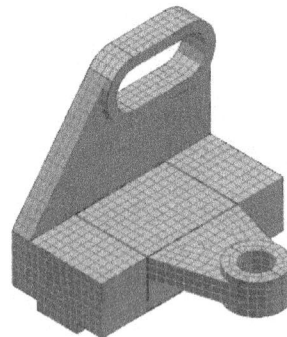

Figure 5-80 Final meshed midsurface model

Hint

1. Copy the **c04_exr02.fem** file from the location *C:\NX Nastran\c04\Exr02* and paste it to *C:\NX Nastran\c05\Exr02*. Change the file name to **c05_exr02.fem**.

2. Open the file in NX.

3. Open the **3D Swept Mesh** dialog box and specify 4 mm element size for CHEXA(8) element type. Also, choose the **Off-Allow Triangles** option from the **Attempt Quad Only** drop-down list and select the **Attempt Free Mapped Meshing** check box.

4. Select the front face of the top feature, as shown in Figure 5-81 and choose the **Apply** button to create swept mesh.

5. Similarly, choose other faces of the model, refer to Figures 5-82 through 5-88.

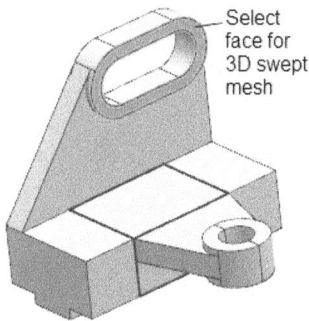

Figure 5-81 *Select face to create mesh*

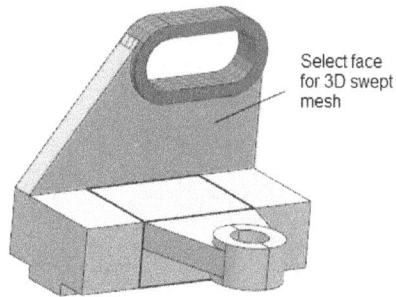

Figure 5-82 *Select front face to create mesh*

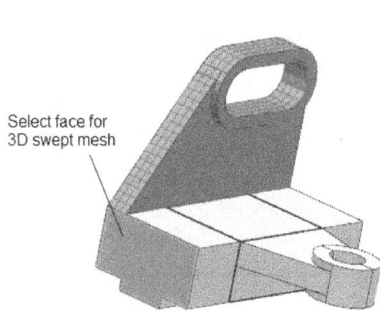

Figure 5-83 *Side face to be selected to create mesh*

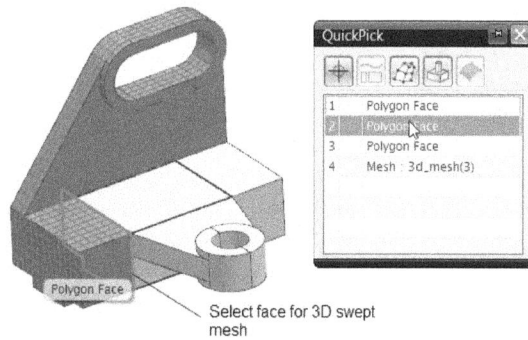

Figure 5-84 *Inner face to be selected to create mesh*

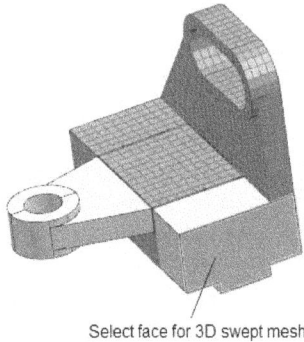

Figure 5-85 *Another side face to be selected to create mesh*

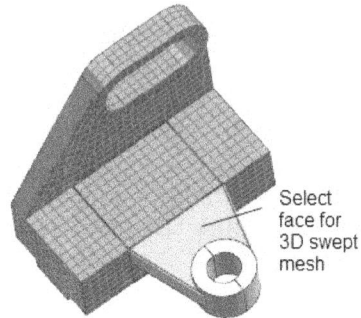

Figure 5-86 *Top face to select to create mesh*

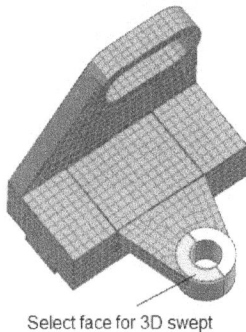

Figure 5-87 *Top circular face to be selected to create mesh*

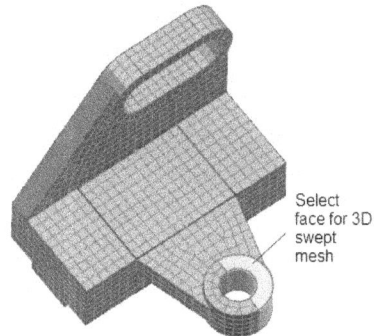

Figure 5-88 *Another top circular face to be selected to create mesh*

Note

*For selecting inner face of the model, you have to invoke the **QuickPick** dialog box, refer to Figure 5-84. To do so, right-click on the face attached to inner face and choose the **Select from list** option from the menu; the **QuickPick** dialog box is displayed. You can use this dialog box to choose the inner face of the model, refer to Figure 5-84.*

Answers to Self-Evaluation Test

1. 3D Sweep Between, 2. 1D elements, **3. Subdivision, Paver, 4. Destination Collector, 5. Solid from Shell Mesh, 6.** T, **7.** T, **8.** F, **9.** T, **10.** T

Chapter 6

Meshing-II

Learning Objectives

After completing this chapter, you will be able to:

- *Determine element quality*
- *Check element edges*
- *Check normals of 2D elements*
- *Check for duplicate nodes and duplicate elements*
- *Check element material orientation*
- *Adjust node proximity to CAD geometry*

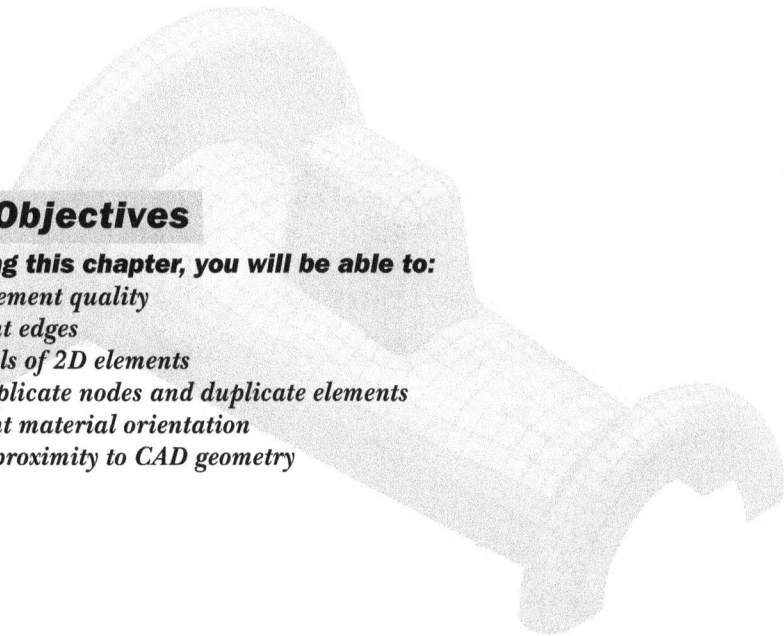

INTRODUCTION

In the previous chapters, you learned to create mesh. After creating the mesh, you need to analyze the quality of the mesh created. To create a proper meshed model, the meshing must clear some specific criteria. You will learn more about these criteria and the tools to check the created element/meshed model in this chapter.

MESH ANALYSIS

After the mesh creation, you need to check the quality of created mesh. To check the mesh quality you need to analyze the element quality and free and non-manifold elements' edges, check the direction of 2D elements, check and merge duplicate nodes and elements, check the material orientation of the elements and node adjustment, and so on. The tools to analyze the mesh are discussed next.

Element Quality

Ribbon: Home > Checks and Information > Element Quality
Menu: Analysis > Finite Element Model Check > Element Quality

This tool is used to check the created elements on the basis of their geometry. The geometry includes size of edges of element, jacobian, skew, warp values, and so on. To check element quality, choose **Menu > Analysis > Finite Element Model Check > Element Quality** from the **Top Border Bar**; the **Element Quality** dialog box will be displayed, as shown in Figure 6-1. You can use the options available in the dialog box for checking the elements/mesh quality. The options and rollouts available in the dialog box are discussed next.

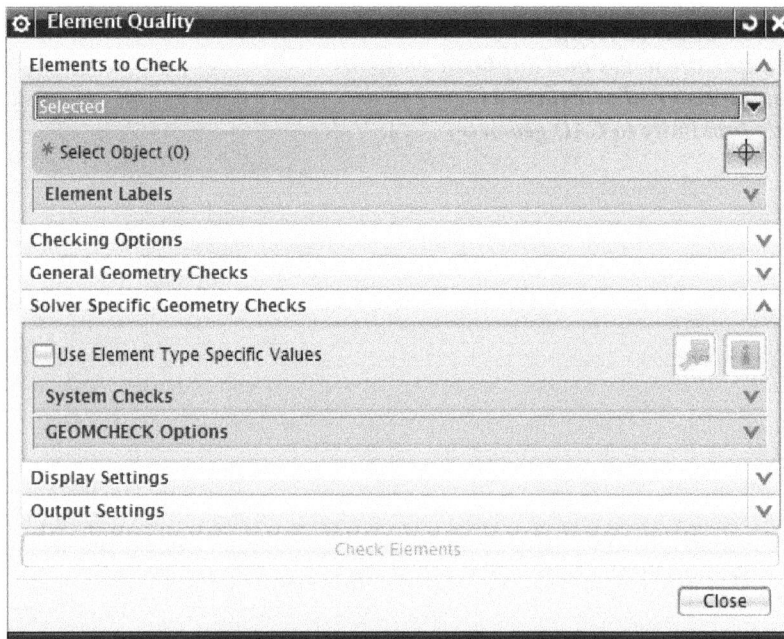

Figure 6-1 The *Element Quality* dialog box

Elements to Check

By using the options available in this rollout, you can select the element edges that need to be checked. There are two options available in the drop-down list of this rollout: **Selected** and **Displayed**. The **Selected** option is used to select user-defined set of elements or mesh while the **Displayed** option is used to select all displayed elements or mesh. You can also select the elements by specifying their labels in the **Labels** edit box of the **Element Labels** sub-rollout. Note that this sub-rollout will be available only when you select the **Selected** option from the drop-down list.

Checking Options

In this rollout, two radio buttons are available: **Warning and Error Limits** and **Error Limits Only**. By default, the **Warning and Error Limits** radio button is selected. While checking the quality, the software will display the warning as well as the error elements that are not following the specified criteria. On selecting the **Error Limits Only** radio button, software will only display the error elements that are not following the specified criteria while checking.

General Geometry Checks

In this rollout, you can define the criteria for the element size. Using the **Calculate Element Edge Size** button, you can find out the maximum and minimum edge length of the selected elements of the model. When you choose this button, the **Information** window will be displayed, as shown in Figure 6-2. The window displays the maximum and minimum sizes of the elements. In the rollout, two check boxes are also available: **Maximum Size >** and **Minimum Size <**. When you select these check boxes, the **Warning Limit** and **Error Limit** edit boxes will be activated. You can specify the size of elements in these edit boxes to define the elements criteria.

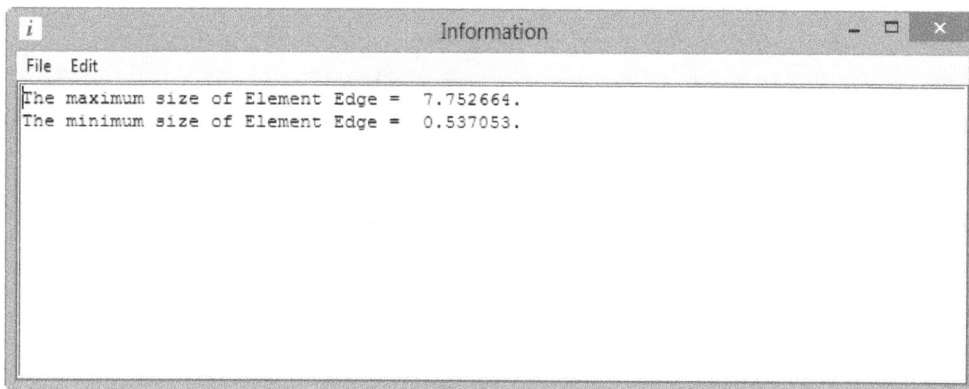

*Figure 6-2 The **Information** window with size of edges*

Solver Specific Geometry Checks

In this rollout, options are available for defining software specific criteria for elements size. When you select the **Use Element Type Specific Values** check box, the **Edit Element Type Specific Values** and **List Element Type Specific Values** buttons will be available. When you choose the **Edit Element Type Specific Values** button, the respective quality checks dialog box will be displayed. For example, in case of NX Nastran, the **NX Nastran Quality Checks** dialog box will be displayed, as shown in Figure 6-3. You can specify the software specific element's criteria in this dialog box. When you choose the **List Element Type Specific Values** button, the **Information** window will be displayed that shows the defined element criterion, as shown in Figure 6-4.

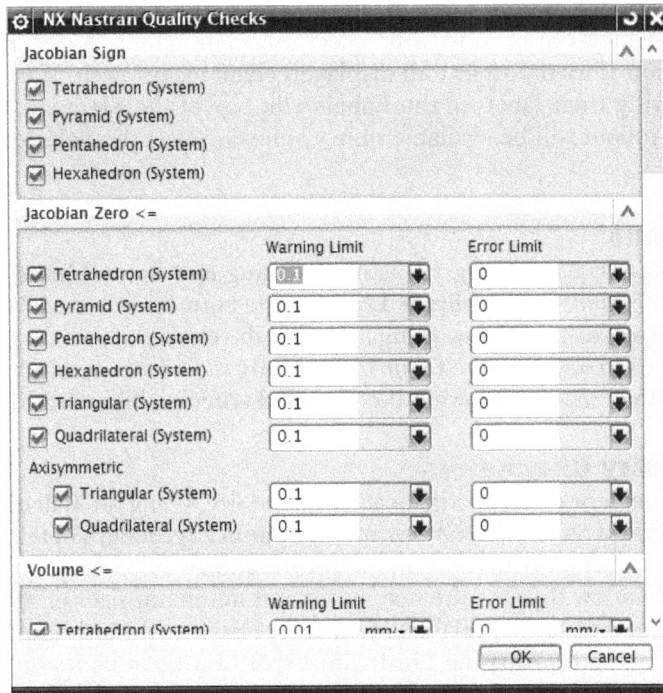

*Figure 6-3 The **NX Nastran Quality Checks** window*

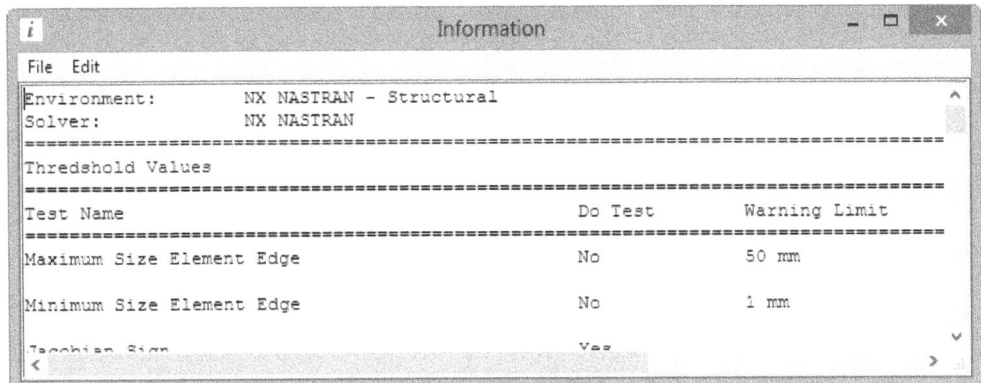

*Figure 6-4 The **Information** window with the element criteria*

You can also specify common criteria to check elements in the edit boxes corresponding to the options that will be available when the **Use Element Type Specific Values** check box is cleared.

Display Settings

In this rollout, options are available for controlling the display of error and failed elements. You can change the color of error and failed elements to be displayed using the respective color swatches. When you select the **Show Element Labels** check box, the labels of the respective violated elements will be displayed. Using the **Hide Input Meshes** button, you can hide the meshes that qualify the specified values.

Output Settings

In this rollout, various options are available for generating an output group or report of the relevant selected elements criteria. For example, if you have selected the **Failed and Warning** option from the **Output Group Elements** drop-down list, a group of failed elements will be created in **Simulation Navigator**, if that exist. Similarly, if you have selected the **Failed and Warning** option from the **Report** drop-down list, an information window will be displayed with the data of failed and warning elements.

The **Check Elements** button available at the bottom of the dialog box is used to check the elements quality after defining the criteria of all elements.

Element Edges

Ribbon: Home > Checks and Information > More > Checks > Element Edges
Menu: Analysis > Finite Element Model Check > Element Edges

The **Element Edges** tool is used to check the created free and non-manifold elements' edges. To check the element quality, choose **Menu > Analysis > Finite Element Model Check > Element Edges** from the **Top Border Bar**; the **Element Edges** dialog box will be displayed, as shown in Figure 6-5. You can use the options available in this dialog box to analyze the element of the created free or overlapped edges of the elements. The options and rollouts available in the dialog box are discussed next.

Elements to Check

In this rollout, two options are available in a drop-down list: **Selected** and **Displayed**. The **Selected** option is used to select a set of elements or mesh while the **Displayed** option is used to select all displayed elements or mesh. You can also select the elements by specifying the label in the **Labels** edit box available in the **Element Labels** sub-rollout of the **Elements to Check** rollout.

Figure 6-5 The Element Edges dialog box

Elements Edge Checks

In this rollout, two check boxes are available: **Free Edges** and **Non-Manifold Edges**. The **Free Edges** check box is used to display all free element edges which are not connected to any other element and exist as separate entities in the model. The **Non-Manifold Edges** check box is used to display the elements edges that share the faces of more than two elements.

Display Settings

Using this rollout, you can change the settings of the free and non-manifold edges to be displayed.

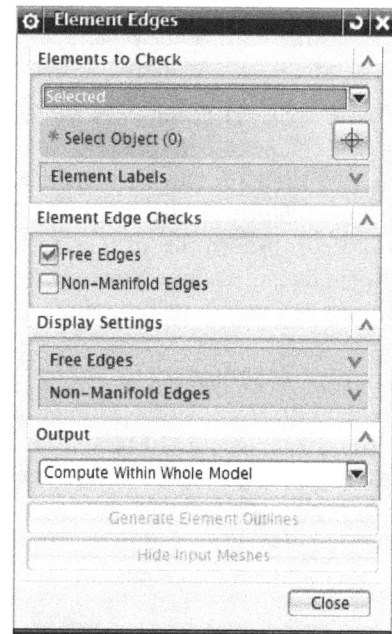

Two sub-rollouts are available in this rollout: **Free Edges** and **Non-Manifold Edges**. You can change the color, font, and width of the displayed edges using these sub-rollouts. Note that the options in these sub-rollouts will be activated only if you have selected the respective check box from the **Elements Edge Checks** rollout.

Output

Using this rollout, you can control the field in which the software will search for the free and non-manifold edges. For example, if you have selected the **Compute Within Visible Model** option from the drop-down list of this rollout, the software will find the deformed edges within meshes that are currently displayed. Whereas, if you have selected the **Compute Within Whole Model** option, the software will find the deformed edges within all meshes including hide meshes of the model.

The **Generate Element Outlines** button is used to highlight the free or non-manifold element edges. The **Hide Input Meshes** button is used to hide the selected mesh or mesh edges that qualify the software criteria so that the failed elements can stand out prominently.

2D Element Normals

Ribbon: Home > Checks and Information > More > Checks > 2D Element Normals
Menu: Analysis > Finite Element Model Check > 2D Element Normals

The **2D Element Normals** tool is used to check and define the normal direction of 2D elements. To check and define element direction, choose **Menu > Analysis > Finite Element Model Check > 2D Element Normals** from the **Top Border Bar**; the **2D Element Normals** dialog box will be displayed, as shown in Figure 6-6. Two options are available in the drop-down list of the **Method** rollout to check and define direction of elements: **Check Normals** and **Orient Normals Using Seed Element**. The options and rollouts available in the dialog box are discussed next.

Check Normals

When you select this option from the **Method** rollout, two options become available in the drop-down list of the **Element to Check** rollout: **Selected** and **Displayed**. The **Selected** option is used to select user-defined set of elements or mesh while the **Displayed** option is used to select all displayed elements or mesh. You can also select the element by specifying the label in the **Labels** edit box in the **Element Labels** sub-rollout of the **Element to Check** rollout.

The **Display Normals** button in the **2D Element Normals** dialog box is used to display the normal direction of the selected elements.

Using the **Reverse Normals** button, you can reverse the direction of the elements.

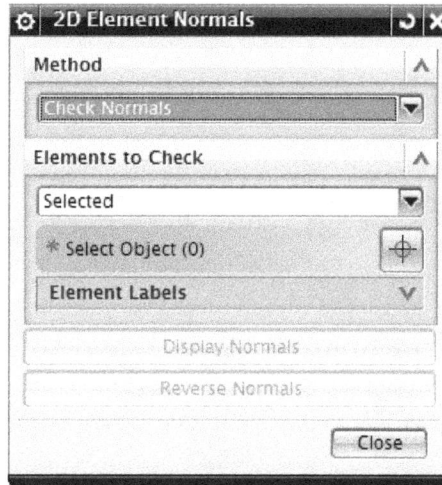

*Figure 6-6 The **2D Element Normals** dialog box*

Orient Normals Using Seed Elements

When you select this option from the **Method** rollout, the **Seed Element (0)** option is highlighted in the **Seed Element** rollout and you are prompted to select the first reference element. The selected element would act as the reference directional element. You can also show the reverse direction of the selected element by selecting the **Reverse Normal** check box.

For displaying the normal direction of other elements with reference to the seed element, three options are available in the drop-down list of the **Scope** rollout: **Connected Elements in Seed Mesh**, **All Visible Connected 2D Elements**, and **Selected Connected Elements**.

Connected Elements in Seed Mesh

Using this option, you can check and define normal direction of all the elements that are connected to the selected seed element.

All Visible Connected 2D Elements

Using this option, you can check and define normal direction of all the currently displayed 2D elements.

Selected Connected Elements

Using this option, you can check and define normal direction of the selected elements. When you select this option, the **Select Element (0)** option will be highlighted in the **Connected Elements** sub-rollout. Next, you need to select the elements whose direction you want to show or change. You can also select the elements by specifying the label in the **Labels** edit box in the **Element Labels** sub-rollout available in the **Connected Elements** sub-rollout.

After selecting all parameters, choose the **Align Normals** button; the normal of the selected elements will match the normal of the seed element. The meshed model with 2D normals is shown in Figure 6-7.

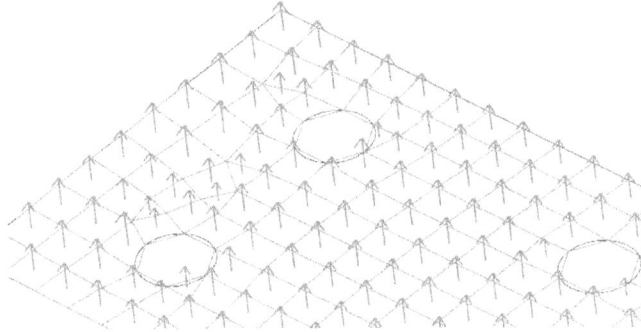

Figure 6-7 *2D Normals of the mesh*

Duplicate Nodes

Ribbon: Home > Checks and Information > More > Checks > Duplicate Nodes
Menu: Analysis > Finite Element Model Check > Duplicate Nodes

The **Duplicate Nodes** tool is used to check and merge the duplicate nodes in the meshed geometry. To check the element quality, choose **Menu > Analysis > Finite Element Model Check > Duplicate Nodes** from the **Top Border Bar**; the **Duplicate Nodes** dialog box will be displayed, as shown in Figure 6-8. The options and rollouts available in the dialog box are discussed next.

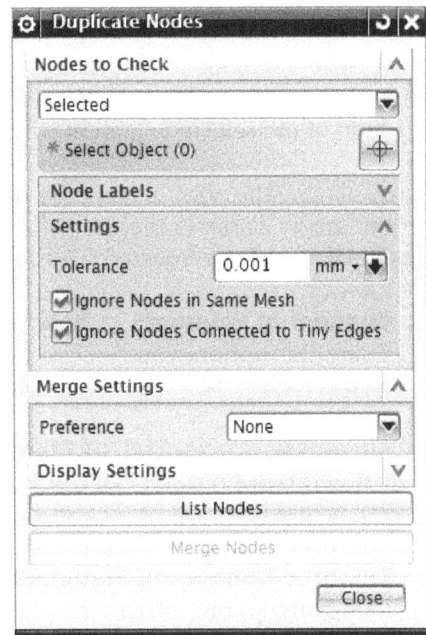

Nodes to Check

In this rollout, two options are available in the drop-down list: **Selected** and **Displayed**. The **Selected** option is used to select the user-defined set of nodes, while the **Displayed** option is used to select all displayed nodes. You can also select the nodes by specifying the label in the **Labels** edit box available in the **Node Labels** sub-rollout of the **Nodes to Check** rollout.

You can set the tolerance value in the **Tolerance** edit box of the **Settings** sub-rollout of the **Nodes to Check** rollout. Using this value, the software will differentiate between two nodes, whether they are duplicate or not. In the **Settings** sub-rollout of the **Nodes to Check** rollout,

Figure 6-8 *The **Duplicate Nodes** dialog box*

two check boxes are available: **Ignore Nodes in Same Mesh** and **Ignore Nodes Connected to Tiny Edges**. If you select the **Ignore Nodes in Same Mesh** check box, the software will check duplicate nodes only between different meshes. And, if you clear this check box, the software will check for duplicate nodes regardless of whether they belong to the same mesh or not. If you select the **Ignore Nodes Connected to Tiny Edges** check box, the software will ignore the nodes that are attached with an edge having a tolerance value less than the specified one.

Merge Settings

In this rollout, various options are available that are used to merge the duplicate nodes. In the **Preference** drop-down list of this rollout, five options are available to merge nodes: **None**, **Retain High Node Label**, **Retain Low Node Label**, **Retain Selected**, and **Merge Selected**. If you select the **None** option, software will merge all the nodes without considering the numbering of nodes. The **Retain High Node Label** option is used to retain the nodes with high label and merge only the low label nodes. The **Retain Low Node Label** option is used to retain the nodes with low label and merge only the high label nodes. The **Retain Selected** option is used to retain the selected nodes. The **Merge Selected** option is used to merge the selected nodes. You can select the nodes using the **Select Node (0)** option which will be available when the **Retain Selected** and **Merge Selected** options are chosen.

Display Settings

The options available in this rollout are used for displaying the duplicate nodes. If you select the **Show Duplicate Nodes** check box, the software will display all duplicate nodes in the meshed model. You can show labels of the nodes using the **Merged** and **Retained** check boxes available in this rollout. You can also change the color display of duplicate nodes using the **Retained Nodes**, **Merged Nodes**, or **Unmergeable Nodes** color swatches. You can control the list to be displayed of duplicate nodes using the options available in the **List Node Pairs** drop-down list.

The **List Nodes** button in the dialog box is used to display the list of all mergeable or non-mergeable nodes. When you choose this button, the **Information** window will be displayed with respect to the selected nodes option, as shown in Figure 6-9. You can select the nodes type from the **List Node Pairs** drop-down list of the **Display Settings** rollout.

The **Merge Nodes** button is used to merge the duplicate nodes on the basis of the settings used in the **Merge Settings** rollout.

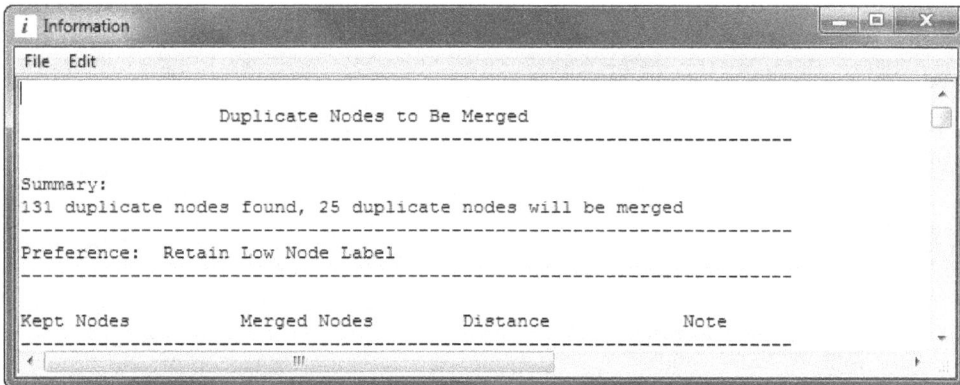

Figure 6-9 *The **Information** window with the selected option*

Duplicate Elements

Ribbon: Home > Checks and Information > More > Checks > Duplicate Elements
Menu: Analysis > Finite Element Model Check > Duplicate Elements

The **Duplicate Elements** tool is used to check the duplicate elements in the meshed geometry. To check duplicate elements, choose **Menu > Analysis > Finite Element Model Check > Duplicate Elements** from the **Top Border Bar**; the **Duplicate Elements** dialog box will be displayed, as shown in Figure 6-10. The options and rollouts available in the dialog box are discussed next.

Elements to Check

In this rollout, two options are available in the drop-down list: **Selected** and **Displayed**. The **Selected** option is used to select the set of elements while the **Displayed** option is used to select all the displayed elements. You can also select the elements by specifying the label in the **Labels** edit box available in the **Element Labels** sub-rollout of the **Elements to Check** rollout.

*Figure 6-10 The **Duplicate Elements** dialog box*

Deletion Settings

In this rollout, you can select the elements to be retained while deleting duplicate elements from the **Preference** drop-down list. There are four options available in this drop-down list: **Retain High Label**, **Retain Low Label**, **Retain Selected**, and **Delete Selected**. The **Retain High Label** option is used to retain the elements with high label and delete low label elements. The **Retain Low Label** option is used to retain the elements with low label and delete high label elements. The **Retain Selected** option is used to retain the selected elements. You can select the elements using the **Select Elements (0)** option which will be available when the **Retain Selected** option is chosen. You can also select elements by specifying the labels in the **Labels** edit box available in the **Element Labels** sub-rollout of this rollout. The **Delete Selected** option is used to delete the selected duplicate elements.

Display Settings

The options in this rollout are used to display duplicate elements. If you select the **Show Duplicate Elements** check box, the software will display all duplicate elements in the meshed model. You can show labels of the duplicate elements using the **Show Element Labels** check boxes in this rollout. You can also change the display settings for the color and line width of duplicate elements using the options available in the rollout.

The **List Elements** button in this dialog box is used to display the list of all duplicate elements. When you choose this button, the **Information** window will be displayed for the list of duplicate elements.

The **Delete Elements** button is used to delete the duplicate elements on the basis of the settings used in the **Deletion Settings** rollout.

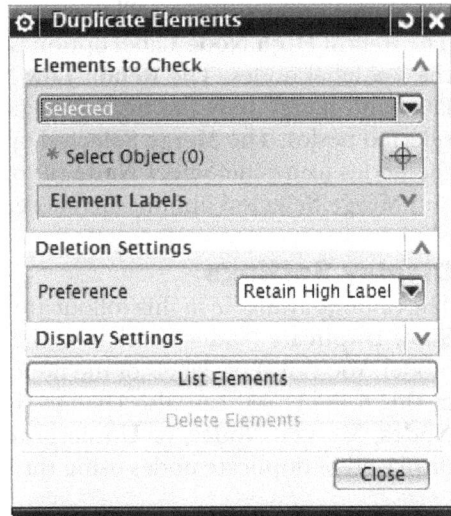

Element Material Orientation

Ribbon: Home > Checks and Information > More > Checks > Element Material Orientation

Menu: Analysis > Finite Element Model Check > Element Material Orientation

The **Element Material Orientation** tool is used to display material orientation of the elements. To view material orientation of the elements, choose **Menu > Analysis > Finite Element Model Check > Element Material Orientation** from the **Top Border Bar**; the **Element Material Orientation** dialog box will be displayed, as shown in Figure 6-11. The options and rollouts available in the dialog box are discussed next.

Elements to Check

In this rollout, two options are available in the drop-down list: **Selected** and **Displayed**. The **Selected** option is used to select the set of elements while the **Displayed** option is used to select all the displayed elements. You can also select the elements by specifying the label in the **Labels** edit box from the **Element Labels** sub-rollout of the **Elements to Check** rollout.

Figure 6-11 The Element Material Orientation dialog box

Display Settings

In this rollout, you can define the display settings of the material orientation for the shell or solid elements. There are four check boxes available to display the material orientation: **Shell Orientation**, **Solid First Direction**, **Solid Second Direction**, and **Solid Third Direction**. These options are discussed next.

Shell Orientation

When you select this check box, the software will display the material orientation of 2D or shell elements. Also, the color swatches corresponding to the shell orientation will be activated. Using this color swatch, you can change the color of arrow displayed for material orientation.

Solid First Direction

When you select this check box, the software will display the first direction of the material coordinate system for 3D or solid elements. Also, the color swatches corresponding to the solid first direction will be activated. Using this color swatch, you can change the color of arrow displayed for material orientation in X direction.

Solid Second Direction

When you select this check box, the software will display the second direction of the material coordinate system for 3D or solid elements. Also, the color swatches corresponding to the solid second direction will be activated. Using this color swatch, you can change the color of arrow displayed for material orientation in Y direction.

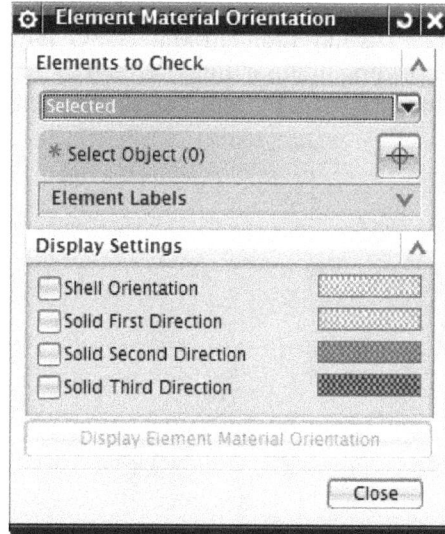

Solid Third Direction
When you select this check box, the software will display the third direction of the material coordinate system for 3D or solid elements. Also, the color swatches corresponding to the solid third direction will be activated. Using this color swatch, you can change the color of arrow displayed for material orientation in Z direction.

The **Display Element Material Orientation** button is used to display the material orientation according to the settings defined in the **Display Settings** rollout. Figure 6-12 shows the meshed midsurface model with the shell orientation.

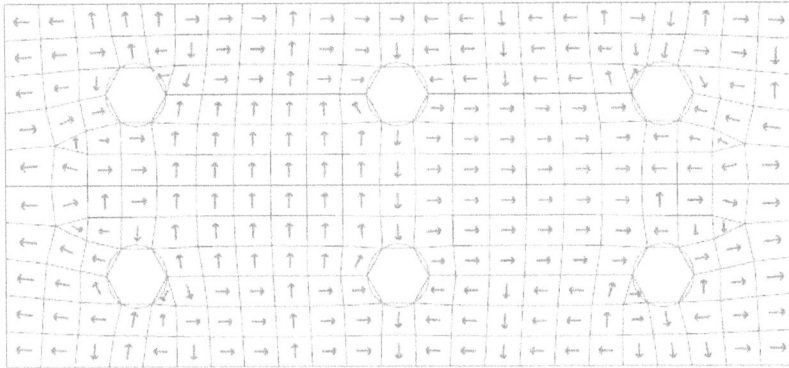

Figure 6-12 Shell orientation of a meshed midsurface model

Adjust Node Proximity to CAD

Ribbon: Home > Checks and Information > More > Checks > Adjust Node Proximity to CAD
Menu: Analysis > Finite Element Model Check > Adjust Node Proximity to CAD

The **Adjust Node Proximity to CAD** tool is used for node adjustment on the CAD geometry within the specified tolerance limit. To adjust the nodes on the CAD geometry, choose **Menu > Analysis > Finite Element Model Check > Adjust Node Proximity to CAD** from the **Top Border Bar**; the **Adjust Node Proximity to CAD** dialog box will be displayed, as shown in Figure 6-13. The options and rollouts available in the dialog box are discussed next.

Nodes to Check
In this rollout, two options are available in the drop-down list: **Selected** and **Displayed**. The **Selected** option is used to select the set of nodes while the **Displayed** option is used to select all displayed nodes.

You can set the tolerance value in the **Tolerance** edit box of the **Proximity Settings** sub-rollout of the **Nodes to Check** rollout. Using this value, the software will

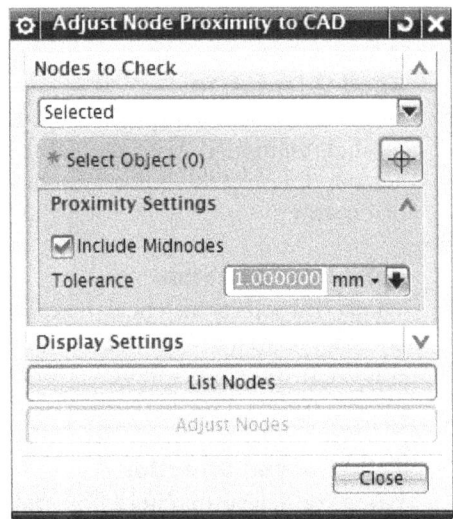

Figure 6-13 The Adjust Node Proximity to CAD dialog box

check the proximity of nodes to relevant model faces of edges. If you select the **Include Midnodes** check box, then the software will also check the proximity of midnode elements.

Display Settings

The options available in this rollout are used to display setting of nodes that failed in proximity check. The **Show Out of Tolerance Nodes** check box is used to display the failed and suggested nodes in the colored point form. You can show labels of the failed nodes using the **Show Node Labels** check box available in this rollout. You can also change the color of failed and suggested nodes using the **Original Node Locations** and **Adjusted Node Locations** color swatches respectively.

The **List Nodes** button in the dialog box is used to display the list of all nodes that failed in proximity check within the specified tolerance limit. When you choose this button, the **Information** window will be displayed for the list of failed nodes.

The **Adjust Nodes** button is used to move the failed nodes to the specified proximity tolerance locations.

> **Note**
> *In NX Nastran, you can check the defined data of analysis anytime. To do so, choose **Menu > Information > Advanced Simulation > Finite Element Model Summary** from the **Top Border Bar**; the **Information** window will be displayed with the summary of analysis, as shown in Figure 6-14.*

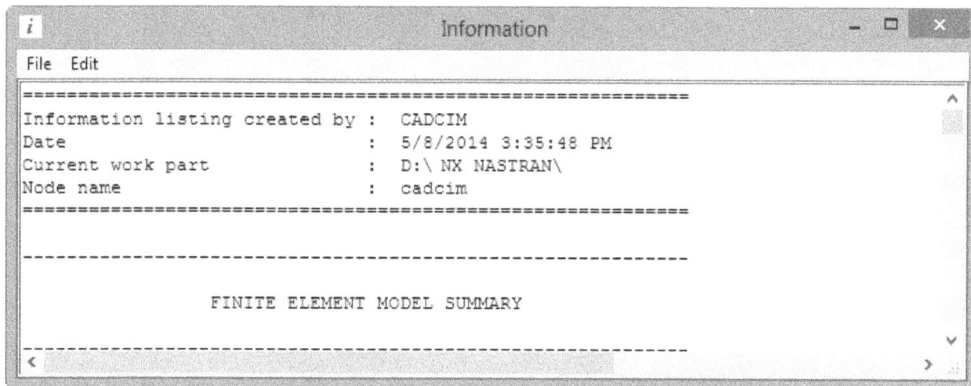

*Figure 6-14 The **Information** window with the list of failed nodes*

TUTORIALS

To perform the tutorials, you need to download the zipped file as *c06_NX_Nastran_input* from the **Input Files** section of the CADCIM website. The complete path for downloading the file is:

Textbooks > CAE Simulation > NX Nastran > NX Nastran 9.0 > Input Files

After the file is downloaded, extract the folder to the location *C:\NX Nastran* and rename it as *c06*.

Tutorial 1

In this tutorial, you will import the extracted FEM file into NX Nastran, as shown in Figure 6-15. After importing the part file, you will check the shell orientation of the model.

(Expected time: 20 min)

The following steps are required to complete this tutorial:

a. Start NX and then open the FEM file in NX.
b. Check the shell orientation of the model.
c. Save and close the model.

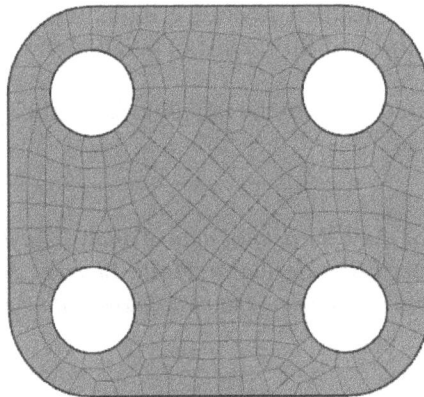

Figure 6-15 Imported FEM model

Starting NX and Opening the FEM File

1. Start NX 9.0 by double-clicking on the shortcut icon of NX 9.0 available on the desktop of your computer.

2. To open an FEM file, choose **File > Open** from the **Ribbon**; the **Open** dialog box is displayed. Select the **c06tut01.fem** file from *C:\NX Nastran\C06\Tut01* and choose the **OK** button; the file is opened in the FEM environment.

Checking Shell Orientation

Next, you need to check the material orientation.

1. For checking the shell orientation, choose **Menu > Analysis > Finite Element Model Check > Element Material Orientation** from the **Top Border Bar**; the **Element Material Orientation** dialog box is displayed, as shown in Figure 6-16.

2. Next, select the **Displayed** option from the drop-down list in the **Elements to Check** rollout. Next, select the **Shell Orientation** check box from the **Display Settings** rollout.

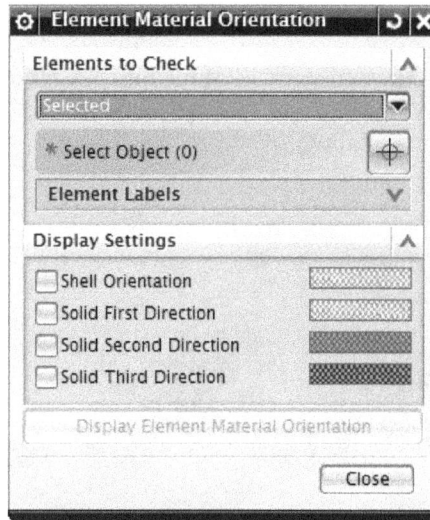

Figure 6-16 The **Element Material Orientation** *dialog box*

3. Choose the **Display Element Material Orientation** button from the bottom of the dialog box; the material orientation of the elements in the FEM model is displayed, as shown in Figure 6-17.

4. Choose the **Close** button to exit the dialog box.

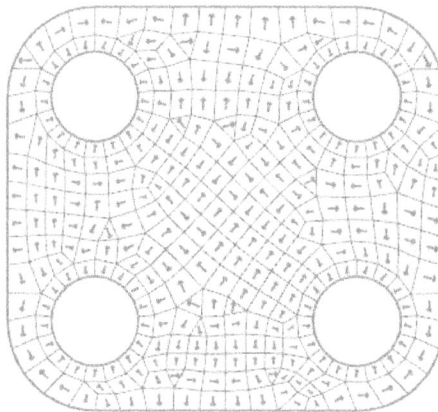

Figure 6-17 Material orientation of FEM model

Saving and Closing the Model

1. Choose **File > Save > Save** from the **Ribbon** to save the model.

2. Choose **File > Close > All Parts** from the **Ribbon** to close all parts.

Note
You will learn to edit the orientation of the FEM model created in this tutorial in later chapter of this textbook.

Tutorial 2

In this tutorial, you will import the extracted FEM file into NX Nastran, as shown in Figure 6-18. After importing the part file, you will check and delete or merge duplicate nodes or elements. You will also check the elements quality.   **(Expected time: 30 min)**

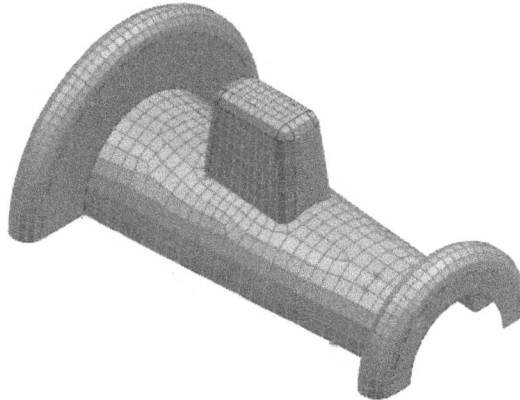

Figure 6-18 *Imported FEM model*

The following steps are required to complete this tutorial:

a.   Start NX and then open the FEM file in NX.
b.   Check the duplicate nodes and merge them.
c.   Check the duplicate elements and delete them.
d.   Check the quality of elements by specifying values, refer to Table 1.
e.   Save and close the model.

Table 1 *Values to be specified for checking elements*

GEOMCHECK Option	Warning Limit	Error Limit
Aspect Ratio	10	100
Skew Angle	45°	30°
Interior Angle (Maximum)	135°	150°
Interior Angle (Minimum)	45°	30°
Warp (Quad)	0.01	0.05
Warp (Solid)	0.866	0.707

Starting NX and Opening the FEM File

1.   Start NX 9.0 by double-clicking on its shortcut icon available on the desktop of your computer.

2.   To open an FEM file, choose **File > Open** from the **Ribbon**; the **Open** dialog box is displayed. Select the **c06tut02_input.fem** file from *C:\NX Nastran\C06\Tut02* and then choose the **OK** button; the file is opened in the FEM environment.

Note

In this model, duplicate nodes and elements are created manually for better understanding of the duplicates. However, the software creates duplicate nodes/elements in some specific meshing operation.

Checking Duplicate Nodes

Next, you need to check duplicate nodes.

1. For checking duplicate nodes, choose **Menu > Analysis > Finite Element Model Check > Duplicate Nodes** from the **Top Border Bar**; the **Duplicate Nodes** dialog box is displayed, as shown in Figure 6-19.

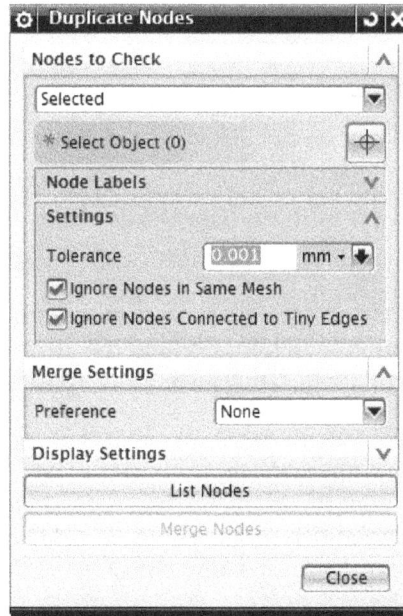

*Figure 6-19 The **Duplicate Nodes** dialog box*

2. Select the **Displayed** option from drop-down list available in the **Nodes to Check** rollout; the duplicate nodes in the model are highlighted.

3. Now, select the **Retain Low Node Label** option from the **Preference** drop-down list of the **Merge Settings** rollout and retain the default settings in rest of the rollouts of the dialog box.

4. Next, choose the **Merge Nodes** button of the dialog box; the seven highest label nodes are merged in the model. Close the dialog box to exit.

Tip

*You can also choose the **List Nodes** button to view the detailed information about the duplicate nodes. When you choose this button, the **Information** window is displayed, refer to Figure 6-20.*

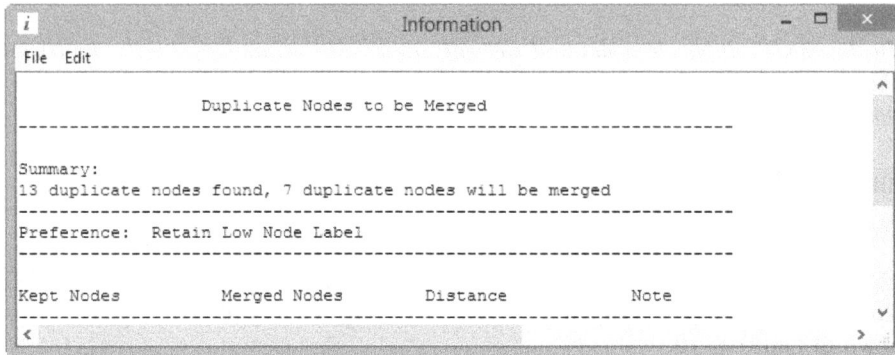

*Figure 6-20 The **Information** window with details of duplicae nodes*

Checking Duplicate Elements

Next, you need to check duplicate elements.

1. For checking duplicate elements, choose **Menu > Analysis > Finite Element Model Check > Duplicate Elements** from the **Top Border Bar**; the **Duplicate Elements** dialog box is displayed, as shown in Figure 6-21.

2. Select the **Displayed** option from drop-down list available in the **Elements to Check** rollout; the duplicate elements in the model are highlighted. To view the labels of the duplicate elements, select the **Show Element Labels** check box in the **Display Settings** rollout.

3. Next, select the **Retain Low Label** option from the **Preference** drop-down list and retain the remaining default settings in the rollout.

4. Now, choose the **Delete Elements** button from the dialog box; the six highest label nodes are deleted from the model. Close the dialog box.

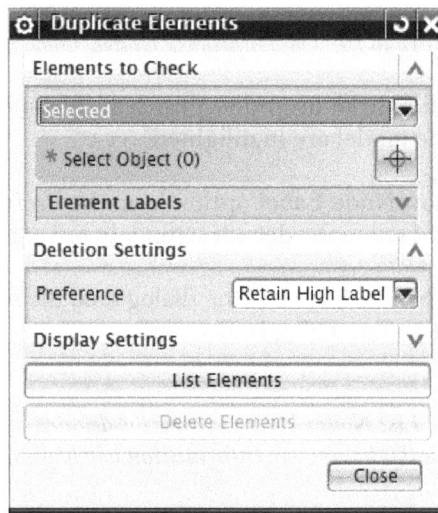

*Figure 6-21 The **Duplicate Elements** dialog box*

Checking Elements Quality
Next, you need to check the elements quality.

1. For checking elements quality, choose **Menu > Analysis > Finite Element Model Check > Elements Quality** from the **Top Border Bar**; the **Elements Quality** dialog box is displayed, as shown in Figure 6-22.

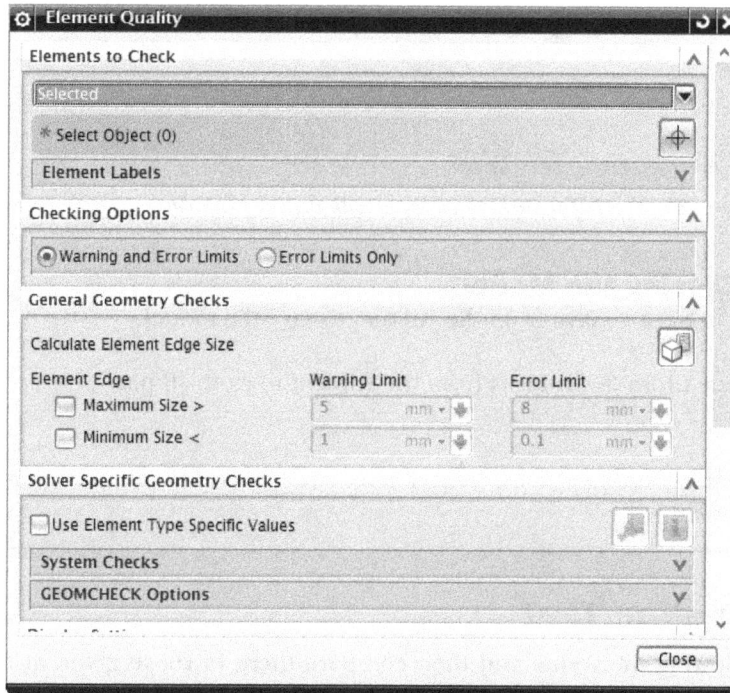

*Figure 6-22 The **Elements Quality** dialog box*

2. Select the **Displayed** option from drop-down list available in the **Elements to Check** rollout; all the elements in the model are selected.

3. Specify the geometry checks values in the respective edit boxes of the **GEOMCHECK Options** sub-rollout of the **Solver Specific Geometry Checks** rollout (for values refer to Table 1).

4. Also, clear the **Taper >**, **Edge Point <**, and **Element Offset >** check boxes from the **GEOMCHECK Options** sub-rollout of the **Solver Specific Geometry Checks** rollout.

5. Next, select the **Failed** option from the **Output Group Element** drop-down list and the **All** option from the **Report** drop-down list in the **Output Settings** rollout. keep the default settings same in the rest of the rollouts of the dialog box.

6. Choose the **Check Elements** button of the dialog box; the Status Bar is displayed with the message **1 failed elements, 47 warning elements** and detailed information of the elements is displayed in the **Information** window, as shown in Figure 6-23. Close the dialog box and information window.

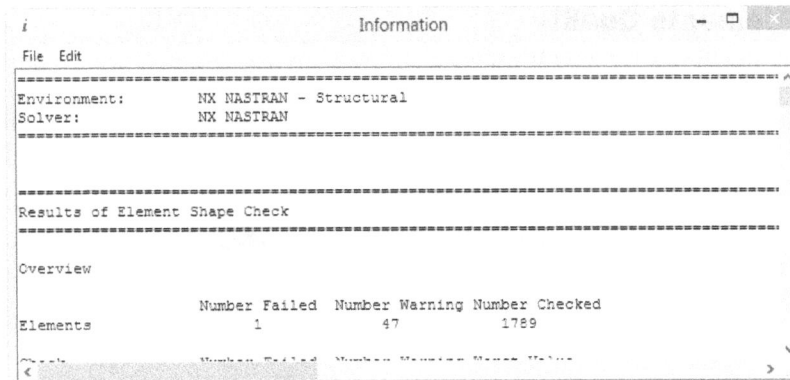

Figure 6-23 *The **Information** window with detailed information of the elements*

Saving and Closing the Model

1. Choose **File > Save > Save** from the **Ribbon** to save the model.

2. Choose **File > Close > All Parts** from the **Ribbon** to close all parts.

Note
You will learn to edit the failed element of the model created in this tutorial in later chapter of this textbook.

Self-Evaluation Test

Answer the following questions and then compare them to those given at the end of this chapter:

1. The _____ tool is used to display material orientation of the elements.

2. The _____ tool is used for node adjustment on the CAD geometry.

3. The **Element Quality** tool is used to check the created elements on the basis of their geometry. (T/F)

4. The **Element Edges** tool is used to analyze the element of the created free or overlapped edges of the elements. (T/F)

5. The **2D Element Normals** tool is used to check and define the normal direction of 2D elements. (T/F)

Review Questions

Answer the following questions:

1. Which of the following tools is used to check and merge the duplicate nodes in the meshed geometry?

   (a) **Element Quality**   (b) **Element Edges**
   (c) **2D Element Normals**   (d) **Duplicate Nodes**

2. Which of the following tools is used to check the duplicate elements in a meshed geometry?

   (a) **Duplicate Nodes**   (b) **Duplicate Elements**
   (c) **Element Quality**   (d) **Element Material Orientation**

3. Which of the following tools is used to view the shell orientation of the 2D meshed model?

   (a) **Element Material Orientation**   (b) **2D Element Normals**
   (c) **Element Quality**   (d) None of these

4. The **Finite Element Model Summary** tool is used to check the defined data of analysis of the model. (T/F)

5. The **Check Normals** and **Orient Normals Using Seed Element** methods are used to check and define direction of elements. (T/F)

EXERCISE

To perform the exercises, you need to extract the *c06_NX_Nastran_input.zip* file that was downloaded at the beginning of the tutorials.

Exercise 1

In this exercise, you will import the FEM file into NX Nastran, as shown in Figure 6-24. After importing the part file, you will check and delete or merge duplicate nodes or elements. You will also check the quality of elements.   **(Expected time: 30 min)**

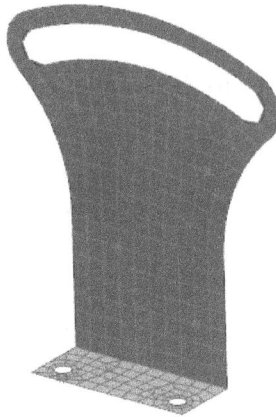

Figure 6-24 Imported FEM model

Note

In this model, duplicate nodes and elements have been created manually for understanding of the duplicates. However, the software creates duplicate nodes or elements in some specific meshing operation.

Hint

1. Start NX and open the **c06_exr01_input.fem** file in NX.
2. Check the duplicate nodes and merge them by selecting the **Retain Low Node Label** option from the **Preference** drop-down list.
3. Check the duplicate elements and delete them for the **Retain Low Label** option from the **Preference** drop-down list.
4. Check the elements quality by selecting the **Use Element Type Specific Values** check box of the **Solver Specific Geometry Checks** rollout.
5. Save and close the model.

Answers to Self-Evaluation Test
1. Element Material Orientation, 2. Adjust Node Proximity to CAD, 3. T, 4. T, 5. T

Chapter 7

Meshing-III

Learning Objectives

After completing this chapter, you will be able to:

- *Edit nodes*
- *Edit elements*
- *Understand mesh refinement*
- *Create mesh associated data for 0D, 1D, 2D, and 3D elements*

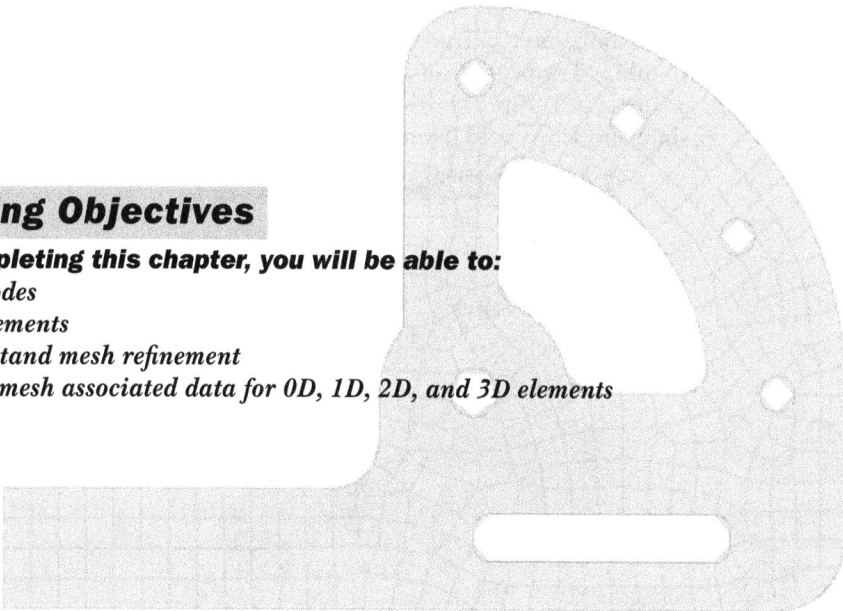

INTRODUCTION

In the previous chapter, you learned to analyze a created mesh. The created mesh generally contains many types of errors. In this chapter, you will learn to refine and edit these mesh errors. You can edit or refine the mesh by using nodes editing, elements editing, or some mesh refining tools.

EDITING NODES

You can edit the nodes position using the node editing tools available in the **Menu** in the **Top Border Bar**. The tools to edit the created node are discussed next.

Translate Tool

Ribbon:	Nodes and Elements > Nodes > Translate
Menu:	Edit > Node > Translate

Using the **Translate** tool, you can translate the position of the nodes. To do so, choose **Menu > Edit > Node > Translate** from the **Top Border Bar**; the **Node Translate** dialog box will be displayed, as shown in Figure 7-1. Two options are available in the drop-down list of the **Type** rollout: **Translate** and **Copy and Translate**. These options are discussed next.

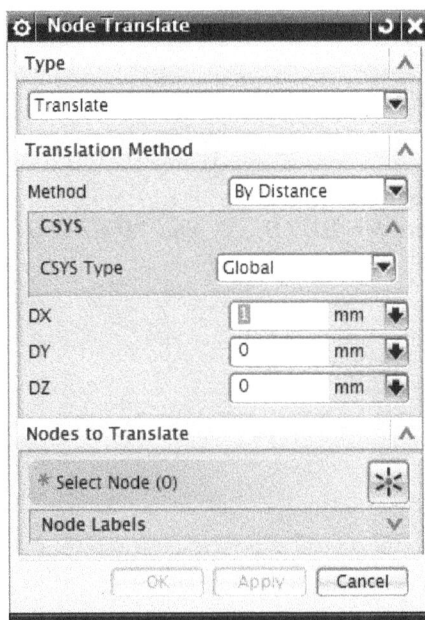

*Figure 7-1 The **Node Translate** dialog box*

Translate

By using this option, you can move the nodes from their position. By default, this option is selected in the **Type** drop-down list and also two corresponding rollouts are available in the dialog box: **Translation Method** and **Nodes to Translate**.

You can use the **Translation Method** rollout to specify the translate method. There are six methods available in the **Method** drop-down list of this rollout: **By Distance**, **Along Direction**,

Align Vectors, **Point to Point**, **Scale Model**, and **By Field**. The other options of this rollout will change according to the method chosen from the drop-down list.

The **Select Node (0)** option is highlighted in the **Nodes to Translate** rollout and you are prompted to select the nodes to translate. You can also select a node by specifying the node label in the **Labels** edit box of the **Node Labels** sub-rollout of this rollout.

Copy and Translate

By using this option, you can create a copy of the nodes or translate them from their position. When you select this option, four rollouts will be available in the dialog box: **Translation Method**, **Nodes to Translate**, **Number of Copies**, and **Specify Label**.

You can specify the number of nodes to be created in the **Specify Number** edit box in the **Number of Copies** rollout.

The **Specify Label** rollout is used to define the label of the translated nodes.

After specifying all the settings, choose the **OK** button to translate the node.

Rotate Tool

| **Ribbon:** | Nodes and Elements > Nodes > Rotate |
| **Menu:** | Edit > Node > Rotate |

Using the **Rotate** tool, you can rotate the nodes. To do so, choose **Menu > Edit > Node > Rotate** from the **Top Border Bar**; the **Node Rotate** dialog box will be displayed, as shown in Figure 7-2. To rotate nodes, two options are available in the drop-down list of the **Type** rollout: **Rotate** and **Copy and Rotate**. These options are discussed next.

Rotate

This option is selected by default in the **Rotate** rollout. As a result, three rollouts are available in the dialog box: **Nodes to Rotate**, **Axis of Rotation**, and **Rotation Angle**.

The **Select Node (0)** option is highlighted in the **Nodes to Rotate** rollout. Therefore, you will be prompted to select the nodes to rotate. You can also select nodes by specifying the node label in the **Labels** edit box available in the **Node Labels** sub-rollout of this rollout.

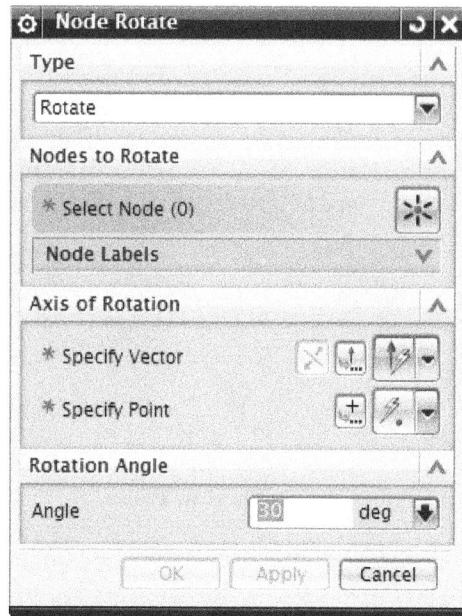

Figure 7-2 The Node Rotate dialog box

Using the **Axis of Rotation** rollout, you can specify the axis around which the node is to be rotated.

You can specify the node rotation angle in the **Angle** edit box available in the **Rotation Angle** rollout.

Copy and Rotate

By using this option, you can rotate and create copy of the nodes. When you select this option, five rollouts will be available in the dialog box: **Nodes to Rotate**, **Number of Copies**, **Axis of Rotation**, **Rotation Angle**, and **Specify Label**. By default, the **Select Node(0)** option is highlighted in the **Nodes to Rotate** rollout and you need to select the nodes to be rotated.

While rotating the nodes, you can also specify the number of nodes to be copied in the **Specify Number** edit box available in the **Number of Copies** rollout.

Using the **Axis of Rotation** rollout, you can specify the axis around which the nodes are to be rotated.

You can specify the nodes rotation angle in the **Angle** edit box available in the **Rotation Angle** rollout. Two radio buttons are also available in this rollout: **Per Copy** and **Total**. If you select the **Per Copy** radio button, each node will be created at the specified angle. If you select the **Total** radio button, all the nodes will be created within the specified angle.

After specifying all the settings, choose the **OK** button to rotate the nodes.

Reflect Tool

Ribbon: Nodes and Elements > Nodes > More > Edit > Reflect
Menu: Edit > Node > Reflect

On using the **Reflect** tool, you can reflect the nodes across a plane. To do so, choose **Menu > Edit > Node > Reflect** from the **Top Border Bar**; the **Node Reflect** dialog box will be displayed, as shown in Figure 7-3. There are two options available in the drop-down list of the **Type** rollout: **Reflect** and **Copy and Reflect**. These options are discussed next.

Reflect

Using this option, you can reflect the nodes across a plane. This option is selected by default. As a result, two rollouts became available in this dialog box: **Nodes to Reflect** and **Reflection Plane**.

The **Select Node (0)** option is highlighted in the **Nodes to Reflect** rollout; therefore you will be prompted to select the nodes to reflect. You can also select a node by specifying the node label in the **Labels** edit box available in the **Node Labels** sub-rollout of this rollout.

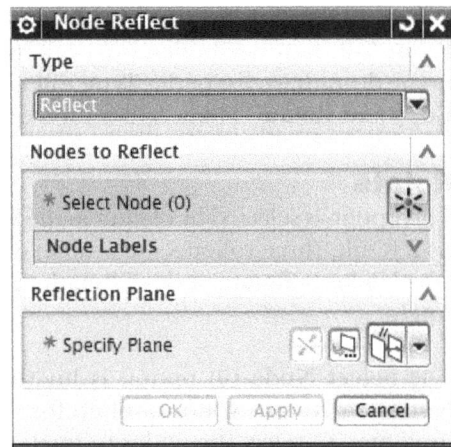

*Figure 7-3 The **Node Reflect** dialog box*

Using the **Reflection Plane** rollout, you can specify the plane across which the node is to be reflected.

Copy and Reflect

Using this option, you can reflect and create a copy of the nodes. When you select the **Copy and Reflect** option from the drop-down list, three rollouts will be available: **Nodes to Reflect**, **Reflection Plane**, and **Specify Label**.

The functions of these rollout, are same as discussed earlier.

After specifying all the settings, choose the **OK** button to reflect the node.

Project Tool

Ribbon: Nodes and Elements > Nodes > More > Edit > Project
Menu: Edit > Node > Project

Using the **Project** tool, you can project the nodes on the selected edge or face. To do so, choose **Menu > Edit > Node > Project** from the **Top Border Bar**; the **Node Project** dialog box will be displayed, as shown in Figure 7-4. In the drop-down list of the **Type** rollout, two options are available: **Project** and **Copy and Project**. These options are discussed next.

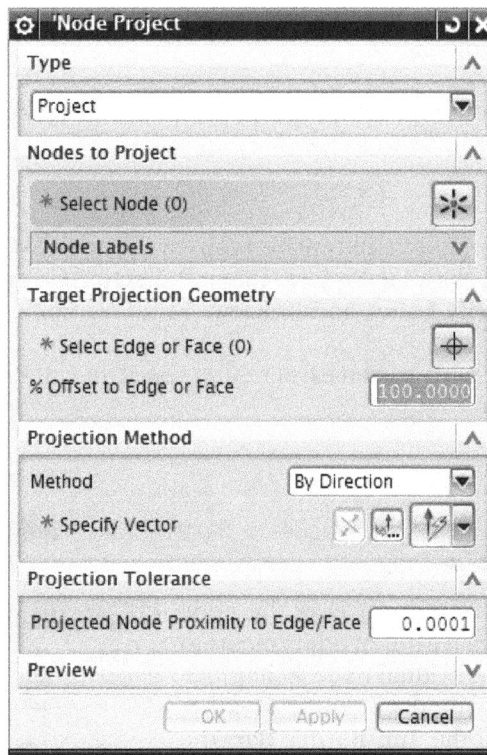

*Figure 7-4 The **Node Project** dialog box*

Project

On using the **Project** option, you can project the nodes on the selected edge or face. This option is selected by default in the rollout and five corresponding rollouts are also available in the dialog box: **Nodes to Project**, **Target Projection Geometry**, **Projection Method**, **Projection Tolerance**, and **Preview**.

The **Select Node (0)** option is highlighted in the **Nodes to Project** rollout. As a result, you will be prompted to select the nodes to project. You can also select a node by specifying the node label in the **Labels** edit box of the **Node Labels** sub-rollout of this rollout.

The **Select Edge or Face (0)** option in the **Target Projection Geometry** rollout is used to select the edge or face to project the node. You can also specify the distance percentage in the **% Offset to Edge or Face** edit box of this rollout using which the node will be projected. For example, if you specify 50%, the software will project the nodes to a halfway between the selected target geometry.

Using the options in the **Projection Method** rollout, you can define the method of projecting nodes from the **Method** drop-down list. There are two options available in this drop-down list: **By Direction** and **By Shortest Distance**. The **By Direction** option is used to project the nodes along a specified vector while the **By Shortest Distance** option is used to project the nodes at the path closest to the target geometry. When you select the **By Direction** option from the drop-down list, then the **Specify Vector** option will become available using which you can define the vector for projecting the nodes.

The **Projection Tolerance** rollout is used to specify projection tolerance. You can specify tolerance in the **Projected Node Proximity to Edge/Face** edit box.

Copy and Project
By using this option, you can project and create a copy of the nodes. When you select this option, six rollouts will be available: **Nodes to Project**, **Target Projection Geometry**, **Projection Method**, **Projection Tolerance**, **Specify Label**, and **Preview**.

These rollouts work similar as discussed earlier. After specifying all the settings, choose the **OK** button to project the node.

Drag Tool

Ribbon:	Nodes and Elements > Nodes > More > Edit > Drag
Menu:	Edit > Node > Drag

Using the **Drag** tool, you can drag the node from its position to make minor adjustments to the location of an individual node to improve the quality of the element. To do so, choose **Menu > Edit > Node > Drag** from the **Top Border Bar**; the **Node Drag** dialog box will be displayed, as shown in Figure 7-5. In the drop-down list of the **Type** rollout, two options are available: **Along Geometry** and **Off Geometry**. These options are discussed next.

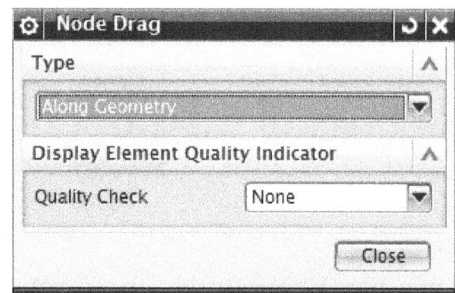

*Figure 7-5 The **Node Drag** dialog box*

The **Along Geometry** option is used to drag a node along its relevant geometry, such as edge or face while the **Off Geometry** option is used to drag a node which is not associated to any geometry.

The options in the **Quality Check** drop-down list of the **Display Element Quality Indicator** rollout are used to check the quality of the elements connected with the dragged node. You can select the type of quality check data to be displayed from this drop-down list.

Align Tool

Ribbon: Nodes and Elements > Nodes > More > Edit > Align
Menu: Edit > Node > Align

Using the **Align** tool, you can align a node between two nodes. To do so, choose tool **Menu > Edit > Node > Align** from the **Top Border Bar**; the **Node Align** dialog box will be displayed, as shown in Figure 7-6. Two rollouts are available in this dialog box: **Define Alignment Line** and **Select Nodes to Align**. These rollouts are discussed next.

Define Alignment Line
The options in this rollout are used to select nodes for aligning. The **Select first node (0)** option is used to select the first node for alignment and the **Select second Node (0)** option is used to select the second node for alignment. After selecting the nodes, an alignment path will be displayed.

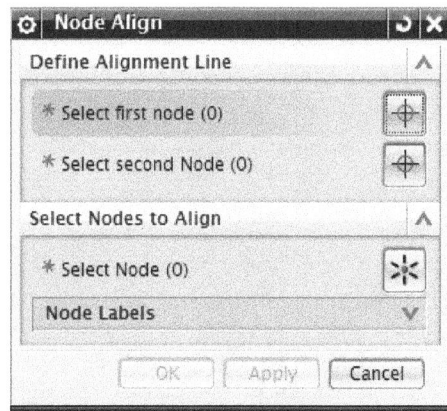

*Figure 7-6 The **Node Align** dialog box*

Select Nodes to Align
The **Select Node (0)** option in this rollout is used to select the node which needs to be aligned. You can also select the node by specifying the node label in the **Labels** edit box of the **Node Labels** sub-rollout of this rollout.

After selecting all the nodes, choose the **OK** button to align the node.

Assign Nodal Coordinate System Tool

Ribbon: Nodes and Elements > Nodes > More > Edit > Assign Nodal Coordinate System
Menu: Edit > Node > Assign Nodal Coordinate System

Using the **Assign Nodal Coordinate System** tool, you can assign nodal coordinate system to the selected nodes or geometry. The created nodal coordinate system may be used while defining mesh. To assign nodal coordinate system to nodes, choose **Menu > Edit > Node > Assign Nodal Coordinate System** from the **Top Border Bar**; the **Assign Nodal CSYS** dialog box will be displayed, as shown in Figure 7-7. You can create displacement type of nodal coordinate system for the structural degrees of freedom or reference type nodal coordinate system for solvers.

Modify Label Tool

Ribbon: Nodes and Elements > Nodes > More > Edit > Label
Menu: Edit > Node > Modify Label

Using the **Modify Label** tool, you can modify the label of a created node. To do so, choose **Menu > Edit > Node > Modify Label** from the **Top Border Bar**; the **Node Modify Label** dialog box will be displayed, as shown in Figure 7-8. There are two rollouts available in this dialog box: **Nodes to Modify** and **New Label**. These rollouts are discussed next.

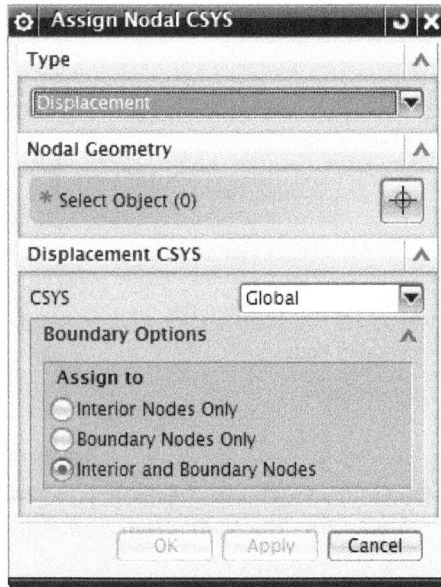

Figure 7-7 The **Assign Nodal CSYS** *dialog box*

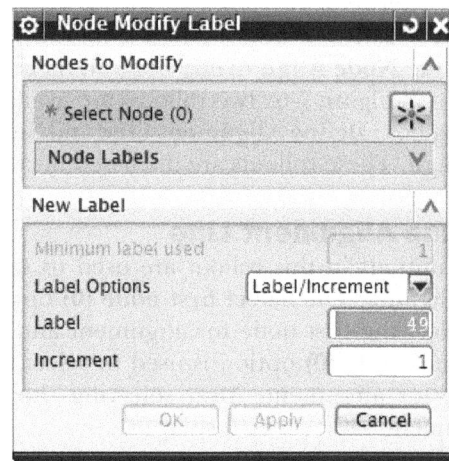

Figure 7-8 The **Node Modify Label** *dialog box*

Nodes to Modify

The **Select Node (0)** option is highlighted in this rollout. As a result, you will be prompted to select the nodes to modify nodes label. You can also select a node by specifying the node label in the **Labels** edit box of the **Node Labels** sub-rollout of this rollout.

New Label

The options in this rollout are used to specify the label for the selected node. For assigning a new label, two options are available in the **Label Options** drop-down list of this rollout: **Label/Increment** and **Offset**. The **Label/Increment** option is used to modify a label by specifying starting label and increment between the labels for the selected nodes. While using the **Offset** option, you can offset the current label for the nodes by a specified value.

After selecting the nodes, choose the **OK** button to assign a new label to the node. Note that, if a label for nodes is already available, then a warning message will be displayed to change the value.

Modify Coordinates Tool

Ribbon: Nodes and Elements > Nodes > More > Edit > Coordinates
Menu: Edit > Node > Modify Coordinates

Using the **Modify Coordinates** tool, you can modify the position of a created node by changing its coordinate values relative to respective coordinate system. To do so, choose **Menu > Edit > Node > Modify Coordinates** from the **Top Border Bar**; the **Node Modify Coordinates** dialog box will be displayed, as shown in Figure 7-9. There are three rollouts available in this dialog box: **CSYS**, **Nodes to Move**, and **New Location**. These rollouts are discussed next.

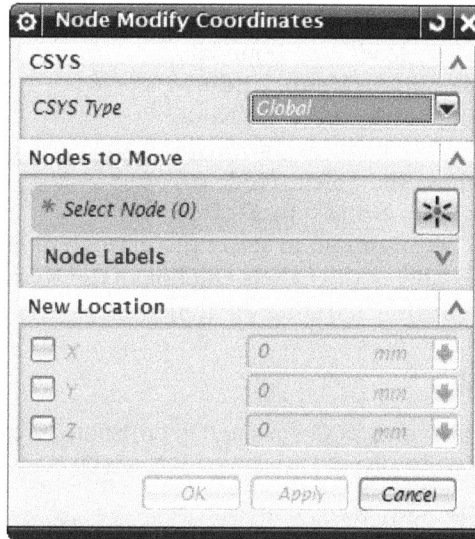

Figure 7-9 The Node Modify Coordinates dialog box

CSYS

In this rollout, you can specify the type of coordinate system to move the nodes. There are four types of coordinate systems available in the **CSYS Type** drop-down list of the **CSYS** rollout: **Global**, **Cartesian**, **Cylindrical**, and **Spherical**. You can create nodes using these options. On selecting the **Cartesian**, **Cylindrical**, or **Spherical** option, the **Specify CSYS** option will also be displayed for selecting or creating the respective coordinate system.

Nodes to Move

The **Select Node (0)** option is highlighted in this rollout. As a result, you will be prompted to select the node to move. You can also select the node by specifying the node label in the **Labels** edit box of the **Node Labels** sub-rollout of this rollout.

New Location

The options available in this rollout depend upon the options selected from the **CSYS Type** drop-down list. To change the position of the node, select the respective axis check boxes from this rollout; the edit boxes will become activated to enter new coordinate values. Specify the values in the edit boxes.

After changing the coordinates of the nodes, choose the **OK** button to move the nodes according to the specified coordinate values.

Delete Tool

Ribbon: Nodes and Elements > Nodes > More > Edit > Delete
Menu: Edit > Node > Delete

Using the **Delete** tool, you can delete those nodes from the window that are not connected to any element. Note that if you delete a node which is carrying any load(s) or boundary condition(s) then the software will invalidate that load(s) or boundary condition(s). To delete a node, choose **Menu > Edit > Node > Delete** from the **Top Border Bar**; the **Node Delete** dialog box will be displayed, as shown in Figure 7-10. In this dialog box, the **Select Node (0)** option is highlighted in the **Nodes to Delete** rollout. As a result, you will be prompted to select the node to delete. You can also select the nodes by specifying the node label in the **Labels** edit box in the **Node Labels** sub-rollout of this rollout. Choose the **OK** button after selecting the node.

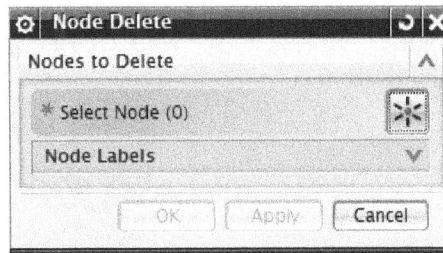

Figure 7-10 *The **Node Delete** dialog box*

Node/Element Tool

Ribbon: Nodes and Elements > Checks and Information > Node/Element
Menu: Information > Advanced Simulation > Node/Element

Using the **Node/Element** tool, you can get information about the selected node or element. To do so, choose **Menu > Information > Advanced Simulation > Node/Element** from the **Top Border Bar**; the **Node/Element Information** dialog box will be displayed, as shown in Figure 7-11. You can also select the **Node** or **Element** option from the drop-down list of the **Type** rollout to get information about the selected Node or Element.

If you select the **Node** option from the **Type** rollout, the **Select Node (0)** option is highlighted in the **Node** rollout. As a result, you will be prompted to select the node. Select the node. Also, options related to the selected node will be displayed in the **Options** rollout. Select the respective check boxes from the **Options** rollout to get the information. The **Label Display** rollout is used to specify the label color. After specifying all settings, choose the **Apply** button; the **Information** window will be displayed with the information of the selected node, refer to Figure 7-12.

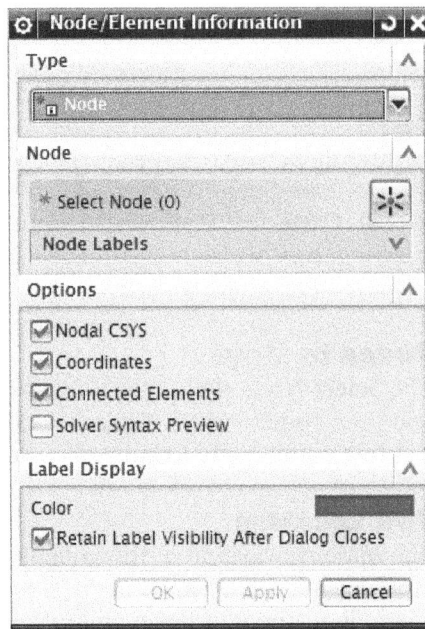

Figure 7-11   *The **Node/Element Information** dialog box*

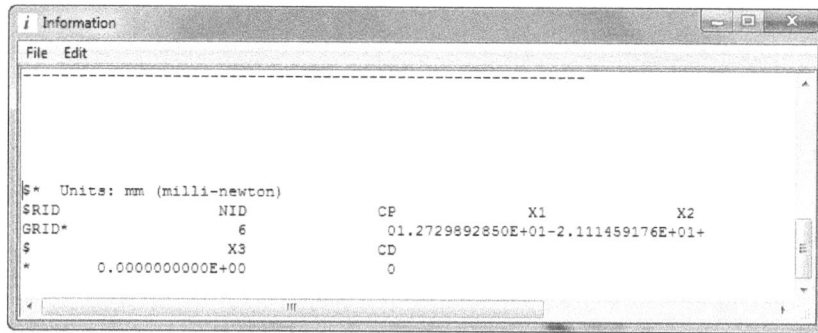

*Figure 7-12 The **Information** window*

Note

Similarly, you can get the information of the elements. Further in this chapter, you will also learn about editing the elements.

Nodal Coordinate System

Ribbon:	Nodes and Elements > Checks and Information > More > Information > Nodal Coordinate System
Menu:	Information > Advanced Simulation > Nodal Coordinate System

Using the **Nodal Coordinate System** tool, you can identify the nodal coordinate system. To do so, choose **Menu > Information > Advanced Simulation > Nodal Coordinate System** from the **Top Border Bar**; the **Identify Nodal CSYS** dialog box will be displayed, as shown in Figure 7-13. Using this dialog box, you can identify displacement coordinate system or reference coordinate system. When you select any node, information related to it will be displayed in the **Coordinate System** rollout.

Move Node Tool

Ribbon:	Nodes and Elements > Nodes > More > Edit > Move
Menu:	Edit > Element > Move Node

The **Move Node** tool is used to move a node to an existing 2D element node. To move the 2D element node, choose **Menu > Edit > Element > Move Node** from the **Top Border Bar**; the **Move Node** dialog box will be displayed, as shown in Figure 7-14. The options available in the dialog box are discussed next.

Selection Steps

In this area, two buttons are available: **Source Node** and **Target Node**. The **Source Node** button is chosen by default. As a result, you will be prompted to select the node to move from its place. After selecting the node to move, choose the **Target Node** button; you will be prompted to select the target node. Select the target node with which the source node will merge.

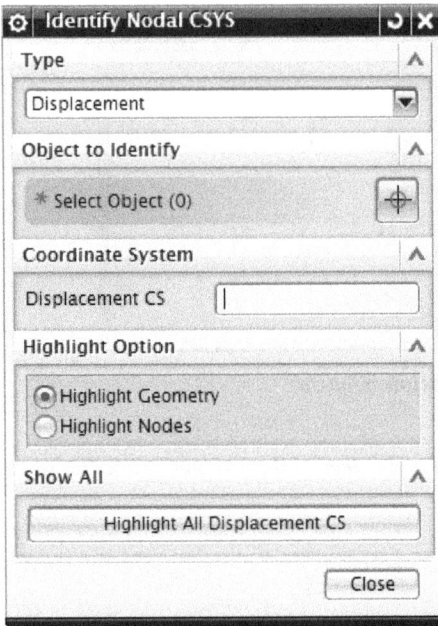

Figure 7-13   The Identify Nodal CSYS
dialog box

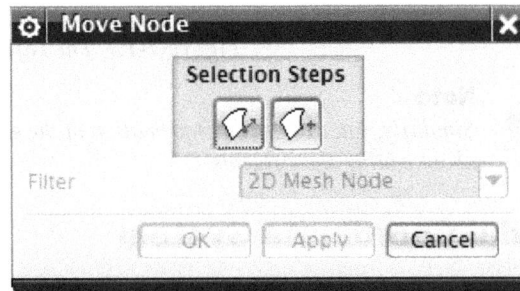

Figure 7-14   The Move Node dialog box

Filter

On choosing the **Target Node** button, the **Filter** drop-down list gets activated. In this drop-down list, three options are available: **2D Mesh Node**, **Node Location on Face**, and **Node Location on Edge**. You can define your existing target node by using these options.

After specifying all settings, choose the **OK** button to move the node from its location.

EDITING ELEMENTS

Similar to nodes editing, you can also edit the elements using the element editing tools. The tools to edit the created elements are discussed next.

Copy and Translate Tool

Ribbon:	Nodes and Elements > Elements > More > Edit > Translate
Menu:	Insert > Element > Copy and Translate

Using the **Copy and Translate** tool, you can create a copy of an element and translate it along a specified vector or orientation. To do so, choose **Menu > Insert > Element > Copy and Translate** from the **Top Border Bar**; the **Element Copy and Translate** dialog box will be displayed, as shown in Figure 7-15. The options and rollouts available in this dialog box are discussed next.

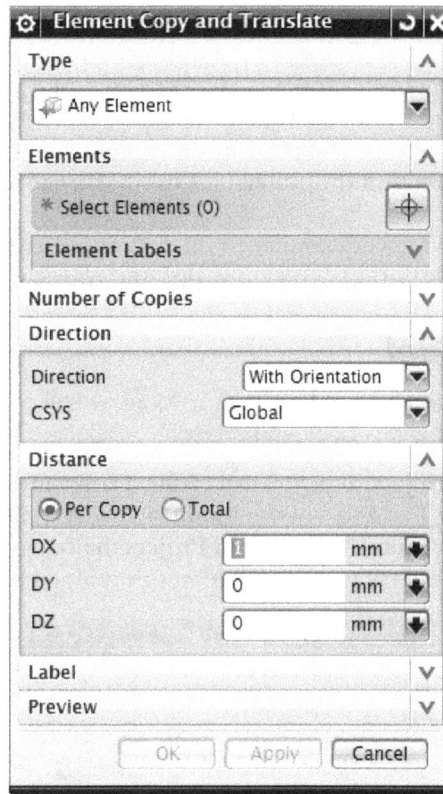

*Figure 7-15   The **Element Copy and Translate** dialog box*

Type

The **Type** drop-down list in this rollout consists of four options: **Any Element, Element Edges, Element Faces**, and **Solid Element Only**. The **Any Element** option is used to copy and translate any 1D, 2D, or 3D element. The **Element Edges** option is used to copy and translate 1D (line) elements. The **Element Faces** option is used to copy and translate 2D (face) elements. The **Solid Elements Only** option is used to copy and translate 3D (solid) elements.

Number of Copies

Using the option in this rollout, you can define the number of copies of the element that are needed to be created .

Direction

Using the options in this rollout, you can define the direction of the selected element to copy and translate. Two options are available in the **Direction** drop-down list of this rollout: **With Orientation** and **Along Vector**. The **With Orientation** option is selected by default and is used to translate the elements by a specified distance from a selected coordinate system. The **CSYS** drop-down list will be activated automatically if the **With Orientation** option is selected by default. You can choose the desired CSYS type from this drop-down list. If you select the **Along Vector** option from the **Direction** drop-down list, you can copy and translate the elements along a vector.

Distance

The options in this rollout are used to specify the distance to copy or translate by using the **Per Copy** or the **Total** radio button of this rollout. The **Per Copy** radio button is used to specify the distance between each element while the **Total** radio button is used to specify the total distance between the elements. The options displayed in this rollout will be changed according to the options selected from the **Direction** drop-down list in the **Direction** rollout.

Other rollouts in this dialog box work similar to the rollouts discussed earlier. After specifying all the settings, choose the **OK** button to create a copy and translation of the selected elements.

Copy and Project Tool

Ribbon: Nodes and Elements > Elements > More > Edit > Project
Menu: Insert > Element > Copy and Project

Using the **Copy and Project** tool, you can create a copy and project selected element on a specified target face. To do so, choose **Menu > Insert > Element > Copy and Project** from the **Top Border Bar**; the **Element Copy and Project** dialog box will be displayed, as shown in Figure 7-16. The options and rollouts available in this dialog box are discussed next.

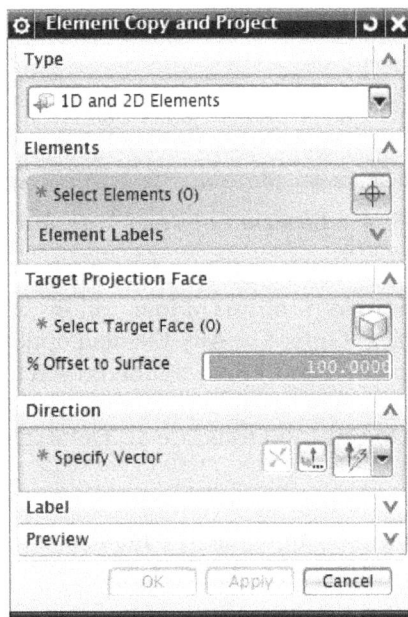

Figure 7-16 *The **Element Copy and Project** dialog box*

Type

In the drop-down list of this rollout, three options are available: **1D and 2D Elements**, **Element Edges**, and **Element Faces**. The **1D and 2D Elements** option is used to copy and project any type of 1D or 2D elements. The **Element Edges** option is used to copy and project 1D (line) elements. The **Element Faces** option is used to copy and project 2D (face) elements.

Target Projection Face

In this rollout, select the **Select Target Face (0)** option; you will be prompted to select the target face on which you want to project the elements. In the **% Offset to Surface** edit box of this rollout, you can specify the percentage of the distance from the target surface. For example, if you specify 50%, the software will project the elements halfway between the original surface and the target surface.

Direction

In this rollout, you can control the direction of the element projection. You have to specify direction vector, if you select the **1D and 2D Elements** or the **Element Edges** option from the **Type** drop-down list. If you select the **Element Faces** option from the **Type** drop-down list, then two radio buttons will be also available in this rollout: **Along Vector** and **Element Normal**. The **Along Vector** radio button is used to project the elements along a specified vector while the **Element Normal** radio button is used to project the elements along element normal direction.

The other rollouts of this dialog box work similar as discussed earlier. After specifying all the settings, choose the **OK** button to create a copy and projection of the selected elements.

Copy and Reflect Tool

Ribbon:	Nodes and Elements > Elements > More > Edit > Reflect
Menu:	Insert > Element > Copy and Reflect

Using the **Copy and Reflect** tool, you can create a copy of the selected element and then reflect it from a specified plane. To do so, choose **Menu > Insert > Element > Copy and Reflect** from the **Top Border Bar**; the **Element Copy and Reflect** dialog box will be displayed, as shown in Figure 7-17. The options and rollouts available in the dialog box are discussed next.

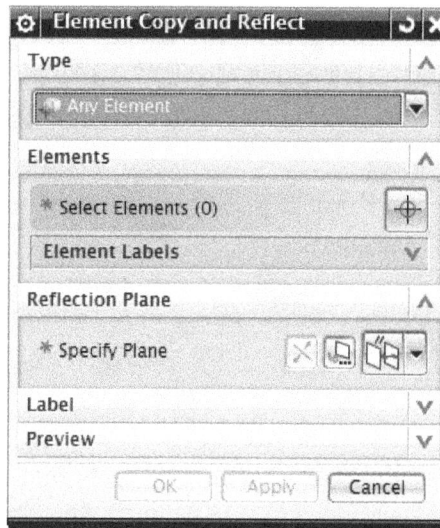

Figure 7-17 The Element Copy and Reflect dialog box

Type

In the drop-down list of the **Type** rollout, four options are available: **Any Element**, **Element Edges**, **Element Faces**, and **Solid Elements Only**. The **Any Element** option is used to copy and reflect any type of 1D, 2D, or 3D element. The **Element Edges** option is used to copy and reflect 1D (line) elements. The **Element Faces** option is used to copy and reflect 2D (face) elements. The **Solid Elements Only** option is used to copy and reflect 3D (solid) elements.

Reflection Plane

In this rollout, you can define the plane along which you want to copy and reflect elements. Select the **Specify Plane** option; you will be prompted to select the plane along which you want to copy and reflect the selected elements. You can also define other planes using the buttons and drop-down list available in this rollout.

Other rollouts of this dialog box work similar as discussed earlier. After specifying all the options, choose the **OK** button to create a copy and then reflect the selected element.

Modify Order Tool

Ribbon: Nodes and Elements > Elements > More > Edit > Order
Menu: Edit > Element > Modify Order

The **Modify Order** tool is used to change the order of the elements. You can also modify the elements to midnode elements or the midnodes elements method from linear to parabolic/parabolic to linear. To do so, choose **Menu >Edit > Element > Modify Order** from the **Top Border Bar**; the **Element Modify Order** dialog box will be displayed, as shown in Figure 7-18. The options in this dialog box can be used for element reordering. The options and rollouts available in this dialog box are discussed next.

Note
This tool works only in Simulation FEM environment.

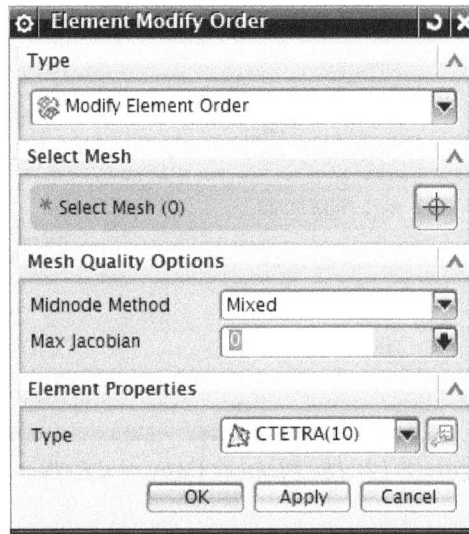

Figure 7-18 *The **Element Modify Order** dialog box*

Type

In the drop-down list of this rollout, two options are available: the **Modify Element Order** and **Modify Midnode Method**. The **Modify Element Order** option is used to modify the element order while the **Modify Midnode Method** option is used to change the midnode elements method.

Select Mesh

In this rollout, the **Select Mesh (0)** option is highlighted. As a result, you will be prompted to select the mesh to change the order of mesh elements. Note that for the **Modify Midnode Method**, you can only select the mesh if the midnode elements are available in the model.

Mesh Quality Options

You can change the midnode method using the options in this rollout. In the **Midnode Method** drop-down list of this rollout, three options are available: **Mixed**, **Curved**, and **Linear**. You can also change the jacobian value using the **Max Jacobian** edit box. This edit box will be available only when you select the **Mixed** option from the **Midnode Method** drop-down list.

Element Properties

This rollout will be available only when you select the **Modify Element Order** option from the drop-down list of the **Type** rollout. The element options in the drop-down list will change according to the mesh selected. You can select the desired element order from the **Type** drop-down list of the **Element Properties** rollout. You can also choose the **Edit Mesh Associated Data** button ⊞ to provide additional information related to the selected element.

After specifying all the options, choose the **OK** button to modify the order of selected elements.

Note

After element reordering, software does not change the element's location or label in the model. It will just change the older elements to newer elements.

Modify Type Tool

| Ribbon: | Nodes and Elements > Elements > More > Properties > Modify Type |
| Menu: | Edit > Element > Modify Type |

The **Modify Type** tool is used to change the element type of the selected mesh. To do so, choose **Menu > Edit > Element > Modify Type** from the **Top Border Bar**; the **Element Modify Type** dialog box will be displayed, as shown in Figure 7-19. You can use the options available in the dialog box for changing the element type of the mesh. The options and rollouts in this dialog box are discussed next.

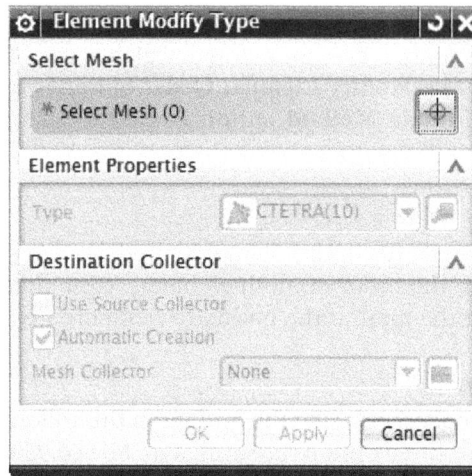

Figure 7-19 The Element Modify Type dialog box

Select Mesh

The **Select Mesh (0)** option is highlighted in this rollout. As a result, you will be prompted to select the mesh to change elements type. On selecting the mesh, the other options in this rollout will be activated.

Element Properties

You can select the desired element type from the **Type** drop-down list of this rollout. Depending upon the mesh selected, the element options will change in the drop-down list. You can also choose the **Edit Mesh Associated Data** button 🔲 to provide additional information related to the element.

Destination Collector

In the **Destination Collector** rollout, the **Use Source Collector** and **Automatic Creation** check boxes are available. The **Use Source Collector** check box is used to change the element type to its original mesh collector. When you select the **Automatic Creation** check box, the software will automatically create the destination collector for the changed element type. In this rollout, the **New Collector** button 🔲 is also available. Using this button, you can define physical properties of the element type and define the new destination location. This location will be active only when both the check boxes will be cleared.

After specifying all the options, choose the **OK** button to modify selected element type.

Split 1D Element Tool

Ribbon: Nodes and Elements > Elements > More > Edit > Split 1D Element
Menu: Edit > Element > Split 1D Element

The **Split 1D Element** tool is used to split the selected 1D element by defining a length or element number. To do so, choose **Menu > Edit > Element > Split 1D Element** from the **Top Border Bar**; the **Split 1D Element** dialog box will be displayed, as shown in Figure 7-20. You can use the options available in the dialog box for splitting 1D elements. The options and rollout available in the dialog box are discussed next.

Elements

In this rollout, the **Select 1D Elements (0)** option is highlighted. As a result, you will be prompted to select 1D element for splitting. You can also select the elements by specifying element label in the **Labels** edit box of the **Element Labels** sub-rollout.

Split Method

In this rollout, the **Split By** drop-down list consists of two methods: **Sub Element Number** and **Sub Element Length**. Depending upon the option selected from the drop-down list, you can define the element number or length in their respective edit boxes.

After specifying all the options, choose the **OK** button to split the 1D elements.

Split Shell Tool

Ribbon:	Nodes and Elements > Elements > More > Edit > Split Shell
Menu:	Edit > Element > Split Shell

The **Split Shell** tool is used to split the shell or 2D elements. You can split the quad elements into a triangle or a triangle into more smaller triangular elements. To split the elements, choose **Menu > Edit > Element > Split Shell** from the **Top Border Bar**; the **Split Shell** dialog box will be displayed, as shown in Figure 7-21. The options and rollouts in this dialog box are discussed next.

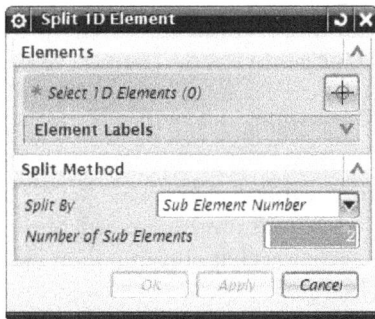

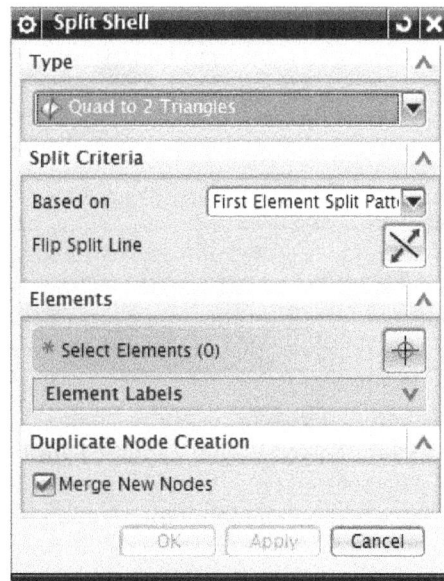

Figure 7-20 The Split 1D Element dialog box   *Figure 7-21 The Split Shell dialog box*

There are eight options available in the drop-down list of the **Type** rollout: **Quad to 2 Triangles**, **Quad to 2 Quads**, **Quad to 4 Quads**, **Quad to 3 Quads**, **Quad to 3 Triangles**, **Triangle to 4 Triangles**, **Split by Line**, and **Triangle to 2 Triangles**. The methods of splitting elements using these options are discussed next.

Quad to 2 Triangles

This option is used to split a quadrilateral element into two triangular elements. When you select this option, four options will be displayed in the **Based on** drop-down list available in the **Split Criteria** rollout. These options are **First Element Split Pattern**, **Element Connectivity**, **Interactive Location**, and **Element Quality**. You can use these options to define the split criteria.

First Element Split Pattern

This option is used to split the element using the first edge of first element selected. You can flip the split line using the **Flip Split Line** button available for this criteria in the rollout.

Element Connectivity

This option is used to split the element by following their connectivity. You can flip the split line using the **Flip Split Line** button available for this criteria in the rollout.

Interactive Location

This option shows the interactive preview of the split line on the element. You can move the cursor to change the split line direction on the selected element.

Element Quality

This option splits those quad elements, whose warp values exceed the values defined in the **Maximum Warp Threshold** edit box.

The **Elements** rollout is used to select the elements. You can also select the elements by specifying the label in the **Labels** edit box of the **Element Labels** sub-rollout of this rollout.

The **Merge New Nodes** check box of the **Duplicate Node Creation** rollout is used to merge the newly created nodes for removing node duplicity.

Quad to 2 Quads

This option is used to split a quadrilateral element into two quadrilateral elements. When you select this option, three options will be displayed in the **Based on** drop-down list of the **Split Criteria** rollout: **First Element Split Pattern**, **Element Connectivity**, and **Interactive Location**. The method to use these options is the same as discussed earlier for **Quad to 2 Triangles**.

Quad to 4 Quads

This option is used to split a quadrilateral element into four quadrilateral elements. When you select this option, only the **Elements** and **Duplicate Node Creation** rollouts will be available in the dialog box. As you select the element, the software will divide elements into four quadrilateral elements. In the **Duplicate Node Creation** rollout, the **Merge New Nodes** check box is available. When you select this check box, the software will merge the created new node while splitting.

Quad to 3 Quads

This option is used to split a quadrilateral element into three quadrilateral elements. When you select this option, the **Interactive Location** option is selected in the **Based on** drop-down list of the **Split Criteria** rollout. Move the mouse to change the split line direction on the selected element for desired splitting.

Quad to 3 Triangles

This option is used to split a quadrilateral element into three triangular elements. When you select this option, two options will be displayed in the **Based on** drop-down list available in the **Split Criteria** rollout: **Interactive Location** and **Element Quality**. The method to use these options is the same as discussed for earlier options.

Triangle to 4 Triangles

This option is used to split a triangular element into four triangular elements. When you select this option, only the **Elements** and **Duplicate Node Creation** rollouts will be available in the dialog box. As you select the element, the software will divide elements into four triangular elements.

Split by Lines

This option is used to split an element by defining a line around the element. When you select this option, beside the **Elements** and **Duplicate Node Creation** rollouts, the **Define Split Line** rollout is also displayed. In this rollout, the **Specify Point** option is highlighted and you will be prompted to select the first point. The point may be on free space or onto the face/edge of the model. You can also define the point using the **Point** dialog box. The **Point** dialog box will be displayed on choosing the **Point Dialog** button or through the **Inferred Point** drop-down list. Similarly, define the second point. After defining both the points, a line will be created between the points. Select the elements and choose the **Apply** button; the elements will be split by the line.

Triangle to 2 Triangles

This option is used to split a triangular element into two triangular elements. When you select this option, two options will be displayed in the **Based on** drop-down list of the **Split Criteria** rollout: **First Element Split Pattern** and **Interactive Location**. You can use these options to define the split criteria in the same manner as discussed previously.

Combine Triangles Tool

Ribbon:	Nodes and Elements > Elements > More > Edit > Combine Triangles
Menu:	Edit > Element > Combine Triangles

The **Combine Triangles** tool is used to combine two triangle shell elements into a single quadrilateral element. To do so, choose **Menu > Edit > Element > Combine Triangles** from the **Top Border Bar**; the **Combine Triangles** dialog box will be displayed, as shown in Figure 7-22. The options and rollouts available in the dialog box are discussed next.

First Element

The **Select Element (0)** option is highlighted in this rollout. As a result, you will be prompted to select first triangle element to combine with triangular element.

Second Element

On selecting the first triangular element, the **Select Element (0)** option is highlighted in this rollout. As a result, you will be prompted to select second triangle element to combine with triangular element.

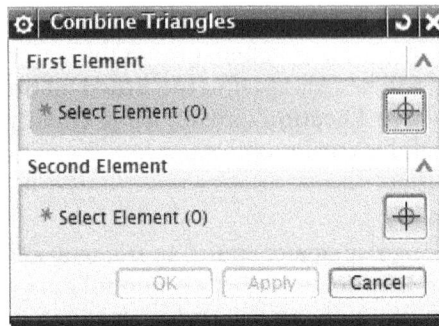

Figure 7-22 The **Combine Triangles** *dialog box*

After selecting both elements, choose the **OK** button to combine the selected triangles and make them a single quadrilateral element.

Modify Label Tool

Ribbon: Nodes and Elements > Elements > More > Edit > Label
Menu: Edit > Element > Modify Label

The **Modify Label** tool is used to change the label of elements. To do so, choose **Menu > Edit > Element > Modify Label** from the **Top Border Bar**; the **Element Modify Label** dialog box will be displayed, as shown in Figure 7-23. The options and rollouts available in the dialog box are discussed next.

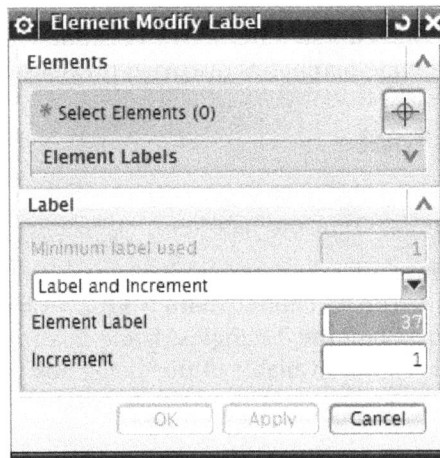

Figure 7-23 The **Element Modify Label** *dialog box*

Elements
The **Select Elements (0)** option is highlighted in this rollout. As a result, you will be prompted to select elements to change their labels.

Label
Using the options of this rollout, you can change the labels of the selected elements. Two options are available in the drop-down list of this rollout: **Label and Increment** and **Offset**. The **Label**

and **Increment** option is used to define a label with a defined increment value for the selected elements. The **Offset** option is used to change the elements label by adding an offset value in their existing label.

Modify Connectivity Tool

Ribbon: Nodes and Elements > Elements > Connectivity
Menu: Edit > Element > Modify Connectivity

The **Modify Connectivity** tool is used to modify the connectivity of the elements. Using this tool, you can modify the connectivity of a single element or group of elements which are attached to a single node. To do so, choose **Menu > Edit > Element > Modify Connectivity** from the **Top Border Bar**; the **Element Modify Connectivity** dialog box will be displayed, as shown in Figure 7-24. To change the connectivity of elements, two options are available in the drop-down list of the **Elements to Modify** rollout: **Single Element** and **Elements Attached to Selected Node**. The methods to modify connectivity using these options are discussed next.

Single Element

This option is used to modify the node connection related to a selected element. When you select the **Single Element** option from the drop-down list, the dialog box will be modified, as shown in Figure 7-25. The options available in the dialog box are discussed next.

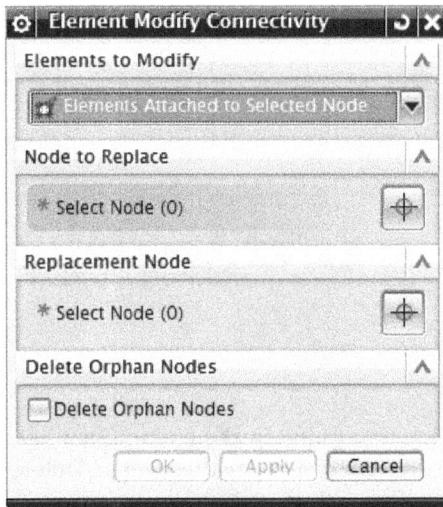

Figure 7-24 *The **Element Modify Connectivity** dialog box*

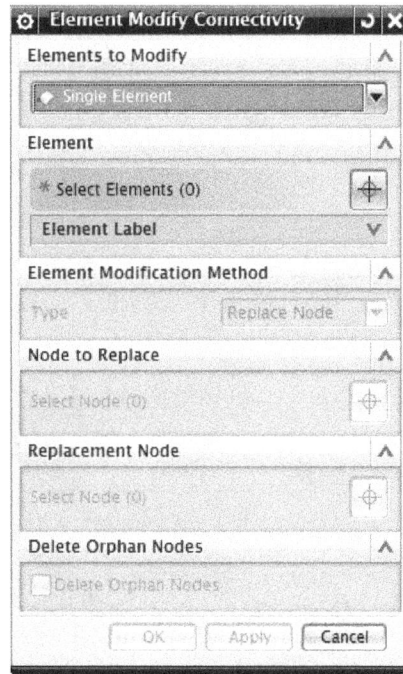

Figure 7-25 *The **Element Modify Connectivity** dialog box for the **Single Element** option*

Element

The **Select Elements (0)** option is highlighted in the **Element** rollout of this dialog box. You will be prompted to select elements to modify their connectivity. You can also select the element by specifying the label in the **Label** edit box of the **Element Label** sub-rollout of this sub-rollout. After selecting the element, the other options in the rollouts will be activated.

Element Modification Method

You can select the type of modification method using the **Type** drop-down list of the **Element Modification Method** rollout. In this drop-down list, the options will change according to the elements selected from the drawing area.

Node to Replace

The option in this rollout is used to select the node to replace. Note that, if you select an outside node from the selected element, a warning message box will be displayed to select the node within the selected element.

Replacement Node

The option in this rollout is used to select the replacement node from the selected node. Note that if you select the node from the selected element, a warning message box will be displayed to select the node outside the selected element.

Delete Orphan Nodes

The **Delete Orphan Nodes** check box of this rollout is used to delete the replace node of the model after choosing the **Apply** button.

Elements Attached to Selected Node

This option is used to modify all elements connected to a single node. When you select this option, the dialog box will be changed, refer to Figure 7-24. The options available in the dialog box are discussed next.

Node to Replace

By using the option of this rollout, you can select the node to replace.

Replacement Node

This rollout is used to select the replacement node from the selected node.

Delete Tool

Ribbon:	Nodes and Elements > Elements > Delete
Menu:	Edit > Element > Delete

The **Delete** tool is used to delete the elements that failed in the element quality check or form some irregularities. To do so, choose **Menu > Edit > Element > Delete** from the **Top Border Bar**; the **Element Delete** dialog box will be displayed, as shown in Figure 7-26.

In this dialog box, the **Select Elements (0)** option is selected by default. As a result, you will be prompted to select elements to delete from the model. Select the element by clicking on it from the model. You can also select the elements by specifying the label in the **Labels** edit box of the **Element Labels** rollout. Choose the **OK** button to delete the selected elements.

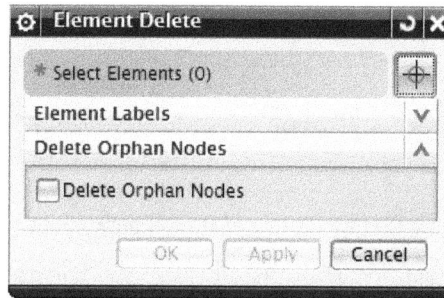

*Figure 7-26 The **Element Delete** dialog box*

Extract Tool

Ribbon: Nodes and Elements > Elements > More > Properties > Extract
Menu: Edit > Element > Extract

The **Extract** tool is used to extract the elements from their original mesh collector. You can use these extracted elements to define the new properties. To do so, choose **Menu > Edit > Element > Extract** from the **Top Border Bar**; the **Element Extract** dialog box will be displayed, as shown in Figure 7-27.

*Figure 7-27 The **Element Extract** dialog box*

In the dialog box, the **Select Elements (0)** option is highlighted in the **Elements** rollout. As a result, you will be prompted to select elements to extract from the mesh. Select the element from the model by clicking on it. You can also select the elements by specifying the label in the **Labels** edit box of the **Element Labels** sub-rollout.

Select the elements from the meshed model and choose the **Apply** button; the selected mesh will be extracted from the model. Figure 7-28 shows the meshed model with extracted elements. The extracted elements are displayed separately in the whole meshed model.

You can view and edit the properties of the extracted element in the **Simulation Navigator**, as shown in Figure 7-29. In the Simulation Navigator, the extracted elements will be saved in a separate node and the remaining mesh will be saved in the locked mesh collector, refer to Figure 7-29.

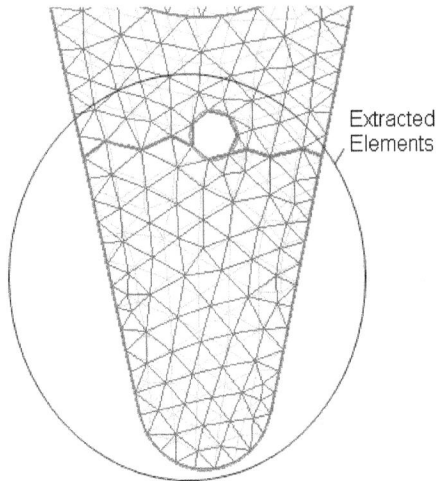

Figure 7-28 *Extracted elements*

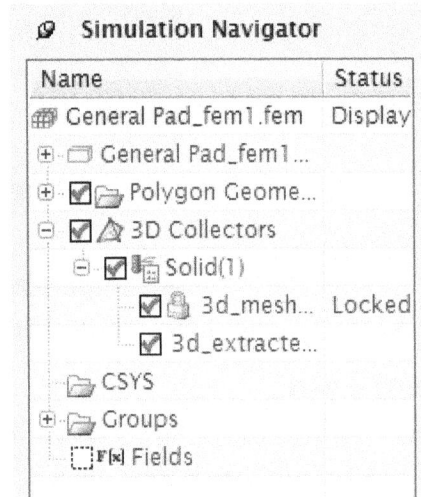

Figure 7-29 *Extracted elements in Simulation Navigator*

Note

After extracting the elements from the mesh, the remaining mesh will be locked. You can unlock this mesh to restore it, but then the extracted elements will vanish. To unlock the mesh, choose **Menu > Edit > Unlock Mesh** *from the* **Top Border Bar**; *the* **Unlock Mesh** *dialog box will be displayed, as shown in Figure 7-30. The* **Mesh** *button is highlighted in this dialog box. Select the locked mesh from the model and choose the* **OK** *button; the meshed node in the Simulation Navigator will turn into updating mode. To update the model, choose* **Menu > Edit > Update** *tool from the* **Top Border Bar**; *the model will be updated and the* **Mesh Update** *message window will be displayed, as shown in Figure 7-31. This message window prompts you to shows the log of the deleted extracted elements. Choose the* **Yes** *button to display the log of the deleted elements; the* **Information** *window will be displayed.*

Note that you can also update the model using **Update** *from the* **Context** *group of the* **Home** *tab in the* **Ribbon**.

Figure 7-30 *The* **Unlock Mesh** *dialog box*

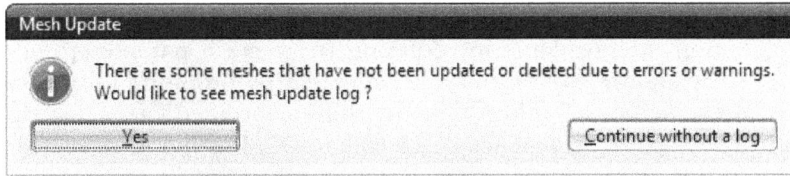

*Figure 7-31 The **Mesh Update** window*

MESH REFINEMENT

After defining the mesh, if any element fails during the analysis, you may have to remesh or refine the mesh at critical areas of the model. The tools used in the process of refining and remeshing the model are discussed next.

2D Local Remesh Tool

Ribbon: Home > Mesh > More > 2D > 2D Local Remesh
Menu: Insert > Mesh > 2D Local Remesh

The **2D Local Remesh** tool is used to remesh the selected 2D mesh/element by defining a size or size factor. To do so, choose **Menu > Insert > Mesh > 2D Local Remesh** from the **Top Border Bar**; the **2D Local Remesh** dialog box will be displayed, as shown in Figure 7-32. You can use the options available in the dialog box for remeshing. The options and rollouts available in the dialog box are discussed next.

Elements to Remesh

In this rollout, the **Select Elements (0)** option is highlighted by default. As a result, you will be prompted to select the mesh/elements to remesh. Select the element from the model by clicking on it. You can also select the elements by specifying the label in the **Labels** edit box of the **Element Labels** sub-rollout of this rollout.

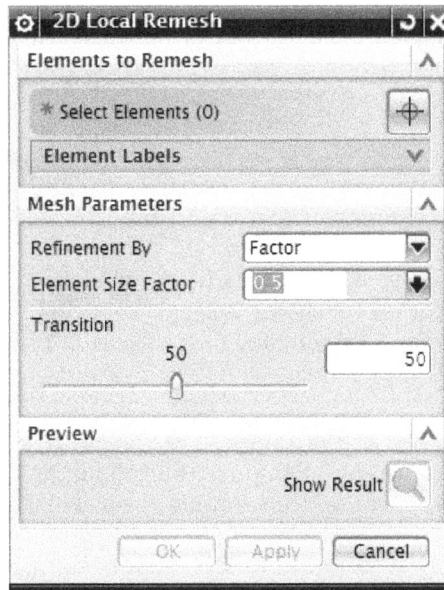

*Figure 7-32 The **2D Local Remesh** dialog box*

Mesh Parameters

In this rollout, options are available for defining the remesh parameters of the selected mesh/elements. There are two options in the **Refinement By** drop-down list: **Factor** and **Size**. By using the **Factor** option, you can resize the mesh by a specified factor. You can specify the factor in the **Element Size Factor** edit box of the rollout. While using the **Size** method, you can define the size of the mesh by specifying element size in the **Element Size** edit box.

The **Transition** slider, available in the rollout, is used to control the rate of size variation while remeshing. The element shape changes randomly for maximum transition value, so software will take less time to remesh. Similarly, use low transition value for high quality remesh.

Preview

By using the **Show Result** button in the **Preview** rollout, you can preview the remesh for the specified settings.

Mesh Control Tool

Ribbon: Home > Mesh > Mesh Control
Menu: Insert > Mesh > Mesh Control

The **Mesh Control** tool is used to refine the selected 2D or 3D mesh by creating density points onto their edges. To do so, choose **Menu > Insert > Mesh > Mesh Control** from the **Top Border Bar**; the **Mesh Control** dialog box will be displayed, as shown in Figure 7-33. For refining the mesh, nine options are available in the drop-down list of the **Density Types** rollout: **Number on Edge**, **Size on Edge**, **Chordal Tolerance on Edge**, **Biasing on Edge**, **Size on Face**, **Weld Row**, **Mapped Hole**, **Fillet**, and **Cylinder**. Some of these options are discussed next.

Number on Edge

This option is used to control the mesh by defining element number on the selected edges of the model. You can create unequal number of elements on different edges. When you select this option, the dialog box will be modified, refer to Figure 7-33. The options and rollouts available in this dialog box are discussed next.

*Figure 7-33 The **Mesh Control** dialog box*

Selection

The **Select Targets (0)** option is highlighted in this rollout. As a result, you will be prompted to select the edges or faces along which you want to control the mesh.

Number on Edge

After selecting the edges, you can define the element numbers on the selected edges in the

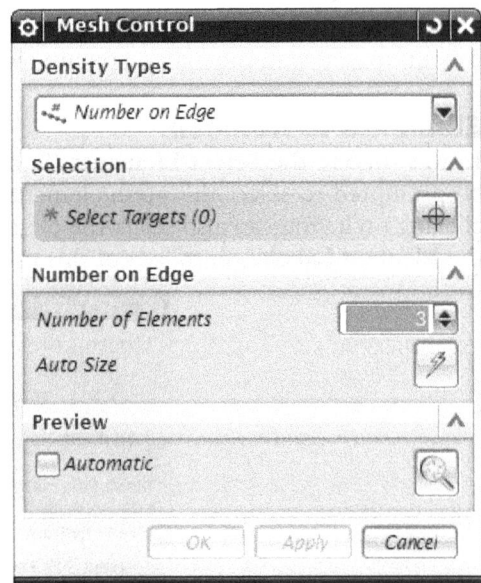

Number of Elements edit box. You can also choose the **Auto Size** button available in this rollout to automatically define the elements on the selected edges.

Preview

The **Preview** button in this rollout is used to preview the defined element's node distribution. Select the **Automatic** check box to distribute the nodes automatically.

> **Note**
> *After applying the element density, a diamond symbol will be created on the selected edge. For viewing the refined mesh, you have to update the model. To do so, choose **Menu > Edit > Update** from the **Top Border Bar**; the model will show the updated mesh. You can also choose the **Update** tool from the **Context** group of the **Home** tab in the **Ribbon**.*

Size on Edge

This option is used to control the mesh by defining elements size on the selected edge of the model. You can create unequal number of elements on different edges. When you select this option, the dialog box will be modified, as shown in Figure 7-34. Some methods using options and rollouts available in this dialog box are discussed next.

Selection

The **Select Targets (0)** option is highlighted in this rollout. As a result, you will be prompted to select the edges or faces along which you want to control the mesh.

Size on Edge

Using the options of this rollout, you can define the elements sizes on the selected edges. The **Location on Edge** drop-down list in the rollout is used to define the position of the elements. This drop-down list consists of four options: **Overall**, **Start**, **End** and **Start and End**. The **Overall** option is used to apply the element density along the entire selected edge. The **Start** option is used to apply the density to the beginning of the edge. The **End** option is used to apply the density to the end of the edge. The **Start and End** option is used to apply the density at the start and end of the edge but not at the middle.

Depending upon the option selected from the **Location on Edge** drop-down list, the name of the size edit box will change. You can enter the element size value in this edit box. You can also choose the **Auto Size** button available in the rollout to automatically define the elements on the selected edge.

Chordal Tolerance on Edge

This option is used to control the mesh by defining chordal tolerance on the selected edge of the model. Using this tolerance value, the software calculates elements by using equations related to the curvature of the selected edge. Note that the software will create higher density on curvature edges while lower density on straight edges. You can create unequal number of elements on different edges. When you select this option from the drop-down list in the **Density Types** rollout, the **Mesh Control** dialog box will be updated and modified, as shown in Figure 7-35. Some of the options of this modified dialog box are discussed next.

Figure 7-34 The **Mesh Control** dialog box on selecting the **Size on Edge** option

Figure 7-35 The **Mesh Control** dialog box on selecting the **Chordal Tolerance on Edge** option

Selection

The **Select Targets (0)** option is highlighted in the **Selection** rollout by default. As a result, you will be prompted to select the edges or faces along which you want to control the mesh.

Chordal Tolerance on Edge

After selecting the edges, you can define the chordal tolerance value on the selected edges in the **Tolerance** edit box. You can also choose the **Auto Size** button available in the rollout to define the tolerance value for the selected edge.

Biasing on Edge

This option is used to control the mesh by using biasing element density along a selected edge. Using this option, you can create dense and coarse nodal distribution along the beginning, end, or center part of the edge. You can create unequal number of elements on different edges. When you select this option, the dialog box will be modified, as shown in Figure 7-36. Some of the options and rollouts available in this dialog box are discussed next.

Selection

The **Select Targets (0)** option is highlighted in this rollout by default. As a result, you will be prompted to select the edges or faces along which you want to control the mesh.

Biasing on Edge

By using the options of this rollout, you can define the element biasing on the selected edges. The **Bias Origin** drop-down list in the rollout is used to define the position of the biasing. This drop-down list consists of three options: **Start of Edge**, **End of Edge**, and **Center of Edge**. The **Start of Edge** option is used to create dense nodal distribution at the start of the edge. The **End of Edge** option is used to create dense nodal distribution at the end of the edge. The **Center of Edge** option is used to create dense nodal distribution at the center of the edge.

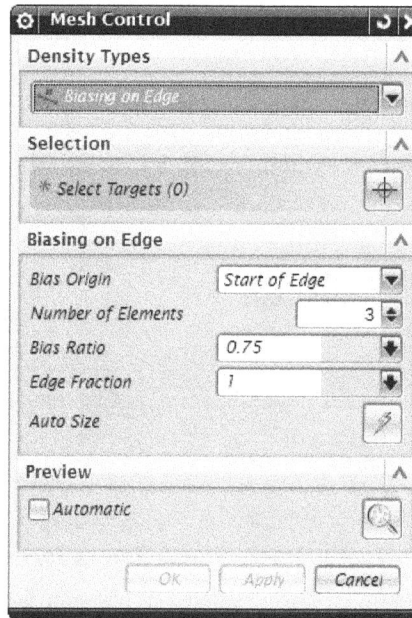

*Figure 7-36 The **Mesh Control** dialog box on selecting the **Biasing on Edge** option*

You can define the element quantity for the selected edges in the **Number of Elements** edit box of this rollout.

The **Bias Ratio** edit box is used to control the fraction of nodes distribution along the selected edge. For example, if you enter **0.75** in this edit box, the software will bias 75% of the edge length and distribute the nodes along 25% of the selected edge.

The **Edge Fraction** edit box available for the **Start of Edge** or **End of Edge** option is used to apply the fraction of start and end location along which the software will do biasing for the defined bias ratio.

You can also choose the **Auto Size** button available in the rollout to define the number of elements, bias ratio, and edge fraction for the selected edge.

Size on Face

This option is used to control the mesh by defining element size on the selected face of the model. When you select this option, the **Mesh Control** dialog box will be modified, as shown in Figure 7-37. Some of the options and rollouts available in this dialog box are discussed next.

Selection

The **Select Targets (0)** option is highlighted in this rollout by default. As a result, you will be prompted to select the faces along which you want to control the mesh.

Size on Face

After selecting the faces, you can define the element size in the **Element Size** edit box. The software will distribute the node around the edges of the selected face. You can also choose the **Auto Size** button available in the rollout to define the elements on the selected edges.

Mapped Hole

This option is used to create structured rows of elements around selected edges of holes or loops. When you select this option, the **Mesh Control** dialog box will be modified, as shown in Figure 7-38. Some of the options and rollouts available in this dialog box are discussed next.

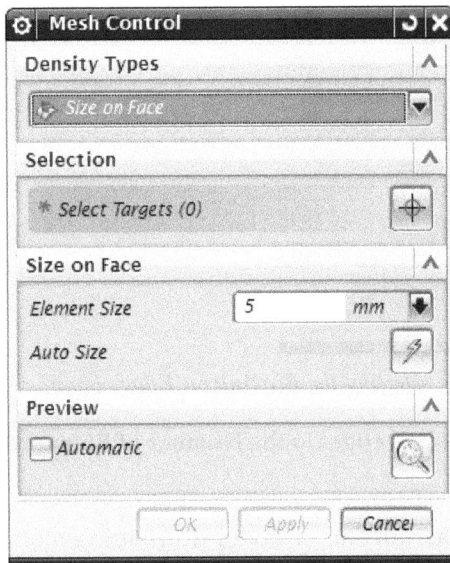

*Figure 7-37 The **Mesh Control** dialog box on selecting the **Size on Face** option*

*Figure 7-38 The **Mesh Control** dialog box on selecting the **Mapped Holes** option*

Selection

The **Select Targets (0)** option is highlighted in this rollout by default. As a result, you will be prompted to select the edges of holes or edges that formed loops along which you want to create the structured elements.

Mapped Holes

The options in this rollout are used to define the layers of elements that will be formed around the selected edges loop. The **Layer Depth** edit box is used to define the distance of the edge element from the selected loop. The **Number of Layers** edit box is used to define the layer numbers to be created between the selected edges and layers.

Spacing on Edge

The options in this rollout are used to control the spacing of elements around the selected edge loop. Three options are available in the **Spacing Type** drop-down list of this rollout: **None**, **By Number**, and **By Size**. The **None** option is used when you auto define the element spacing for the defined layer depth and number of layers. By using the **By Number** option, you can define the number of elements to be created around the edge. The **By Size** option is used to define the size of elements to be created around the edge. You can define the values for the **By Number** and **By Size** options in their relevant edit boxes. You can also choose the **Auto Size** button available in the rollout to define the elements on the selected edges.

Note
*The **Mapped Hole** option is only valid for holes or looped edges. If the selected edges do not make a loop, the **Invalid Loop** window will be displayed after choosing the **Apply** button, as shown in Figure 7-39. You must select the closed loop to create mapping.*

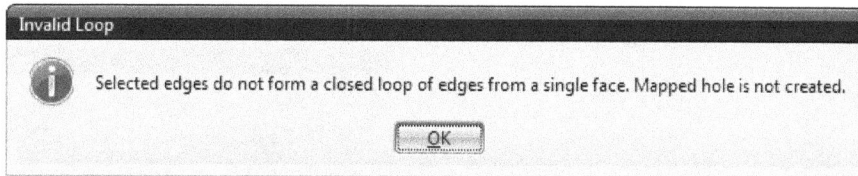

Invalid Loop

Selected edges do not form a closed loop of edges from a single face. Mapped hole is not created.

OK

*Figure 7-39 The **Invalid Loop** window*

Figure 7-40 shows the model with default meshing and Figure 7-41 shows the model after the meshes are controlled around edges.

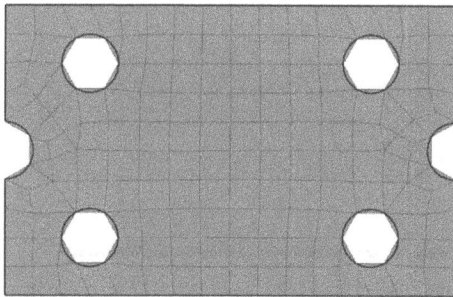

Figure 7-40 Model with default meshing

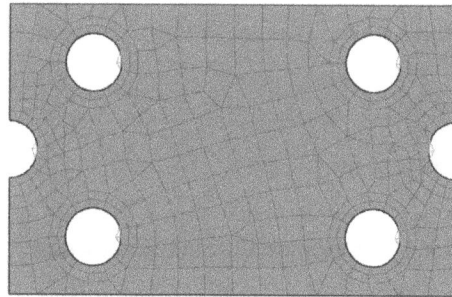

Figure 7-41 Meshed model after mesh controlling

MESH ASSOCIATED DATA

Ribbon: Home > Properties > More > Properties > Mesh Associated Data
Menu: Tools > Mesh Associated Data

The **Mesh Associated Data** dialog box is used to define the properties of the related mesh or elements. You need to first mesh the model and then define the associated data or can also define the associated data while creating mesh using the **Edit Mesh Associated Data** button available in the **Element Properties** rollout of the relevant element or the mesh creation dialog box. Note that the **Mesh Associated Data** dialog box changes according to the element type defined in mesh.

To define the associated data for meshed model, choose **Menu > Tools > Mesh Associated Data** from the **Top Border Bar**; the **Mesh Associated Data** dialog box will be displayed, as shown in Figure 7-42. The **Select Mesh (0)** option is highlighted in this dialog box. You will be prompted to select the mesh. Depending upon the meshing process used in the model, the dialog box gets modified. The methods to define the associated data for different mesh or element type are discussed next.

Note
This tool is available only in the Simulation environment and works in the Simulation FEM file.

Mesh Associated Data for 0D Element

As discussed previously, the 0D elements are used to create concentrated masses on a point. To define the associated data for 0D element/mesh, choose 0D element/mesh from the model; the dialog box will be modified. Figure 7-43 shows the **Mesh Associated Data** dialog box for **CONM1** element type meshed model.

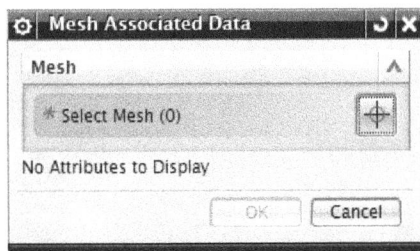

*Figure 7-42 The **Mesh Associated Data** dialog box*

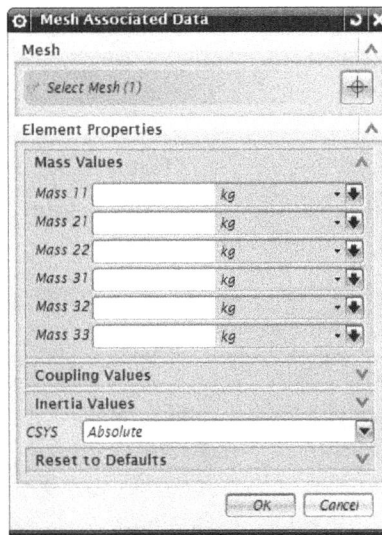

*Figure 7-43 The **Mesh Associated Data** dialog box for **CONM1** (0D) element*

You can specify the mass values in their respective edit boxes and also define the coordinate system of the element. Choose the **OK** button from the dialog box, the data will be stored in respective mesh. Similarly, you can specify mesh associated data for other 0D element/mesh.

Mesh Associated Data for 1D Element

As discussed previously, the 1D elements are used for creating a rod or beam type. You can define orientation or offset of the relevant section around geometry. To define the associated data for 1D element/mesh, choose 1D element/mesh from the model; the **Mesh Associated Data** dialog box will be modified. Figure 7-44 shows the **Mesh Associated Data** dialog box for **CBEAM** element type meshed model.

In the dialog box, two methods are available in the **Method** drop-down list of the **Section Orientation** sub-rollout of the **Element Properties** rollout: **Orientation Vector** and **Orientation Node**. You can define the section orientation by specifying a vector or node. Figure 7-45 shows the model with default orientation of the 1D element. Wherever, Figure 7-46 shows the model after defining the required orientation.

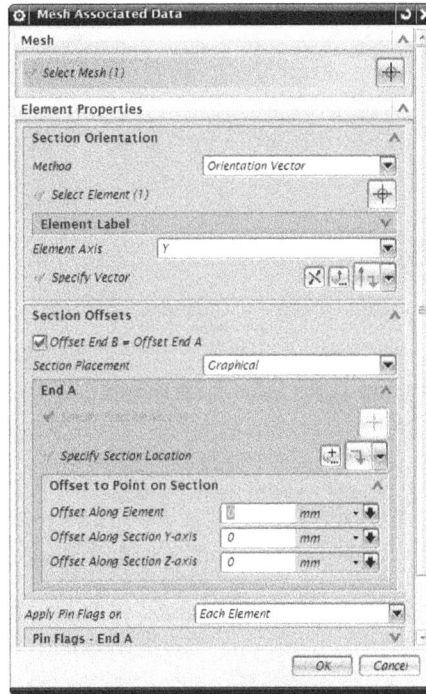

Figure 7-44 The **Mesh Associated Data** dialog box for **CBEAM** (1D) element

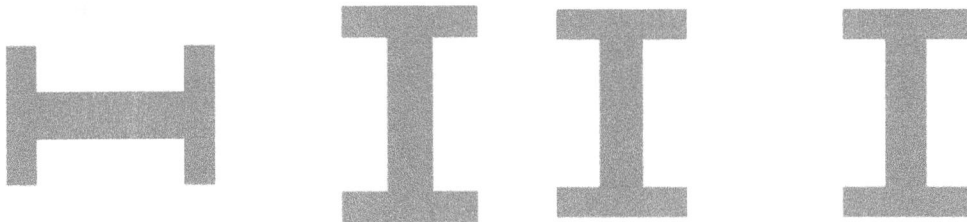

Figure 7-45 Default orientated meshed model   *Figure 7-46* Meshed elements after reorientation

Similarly, you can change the offset value of the elements. The options for offsetting the elements are available in the **Section Offsets** sub-rollout of the **Mesh Associated Data** dialog box. You can also specify the offset values in the edit boxes available in the **Offset to Point on Section** sub-rollout. In the **Section Offsets** sub-rollout, the **Offset End B= Offset End A** check box is selected by default. You can specify different offset values for different ends of the model. Figure 7-47 shows the model with default 1D element. Figure 7-48 shows the model with unequal offset element created on clearing the **Offset End B= Offset End A** check box.

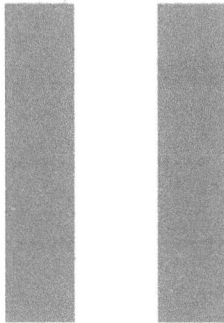

Figure 7-47 *Default meshed model*

Figure 7-48 *Unequal offset element*

Mesh Associated Data for 2D Element

As already discussed, the 2D elements are also known as shell elements. To define the associated data for 2D element/mesh, choose 2D element/mesh from the model; the dialog box will be updated. Figure 7-49 shows the **Mesh Associated Data** dialog box for **CQUAD4** element type meshed model. You can specify the shell offset, material orientation, and thickness specification using this dialog box. The options and rollouts in this dialog are discussed next.

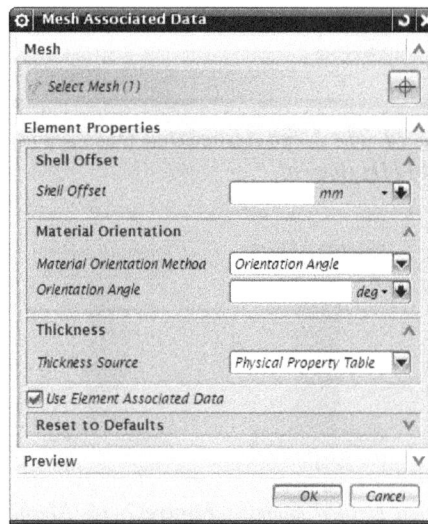

Figure 7-49 *The* **Mesh Associated Data** *dialog box for*
CQUAD4 *(2D) element*

Shell Offset

This option is used to specify the offset value of a shell element. You can specify the offset value in the **Shell Offset** edit box of the **Shell Offset** sub-rollout in the **Element Properties** rollout. This offset value is the distance of material orientation from the shell mesh.

Material Orientation

There are four options available in the **Material Orientation Method** drop-down list of the **Material Orientation** sub-rollout for defining orientation: **MCID**, **Orientation Angle**, **Tangent Curve**, and **Vector**. The following methods can be used to define material orientation.

MCID

When you select the MCID option, the software will orient the material along X-axis of the selected coordinate system. You can select any coordinate system type from the **MCID Type** drop-down list while selecting the **User Defined** option from the **MCID Definition** drop-down list, available in the **Material Orientation** sub-rollout of the **Element Properties** rollout. If you select the **Inherited from Layup** option from the **MCID Type** drop-down list, the material orientation will define the layups nodes. This option is applicable only for ply-based process.

Orientation Angle

This option is used for defining orientation by specifying an angle. You can specify the angle value in the **Orientation Angle** edit box that will be available in the **Material Orientation** sub-rollout of the **Element Properties** rollout.

Tangent Curve

Using this option, you can define the orientation along the selected curve of the model. When you select this option, the **Primary Direction** sub-rollout and the **Rotational Angle** edit box will be activated. The **Select Edge or Curve (0)** option available in the **Primary Direction** sub-rollout is used to select the edge to define the orientation. The **Rotational Angle** edit box is used to specify the angle value along the selected edge to define the orientation.

Vector

This option is used for defining the orientation along a vector. You can specify the vector using the **Vector** dialog box that appears when you choose the **Vector Dialog** button. The **Rotational Angle** edit box in this area is used to specify the angle value along the specified vector.

Thickness

You can specify the thickness on the shell mesh using drop-down list available in this sub-rollout. For defining thickness, three options are available in this sub-rollout: **Physical Property Table**, **Midsurface**, and **Field**. On selecting, the **Physical Property Table** option, the software will specify the thickness of shell mesh defined in associated physical property (mesh collector). On selecting, the **Midsurface** option, the software will specify the geometry thickness to the shell mesh. Whereas on selecting the **Field** option, you have to specify the table field for a variable thickness.

Figure 7-50 shows a meshed shell geometry with default orientation and Figure 7-51 shows the geometry after specifying orientation around the hole.

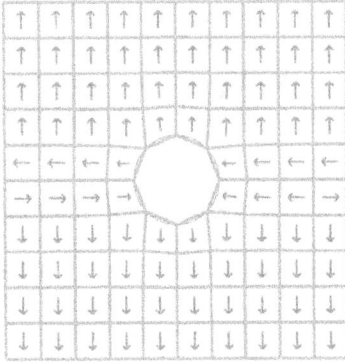

Figure 7-50 *Default mesh orientation*   **Figure 7-51** *Mesh orientation around the hole*

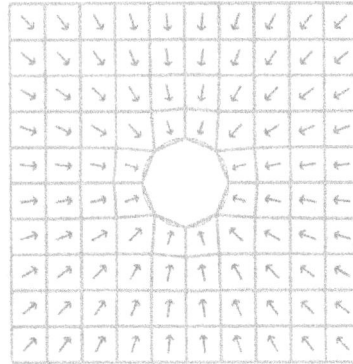

Mesh Associated Data for 3D Element

As discussed previously, the 3D elements are used for solid structure. You can apply material orientation value to these 3D elements. To define the associated data for 3D element/mesh, choose 3D element/mesh from the model; the **Mesh Associated Data** dialog box will be modified. Figure 7-52 shows the **Mesh Associated Data** dialog box for **CTETRA4** element type meshed model. For defining material orientation, six options are provided in the **Material Orientation Method** drop-down list: **Tangent Curve**, **Surface Normal**, **Vector**, **Physical Property Table**, **Spatial Field**, and **Element Free Face Proximity**. The options that can be used for material orientation are discussed next.

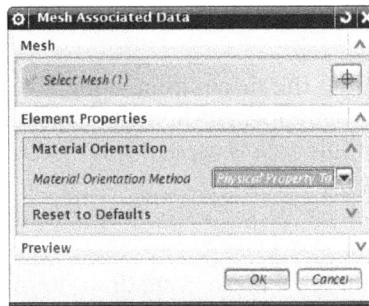

Figure 7-52 *The **Mesh Associated Data** dialog box for **CTETRA4** (3D) element*

Tangent Curve

By using this option, you can define the orientation along the selected curve of the model. When you select this option, the rotational angle edit boxes will be displayed. You can specify the rotational values along the X, Y, and Z axes in these edit boxes.

Surface Normal

By using this option, you can define the orientation using a face as the primary direction and a vector/tangent curve as the secondary direction. When you select this option, the rotational angle edit boxes will also be displayed. You can specify the X, Y, and Z axes rotational values in these edit boxes.

Vector

By using this option, you can define the orientation using a vector in the primary and a vector/curve in the secondary direction. When you select this option, the rotational angle edit boxes will be displayed. You can specify the rotational values along the X, Y, and Z axes in these edit boxes.

Physical Property Table

By using this option, the software will orient the material using the specification defined in the physical property table to which the element/mesh belongs. By default, the default coordinate system is selected in the table.

Spatial Field

This option is used for defining the material orientation in the imported simulation file.

Element Free Face Proximity

By using this option, you can define the orientation of the model using the element face. The software uses the normal from the selected element as the primary direction. When you choose multiple faces of the elements, the software calculates the normals along their combinations. For the secondary direction, you can choose a vector/curve. When you select this option, the rotational angle edit boxes will be displayed. You can specify the rotational value of the X, Y, and Z axes in these edit boxes.

Figure 7-53 shows a meshed 3D geometry with default orientation and Figure 7-54 shows the geometry after defining the orientation around the center of the model.

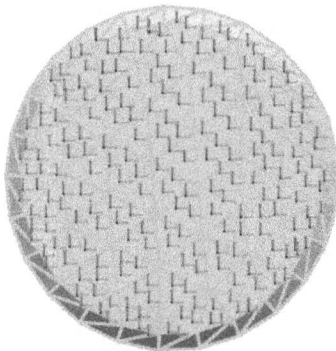

Figure 7-53 Default mesh orientation

Figure 7-54 Mesh orientation around the center of the model

TUTORIALS

Tutorial 1

In this tutorial, you will import the FEM file completed in Tutorial 1 of Chapter 6 into the NX Nastran. After importing the part file, you will orient the material of the model along the top edge. The default material orientation of the model is shown in Figure 7-55.

(Expected time: 30 min)

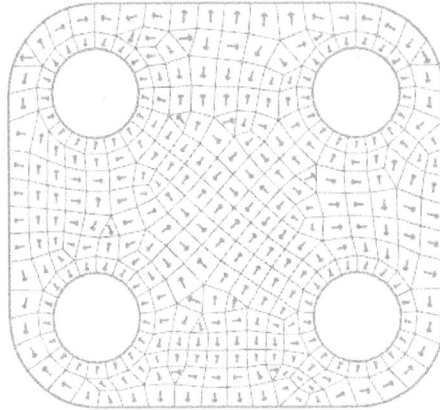

Figure 7-55 Default material orientation of the FEM model

The following steps are required to complete this tutorial:

a. Start NX and open the FEM file in NX.
b. Define the material orientation of the model.
c. Check the defined orientation of the model.
d. Save and close the model.

Starting NX and Opening the FEM File

1. Start NX 9.0 by double-clicking on the shortcut icon of **NX 9.0** on the desktop of your computer.

2. Copy the *c06tut01.fem* file from the location *C:\NX Nastran\c06\Tut01* and paste it to *C:\NX Nastran\c07\Tut01*. Change the file name to *c07tut01.fem*. You can also download this file from *www.cadcim.com*.

> **Note**
> *You must create the Tut01 folder in c07 folder at the location C:\NX Nastran\ as you have created in the earlier chapters.*

3. Choose **File > Open** from the **Ribbon**; the **Open** dialog box is displayed. Choose **FEM Files (*.fem)** from the **Files of type** drop-down list and select the *c07tut01.fem* file from the location you have saved. Choose the **OK** button; the file is opened in the FEM environment.

Defining Material Orientation

Next, you need to define the material orientation.

1. Choose **Menu > Tools > Mesh Associated Data** from the **Top Border Bar**; the **Mesh Associated Data** dialog box is displayed. Select the meshed model from the environment; the dialog box will get modified, refer to Figure 7-56.

2. Select the **Tangent Curve** option from the **Material Orientation Method** drop-down list from the **Material Orientation** sub-rollout of the **Element Properties** rollout.

3. Choose the **Select Edge or Curve (0)** option in the **Primary Direction** sub-rollout; you are prompted to select the edge along which you want to define the material orientation.

4. Select the top edge of the model, refer to Figure 7-57. Choose the **OK** button; the material is oriented along the edge direction.

Figure 7-56 *The Mesh Associated Data dialog box*

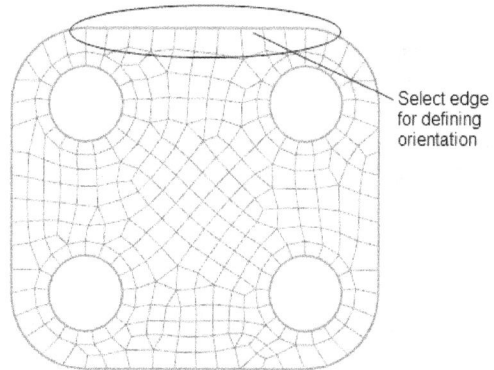

Figure 7-57 *Material orientation of the FEM model*

Checking Material Orientation

Now, you need to check the material orientation of the model.

1. Choose **Menu > Analysis > Finite Element Model Check > Element Material Orientation** from the **Top Border Bar**; the **Element Material Orientation** dialog box will be displayed.

2. Select the **Displayed** option from drop-down list in the **Elements to Check** rollout, refer to Figure 7-58, and select the **Shell Orientation** check box in the **Display Settings** rollout.

3. Choose the **Display Element Material Orientation** button from the dialog box; the FEM model will display the material orientation, as shown in Figure 7-59. Choose the **Close** button to exit from the dialog box.

Figure 7-58 *The Element Material Orientation* dialog box

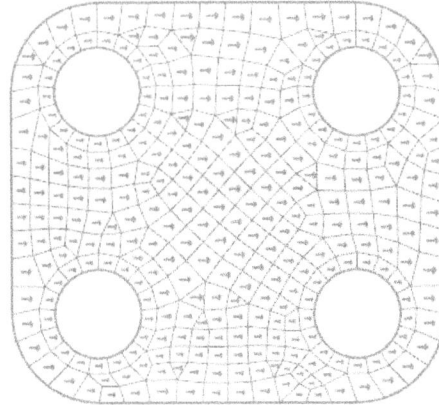

Figure 7-59 Defined material orientation of the FEM model

Saving and Closing the Model

1. Choose **File > Save > Save** from the **Ribbon** to save the model.

2. Choose **File > Close > All Parts** from the **Ribbon**; all displayed parts get closed.

Tutorial 2

In this tutorial, you will import the FEM file edited in Tutorial 2 of Chapter 6 into NX Nastran. After importing the part file, you will edit the failed elements.

(Expected time: 30 min)

The following steps are required to complete this tutorial:

a. Start NX and open the FEM file in NX.
b. Highlight the failed elements.
c. Edit the failed elements and recheck elements quality.
d. Save and close the model.

Starting NX and Opening the FEM File

1. Start NX 9.0 by double-clicking on the shortcut icon of **NX 9.0** on the desktop of your computer.

2. Copy the **c06tut02_input.fem** file from *C:\NX Nastran\c06\Tut02* and paste it to *C:\NX Nastran\c07\Tut02*. Change the file name to *c07tut02.fem*. You can also download the file from *www.cadcim.com*.

Note
You must create the Tut02 folder in c07 folder at the location C:\NX Nastran\ as created in the previous chapter.

3. Choose **File > Open** from the **Ribbon**; the **Open** dialog box is displayed. Choose **FEM Files (*.fem)** from the **Files of type** drop-down list and select the *c07tut02.fem* file from the location. Choose the **OK** button; the file is opened in the FEM environment.

Highlighting the Failed Element

Next, you need to highlight the failed element.

1. Choose **Menu > Analysis > Finite Element Model Check > Elements Quality** from the **Top Border Bar**; the **Elements Quality** dialog box is displayed. Make sure **Failed and Warning** and **All** options are selected in the **Output Group Elements** and **Report** drop-down lists respectively from the **Output Settings** rollout. Select all elements and choose the **Check Elements** button of the dialog box; all failed and warning elements are highlighted in the model, refer to Figure 7-60.

Note
1. *You do not need to refill the elements quality value again in the **Elements Quality** dialog box as you have already filled those values.*

2. *You can also show only the failed elements using the **Output Group** sub-node under the **Groups** node created in the **Simulation Navigator** in the previous chapter. To do so, right-click on the **Output Group** node and choose the **Show Only** option.*

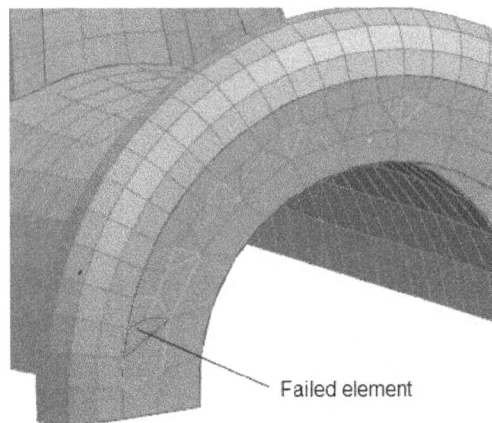

Failed element

Figure 7-60 Failed element in the FEM model

Editing the Failed Element

Now, you need to edit the failed elements. For editing the failed element, you will use the **Split Shell** tool.

1. Choose **Menu > Edit > Element > Split Shell** from the **Top Border Bar**; the **Split Shell** dialog box is displayed.

2. Select the **Quad to 3 Triangles** option from the drop-down list in the **Type** rollout, as shown in Figure 7-61. Now, move the cursor on the failed quad element to divide it into three

triangular elements, refer to Figure 7-62. Select the element to divide it. Choose the **Cancel** button to exit the dialog box.

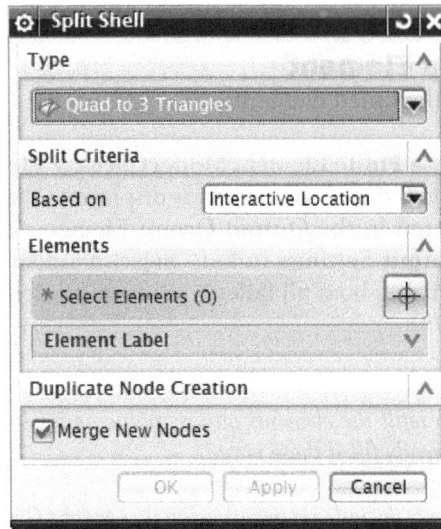

*Figure 7-61 The **Split Shell** dialog box*

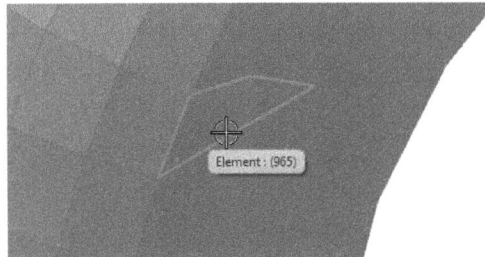

Figure 7-62 Division of failed element

Rechecking the Quality of Elements

1. Recheck the quality of divided elements using the options in the **Elements Quality** dialog box. The failed element is resolved and the message **0 failed elements, 48 warning elements** is displayed in the **Information** window. Close all the windows and then the dialog box to exit.

Saving and Closing the Model

1. Choose **File > Save > Save** from the **Ribbon** to save the model.

2. Choose **File > Close > All Parts** from the **Ribbon**; all displayed parts get closed.

Tutorial 3

In this tutorial, you will import the FEM file shown in Figure 7-63 into NX Nastran and then refine and remesh the elements around the holes and small rounds. You can download this file from *www.cadcim.com*. The complete path for downloading the file is:

Textbooks > CAE Simulation > NX Nastran > NX Nastran 9.0 > Input File

(Expected time: 30 min)

The following steps are required to complete this tutorial:

a. Start NX and open the FEM file in NX.
b. Refine mesh around holes using the **Mesh Control** tool.
c. Remesh around the round edges.
d. Edit rounds of bottom cut.
e. Check elements and edit failed elements.
f. Save and close the model.

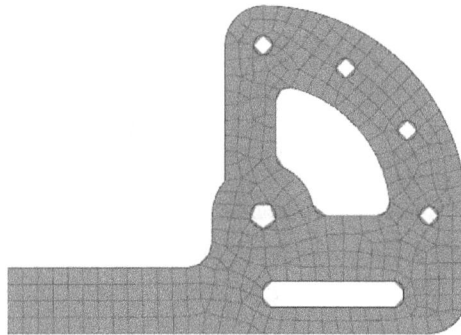

Figure 7-63 *FEM model for remeshing*

Starting NX and Opening the FEM File

First, you need to download the FEM file and then import it.

1. Start NX 9.0 by double-clicking on the shortcut icon of **NX 9.0** on the desktop of your computer.

2. Download the **c07tut03_input** zipped folder file from *www.cadcim.com* and extract the folder to the location *C:\NX Nastran\c07\Tut03*.

Note
You must create the Tut03 folder in c07 folder at the location C:\NX Nastran\ as created in the previous chapter.

3. To open an FEM file, choose **File > Open** from the **Ribbon**; the **Open** dialog box is displayed. Select the **c07tut03_input.fem** file from the location you extracted and choose the **OK** button; the file is opened in the FEM environment.

Mesh Refining Around the Holes

For refining holes, you need to use the **Mapped Hole** option of the **Mesh Control** tool.

1. To refine the mesh around the holes, choose **Menu > Insert > Mesh > Mesh Control** from the **Top Border Bar**; the **Mesh Control** dialog box is displayed. Select the **Mapped Hole** option from the drop-down list in the **Density Types** rollout; the options in the dialog box get modified, refer to Figure 7-64.

*Figure 7-64 The **Mesh Control** dialog box for the **Mapped Hole** option*

2. Enter the values given below in the edit boxes in different rollouts:
   Layer Depth : 14
   Number of Layers : 2
   Spacing Type : By Size
   Element Size : 4

3. Select all the circular edges of the model, refer to Figure 7-65. Choose the **OK** button from the dialog box; a diamond symbol is created on all the edges.

Now, you need to update the mesh to available around the holes.

4. Choose **Menu > Edit > Update** from the **Top Border Bar**; the mesh around the holes is refined, as shown in Figure 7-66.

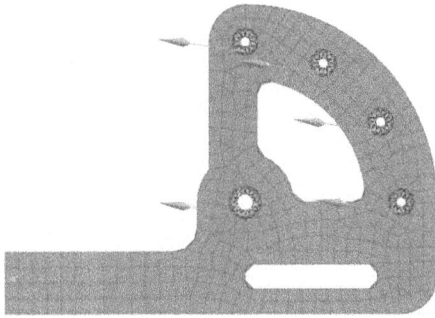

Figure 7-65 Selecting the edges of the holes

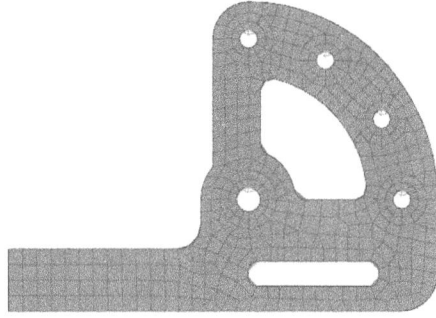

Figure 7-66 Mesh refined around holes

Remeshing Around the Round Edges

After refining the holes, you need to remesh the meshing around the small rounds of the curvature profile of the model.

1. Choose **Menu > Insert > Mesh > 2D Local Remesh** from the **Top Border Bar**; the **2D Local Remesh** dialog box is displayed, refer to Figure 7-67.

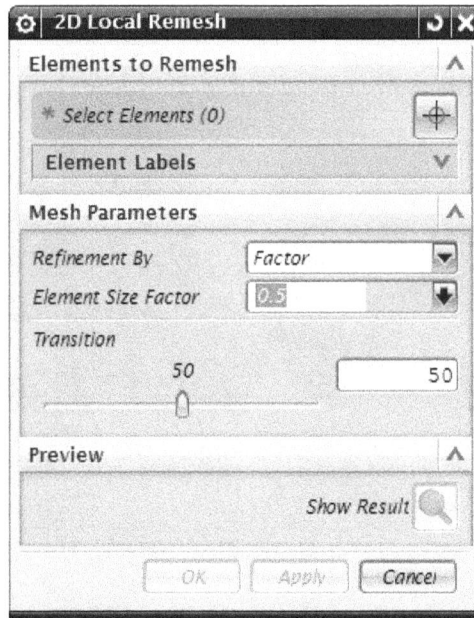

Figure 7-67 The **2D Local Remesh** dialog box

2. Select the **Factor** option from the **Refinement By** drop-down list, if it is not selected by default. Enter **0.5** in the **Element Size Factor** edit box and set the **Transition** to **50**.

3. Select the elements around the round edges of the model, as shown in Figure 7-68, and choose the **OK** button; elements are remeshed, as shown in Figure 7-69.

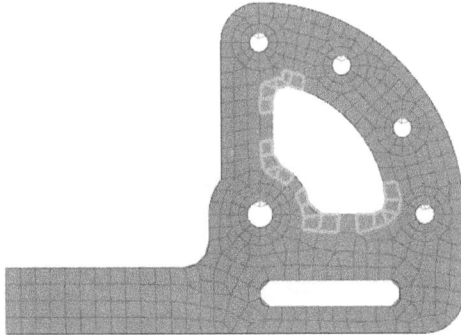

Figure 7-68 *Selecting the elements around the round edges*

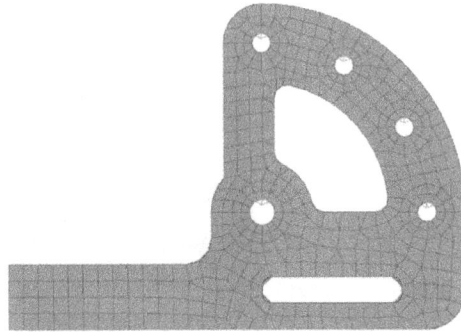

Figure 7-69 *Mesh after remeshing around round edges*

For editing the elements around rounds, you need to delete round's elements. Then create and project nodes around edges of rounds. In the end you need to create elements through nodes.

Deleting Elements

1. To delete the elements, choose **Menu > Edit > Element > Delete** from the **Top Border Bar**; the **Element Delete** dialog box is displayed.

2. Make sure the **Delete Orphan Nodes** check box is selected in the **Delete Orphan Nodes** rollout. Select the elements around the rounds, as shown in Figure 7-70, and choose the **OK** button; all the selected elements are deleted, as shown in Figure 7-71.

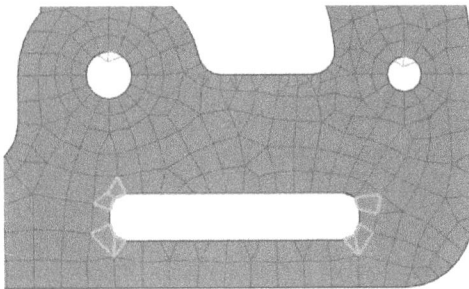

Figure 7-70 *The elements around the rounds selected*

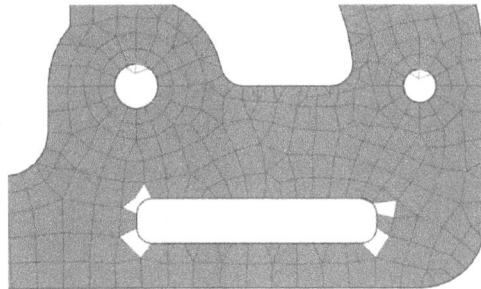

Figure 7-71 *Model after deleting elements*

Creating Nodes Around Rounds

In this section, you need to first create nodes between the existing nodes of the deleted elements.

1. To create nodes between the existing nodes, choose **Menu > Insert > Node > Between Nodes** from the **Top Border Bar**; the **Node Between Nodes** dialog box is displayed.

2. Make sure that the **Between Two Nodes** option is selected in the drop-down list available in the **Type** rollout. Enter **2** in the **Specify Number** edit box of the **Number of Nodes** rollout, refer to Figure 7-72.

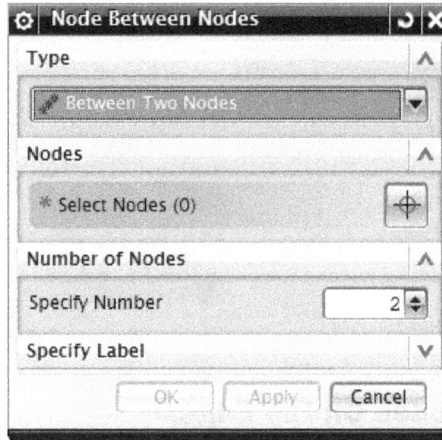

Figure 7-72 *The Node Between Nodes dialog box*

3. Select the two nodes, as shown in Figure 7-73, and choose the **Apply** button; two nodes are created between the selected nodes, as shown in Figure 7-74.

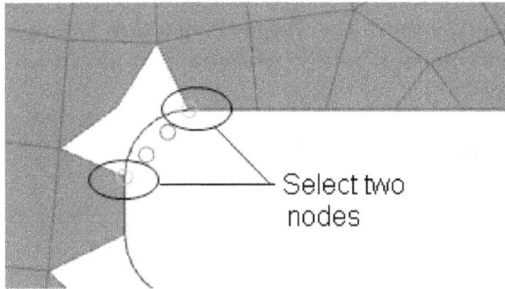

Figure 7-73 *Two nodes selected*

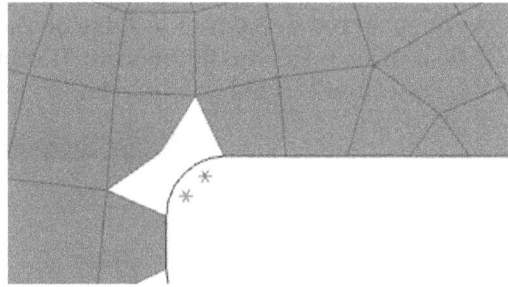

Figure 7-74 *Two nodes created between the selected nodes*

4. Similarly, select two nodes from the upper edge of the deleted element, as shown in Figure 7-75 and choose the **Apply** button with previous settings; two nodes are created between the selected nodes.

5. Similarly, create nodes with the nodes of other deleted elements. The final created nodes around the round edges are shown in Figure 7-76. Choose the **Cancel** button to exit the dialog box.

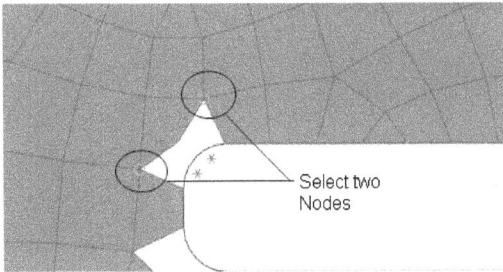

Figure 7-75 Select two nodes

Figure 7-76 Final created nodes

Projecting Created Nodes On the Edges

Projecting the free nodes on the round edges.

1. To project free nodes on the round edges, choose **Menu > Edit > Node > Project** from the **Top Border Bar**; the **Node Project** dialog box is displayed.

2. Make sure that the **Project** option is selected from the drop-down list in the **Type** rollout. Next, enter **100** in **%Offset to Edge or Face** edit box in the **Target Projection Geometry** rollout. Also, select the **By Shortest Distance** option from the **Method** drop-down list in the **Projection Method** rollout, refer to Figure 7-77.

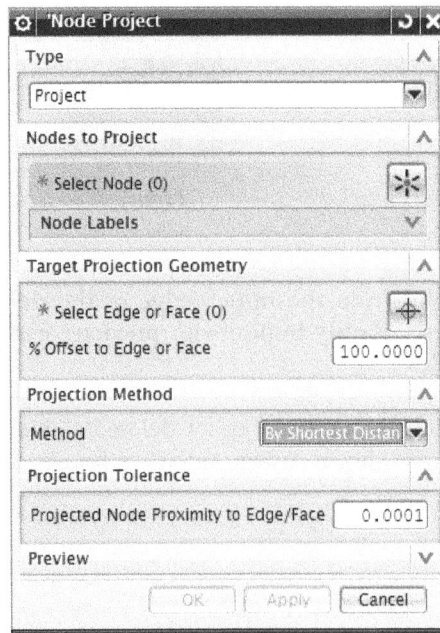

*Figure 7-77 The **Node Project** dialog box*

3. Select the two free nodes to project onto the round edge, as shown in Figure 7-78, and choose the **Apply** button. Both nodes are projected on the edge, as shown in Figure 7-79.

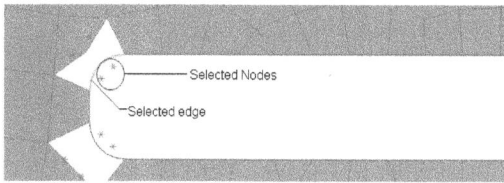

Figure 7-78 *Two nodes and edge selected to project on*

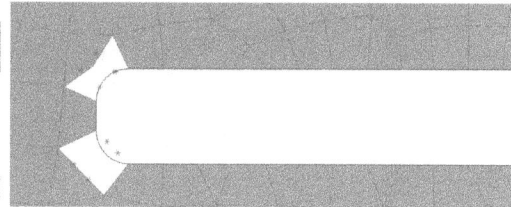

Figure 7-79 *Projected nodes*

4. Similarly, project other free nodes on the rounds. The final projected nodes around the round edges are shown in Figure 7-80. Choose the **Cancel** button to exit the dialog box.

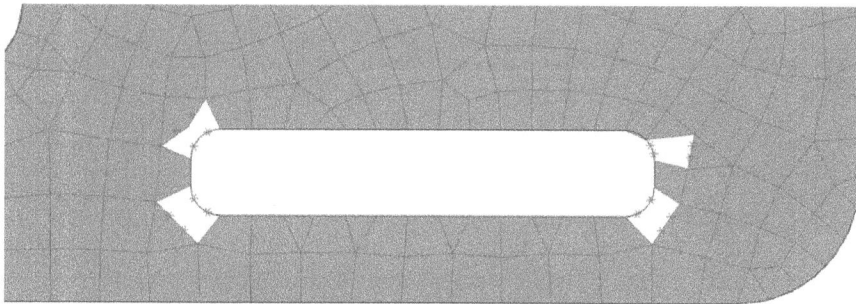

Figure 7-80 *Final projected free nodes*

Creating Elements Using Created Nodes

Now, you need to create elements using created nodes.

1. To create elements, choose **Menu > Insert > Element > Create** from the **Top Border Bar**; the **Element Create** dialog box is displayed.

2. Select the **2D** option from the drop-down list of the **Element Family** rollout and the **CQUAD4** option from the **Type** drop-down list in the **Element Properties** rollout, as shown in Figure 7-81.

3. Select the four nodes in a sequential order; the element is created using these nodes, as shown in Figure 7-82. Note that if you are not able to create an element, as shown in Figure 7-82, then you may need to drag the existing nodes upto the nodes created on the element.

4. Similarly, create other element using the nodes. The final created elements around the round edges are shown in Figure 7-83. Choose the **Close** button to exit the dialog box.

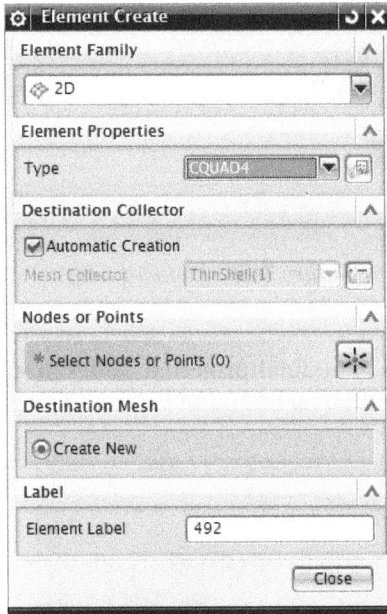

Figure 7-81 The **Element Create** dialog box

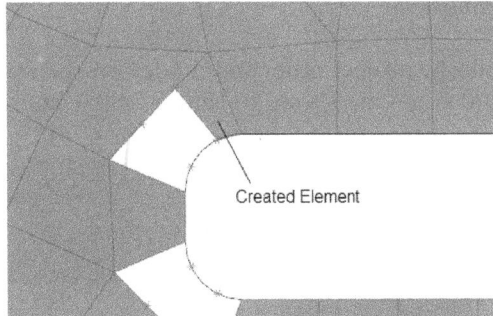

Figure 7-82 *Element created using four nodes*

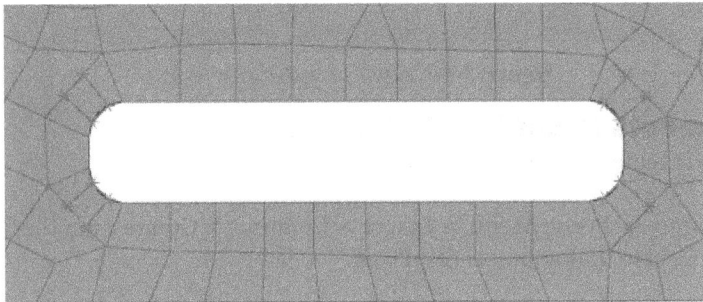

Figure 7-83 *Final created elements around the rounds edges*

Checking Elements Quality

After refining and remeshing, you need to check the quality of the elements in the mesh.

1. Choose **Menu > Analysis > Finite Element Model Check > Elements Quality** from the **Top Border Bar**; the **Elements Quality** dialog box is displayed, as shown in Figure 7-84.

2. Select the **Displayed** option from the drop-down list in the **Elements to Check** rollout; all the elements in the model are selected.

3. Select the **Use Element Type Specific Values** check box available in the **Solver Specific Geometry Checks** rollout.

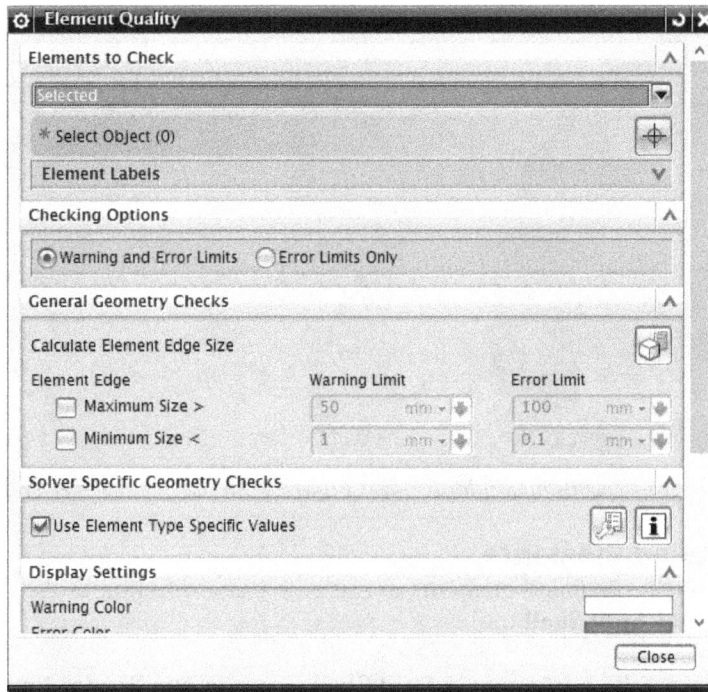

Figure 7-84 The Elements Quality dialog box

4. Select the **Failed** option from the **Output Group Elements** drop-down list and the **All** option from the **Report** drop-down list available in the **Output Settings** rollout. Retain the default settings in the rest of the rollouts in the dialog box.

5. Choose the **Check Elements** button of the dialog box; the status bar displays the number of failed and warning elements and the detailed information is displayed in the **Information** window. Also, the failed elements are highlighted in the model, refer to Figure 7-85, and added to the **1 - OUTPUT GROUP** node under the **Groups** node in the **Simulation Navigator**.

6. Right-click on the **1 - OUTPUT GROUP** node and choose the **Show Only** option from the menu displayed; only the failed elements are displayed and rest of the elements get hidden, refer to Figure 7-86.

Note
*On choosing the **Show Only** option, the meshes check boxes under the **2D collectors** node or in the respective mesh type collectors in the **Simulation Navigator** get cleared. You can display all the meshes again by selecting these check boxes.*

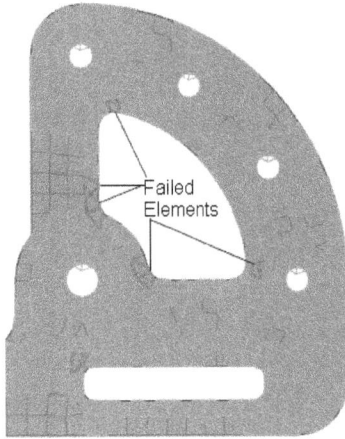

Figure 7-85 *Failed elements highlighted*

Figure 7-86 *Only failed elements displayed*

Editing the Failed Elements

Next, you need to edit the failed elements in this step using different tools. You can split the elements by using the **Split Shell** tool.

1. Choose **Menu > Edit > Element > Split Shell** from the **Top Border Bar**; the **Split Shell** dialog box is displayed.

2. Select the **Quad to 2 Quads** option available in the drop-down list in the **Type** rollout, refer to Figure 7-87. Next, select the **First Element Split Pattern** option from the **Based on** drop-down list in the **Split Criteria** rollout.

3. Select the failed elements and divide them, as shown in Figure 7-88. If needed, choose the **Flip Split Line** button available in the **Split Criteria** rollout to change the direction of division. Next, choose the **OK** button to accept the division. You can also use other tools to resolve the failed elements, if needed.

4. Check the elements quality again; all failed elements get resolved.

Saving and Closing the Model

1. Choose **File > Save > Save** from the **Ribbon** to save the model.

2. Choose **File > Close > All Parts** from the **Ribbon**; all displayed parts get closed.

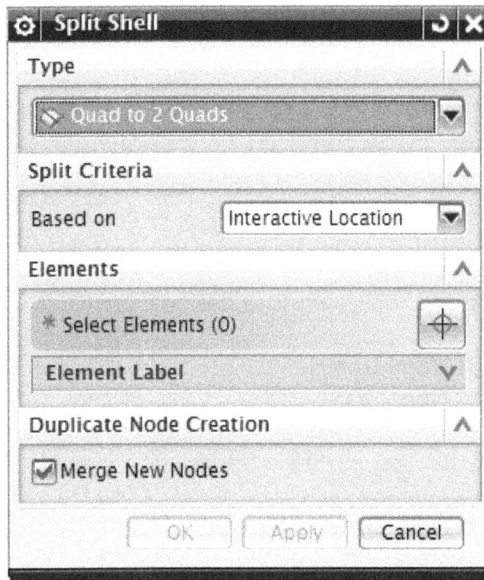

Figure 7-87 The **Split Shell** dialog box

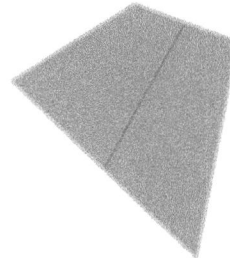

Figure 7-88 Failed element selected for division

Self-Evaluation Test

Answer the following questions and then compare them to those given at the end of this chapter:

1. The _____ tool is used to change the element type of the selected mesh.

2. The _____ tool is used to split the selected 1D element by defining its size.

3. The _____ tool is used to combine two triangle shell elements to convert them into a single quadrilateral element.

4. The _____ dialog box is used to define the material orientation of 2D element or mesh.

5. The _____ is used to remesh the selected 2D mesh/element by defining a size or size factor.

6. The **Node Translate** tool is used to translate the position of the nodes. (T/F)

7. The **Node Rotate** tool is used to rotate the nodes around a given axis. (T/F)

8. The **Node Modify Label** tool is used to modify the label of a created node. (T/F)

9. The **Node Delete** tool is used to delete a node from the window. (T/F)

10. The **Element Modify Order** tool is used to change the order of the elements. (T/F)

Review Questions

Answer the following questions:

1. Which of the following tools is used to project the nodes on selected edge or plane?

    (a) **Node Translate**   (b) **Node Project**
    (c) **Node Rotate**   (d) None of these

2. Which of the following tools is used to align a node between two nodes?

    (a) **Node Align**   (b) **Node Rotate**
    (c) **Node Modify Label**   (d) None of these

3. Which of the following tools is used to drag the nodes from their position?

    (a) **Node Drag**   (b) **Node Modify Label**
    (c) **Node Translate**   (d) None of these

4. Which of the following tools is used to split the 2D elements?

    (a) **Split 1D Element**   (b) **Split Shell**
    (c) **Element Modify Order**   (d) None of these

5. Which of the following tools is used to move a node to an existing 2D element node?

    (a) **Move Node**   (b) **Element Modify Label**
    (c) **Split Shell**   (d) None of these

6. Which of the following tools is used to change the label of elements?

    (a) **Move Node**   (b) **Element Modify Label**
    (c) **Split Shell**   (d) None of these

7. Which of the following tools is used to split the shell or 2D elements?

    (a) **Move Node**   (b) **Element Modify Label**
    (c) **Split Shell**   (d) None of these

8. Which of the following tools is used to refine the selected 2D or 3D mesh by creating density points onto their edges?

    (a) **Mesh Control**   (b) **Element Delete**
    (c) **2D Local Remesh**   (d) None of these

9. The **Node Reflect** tool is used to reflect the nodes across a plane. (T/F)

10. The **Element Modify Connectivity** tool is used to modify the connectivity of the elements. (T/F)

EXERCISE

Exercise 1

In this exercise, you will import the FEM file edited in Exercise 1 of Chapter 6, as shown in Figure 7-89, into NX Nastran. After importing the part file, you will check the element quality and edit the failed elements. **(Expected time: 45 min)**

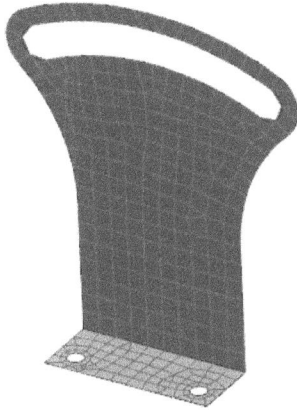

Figure 7-89 *Imported FEM model*

Hint

1. Copy the **c06_exr01_input.fem** file from the location *C:\NX Nastran\c06\Exr01* and paste it onto the *C:\NX Nastran\c07\Exr01* location. Change the file name to **c07_exr01.fem**.
2. Open the file in NX.
3. Check the quality of elements by selecting the **Use Element Type Specific Values** check box available in the **Solver Specific Geometry Checks** rollout of the **Element Quality** check box.
4. Edit the failed elements by dividing them into 2 or 3 triangles or in 2 quad elements.
5. Save and close the model.

Answers to Self-Evaluation Test

1. Modify Type, **2.** Split 1D Element, **3.** Combine Triangles, **4.** Mesh Associated Data, **5.** 2D Local Remesh, **6.** T, **7.** T, **8.** T, **9.** T, **10.** T

Chapter 8

Connections and Contacts

INTRODUCTION

Many times you analyze a load on a part with respect to another part. These parts are connected to each other either through connection like bolts, weld, or through some contacts like gluing. In this chapter, you will learn about these connections and contacts and will apply them on the model or the assembly processed earlier. Therefore, before learning these you will learn to import an assembly in NX Nastran.

ASSEMBLY FEM

You can import an assembly in the Nastran environment by using FEM method in which you can create enhanced working environment for individual parts of the assembly. In Assembly FEM, software creates the ***.afm** file. To create Assembly FEM, first open the assembly in Assembly environment, and then choose **Advanced Simulation** from **File > Applications** from **Ribbon**; the file will open in Advanced Simulation environment. Choose the **Change Window Drop-down** from the **Context** group of the **Home** tab; a flyout will be displayed, refer to Figure 8-1. Choose the **New Assembly FEM** option from this flyout; the **New Part File** window along with the **Blank** name template will be displayed, as shown in Figure 8-2. The name of the assembly file with ***.afm** extension will be displayed in the **Name** edit box of the **New File Name** rollout. You can change the file name or folder location from this rollout. Choose the **OK** button from the **New Part File** window; the **New Assembly FEM** dialog box will be displayed, as shown in Figure 8-3. You can change the solver and analysis type using the options available in drop-down lists in the **Solver Environment** rollout of this dialog box. Choose the **OK** button from the dialog box. You will enter in the **Assembly FEM** environment but the assembly will not be displayed. You can see that all parts are available as Ignored status in the **Simulation Navigator**, as shown in Figure 8-4. To see these parts in the Assembly FEM environment, you have to individually map them with the solver environment. By doing this, you can create individual FEM and idealize the file of each component.

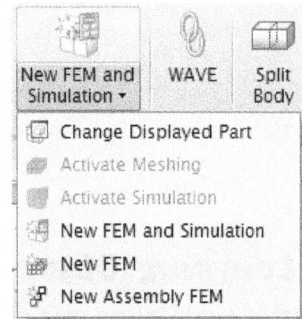

Figure 8-1   The Change Window Drop-down flyout

To map the ignored component, right-click on the component in the **Simulation Navigator**; a shortcut menu will be displayed, as shown in Figure 8-5. Choose the **Map New** option from the shortcut menu; the **New Part File** window will be displayed again, as shown in Figure 8-6. Select the **NX Nastran** option, available under the **Templates** rollout and choose the **OK** button; the **New FEM** dialog box and the **CAD Part** window with preview of the part will be displayed, refer to Figures 8-7 and 8-8. The **New FEM** dialog box is used to create FEM and idealized files as discussed in the previous chapters. Choose the **OK** button from the dialog box; the FEM and idealize files of the part will be created and also the part will be shown in the environment.

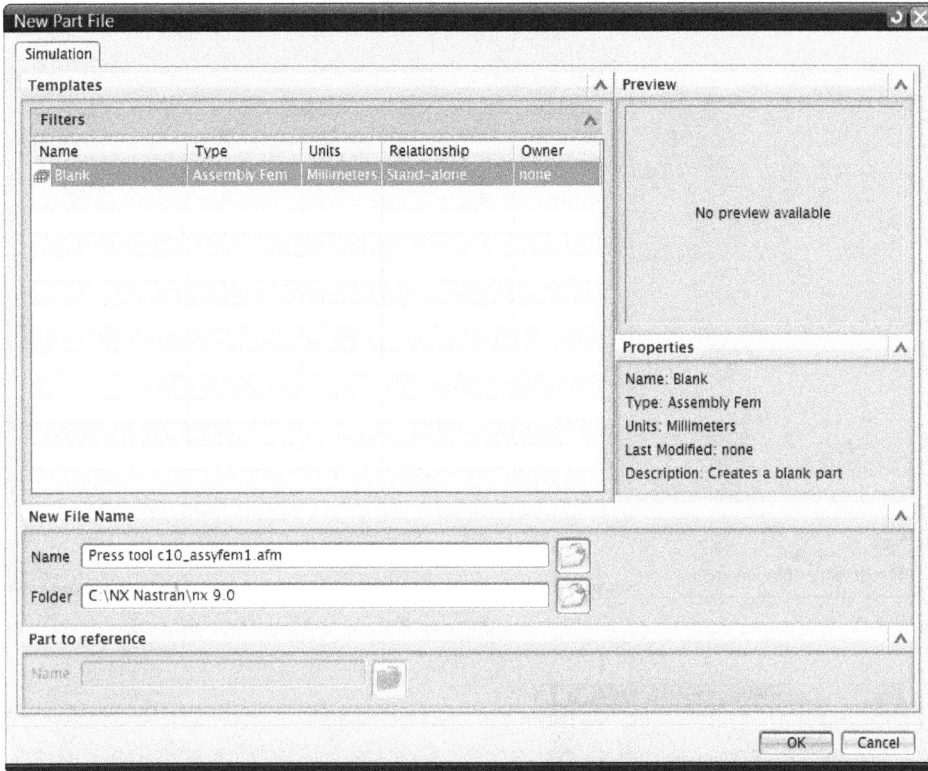

Figure 8-2 The **New Part File** *window*

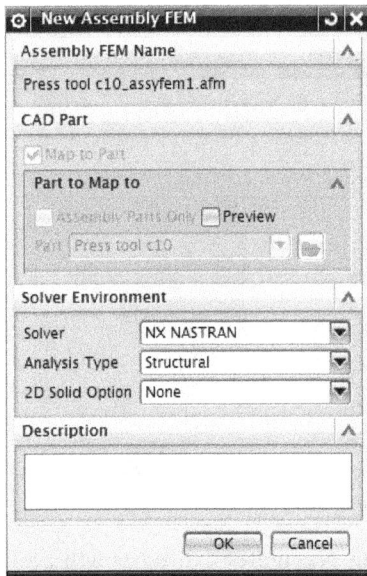

Figure 8-3 The **New Assembly FEM** *dialog box*

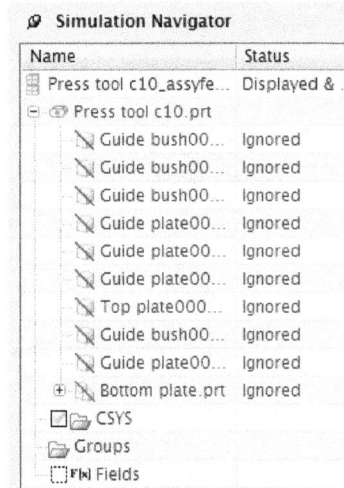

Figure 8-4 *Assembly parts in the* **Simulation Navigator**

Figure 8-5 *Shortcut menu*

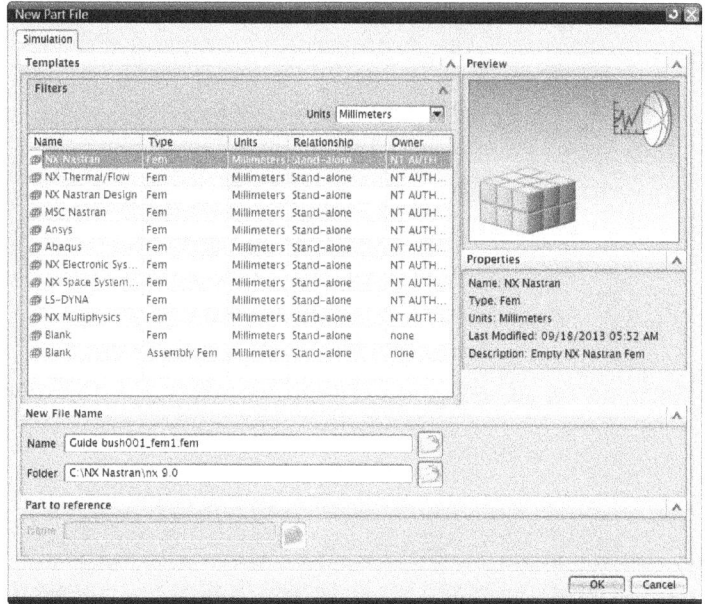

Figure 8-6 The **New Part File** *window*

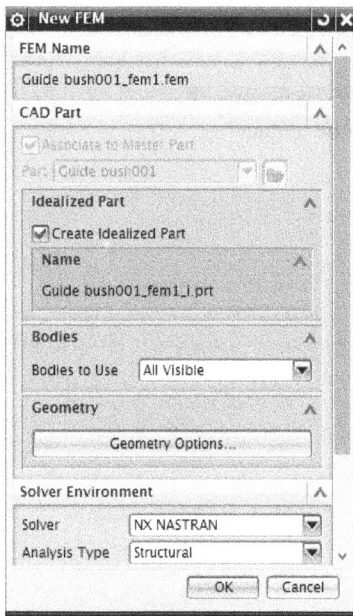

Figure 8-7 The **New FEM** *dialog box*

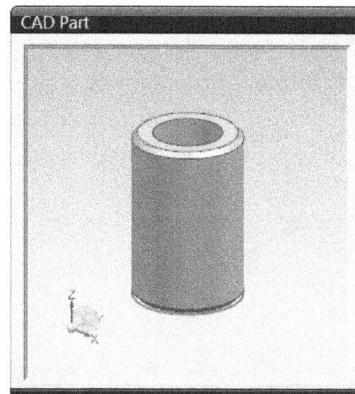

Figure 8-8 The **CAD Part** *preview window*

Similarly, map other components from the **Simulation Navigator**; all parts will be displayed in the Advanced Simulation environment and the status of the parts will be changed in the **Simulation Navigator**, refer to Figure 8-9. For meshing this assembly, you have to make every part, a work part individually. To do so, right-click on the FEM node of the respective file name in the **Simulation Navigator**; a shortcut menu will be displayed, refer to Figure 8-10. Choose the **Make Work Part** option from the shortcut menu; rest of the assembly will become inactive and meshing options will become available in this environment. Create meshing on the part and similarly on other parts of the assembly. To display complete assembly, right-click on the assembly (*.afm node) in the **Simulation Navigator** and choose the **Make Work Part** from the shortcut menu.

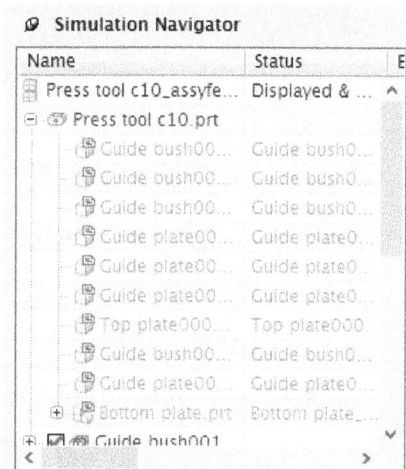

*Figure 8-9 The **Simulation Navigator** with changed status of the parts*

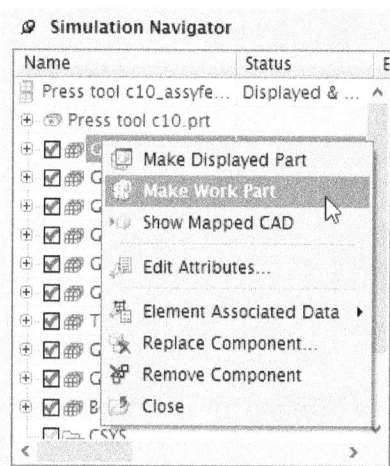

Figure 8-10 Shortcut menu displayed using FEM node

Note

*After meshing all the parts, the labels of the nodes and elements remain inconflict with respect to individual parts. You need to solve these conflict labels. To do so, right click on the *.afm node from the **Simulation Navigator**; a shortcut menu will be displayed, as shown in Figure 8-11. Choose the **Assembly Label Manager** option from the shortcut menu; the **Assembly Label Manager** dialog box will be displayed, as shown in Figure 8-12. You can see that the nodes and elements are showing conflict status in their respective tab under the **Labels** rollout. To resolve the conflicts labels, choose the **Automatically Resolve** button from the **Automatic Label Resolution** rollout; all the conflicts nodes and elements will be resolved. Choose the **OK** button to exit the dialog box.*

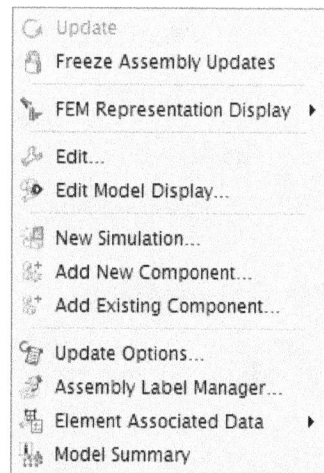

Figure 8-11 Shortcut menu displayed

Tip

*In a meshed assembly, it becomes difficult to differentiate the parts. You can apply different colors to individual meshed components. To do so, choose **Menu > Preferences > Model Display** from the **Top Border Bar**; the **Model Display** dialog box will be displayed. Choose the **Element** tab from the dialog box and the **Mesh** option from the **Color Basis** drop-down list, as shown in Figure 8-13, and choose the **OK** button; the parts will be displayed in different colors.*

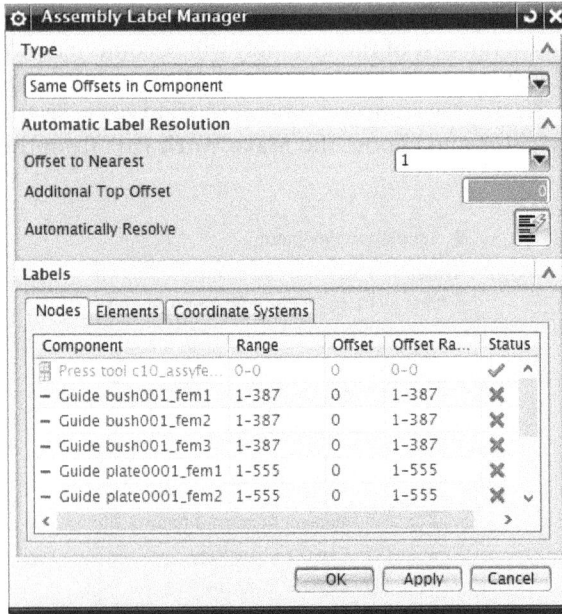

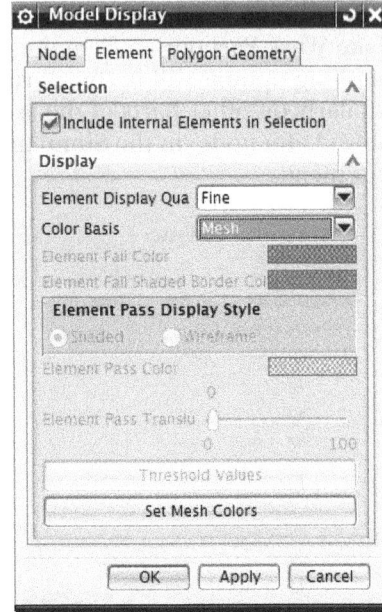

Figure 8-12 The ***Assembly Label Manager*** *dialog box*   *Figure 8-13* The ***Model Display*** *dialog box*

CONNECTIONS

In NX Nastran most of the times, you analyze the parts that are bolt or weld connected in real world. You can create these types of connections in 1D elements. Through these connections, you can also connect multiple sheets or bodies for analysis. The connections and their creation methods in NX Nastran are discussed next.

Note

All the connections are created using 1D elements/mesh. You must define the section and material for each in mesh collector.

1D Connection

Ribbon:	Home > Connections > 1D Connection
Menu:	Insert > Mesh > 1D Connection

The 1D connections are used to connect multiple sheets or bodies in FEM environment. To do so, choose **Menu > Insert > Mesh > 1D Connection** from the **Top Border Bar**; the **1D Connection** dialog box will be displayed, as shown in Figure 8-14. You can create seven types of connections using the options available in the drop-down list under the **Type** rollout; **Node to Node**, **Element Edge to Element Face**, **Point to Point**, **Point to Edge**, **Point to Face**, **Edge to Edge**, and **Edge to Face**. These options are discussed next.

Node to Node

Using this option, you can create connection between selected nodes. When you choose this option, the dialog box will be modified. You can use the rollouts available for this option for creating connections in the following manner.

Source

The **Select Node (0)** option is highlighted in this rollout and you are prompted to select source node for the connection. Select the source node from the drawing area. You can select multiple nodes as source nodes. Also, you can select the nodes by specifying the label of the node in the **Labels** edit box of the **Node Labels** sub-rollout of this rollout.

Figure 8-14 The **1D Connection** *dialog box*

Target

After selecting the source nodes, select the **Node** button in this rollout; you are prompted to select target nodes using which the connection will be created. Select the target nodes from the drawing area. You can select multiple nodes as target. You can also select the target node by specifying the node label in the **Labels** edit box of the **Node Labels** sub-rollout of this rollout.

Connection Element

In this rollout, there are two sub-rollouts: **Elements Properties** and **Destination Collector**. You can select the 1D elements type from the **Type** drop-down list in the **Elements Properties** sub-rollout. The **Edit Mesh Associated Data** button is available next to the drop-down list. Using this, you can provide additional information related to the element.

Using the options in the **Destination Collector** sub-rollout, you can define the physical properties of the elements using the **New Collector** button. If you select the **Automatic Creation** check box, the software will create the element at default location with default properties.

Label

In this rollout, the **Label** edit box is available using which you can assign labels to the connections.

Element Edge to Element Face

Using this option, you can create connection between the edges and the faces of different elements. When you choose this option, the dialog box will be modified, as shown in Figure 8-15. In this dialog box, the **Source** rollout is used to select element edges while the **Target** rollout is used to select element faces. Other rollouts and options in this dialog box are discussed next.

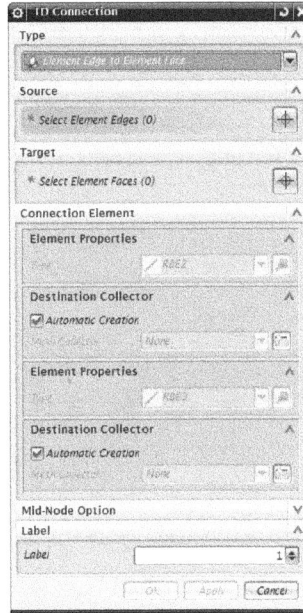

*Figure 8-15   The **1D Connection** dialog box for the* ***Element Edge to Element Face*** *option*

Connection Element

In this rollout, there are four sub-rollouts: two **Elements Properties** and two **Destination Collector**. The **RBE2** element type is selected in the first **Elements Properties** sub-rollout and the **RBE3** element type is selected in the second **Elements Properties** sub-rollout. The software projects **RBE2** element from the source edge to the target face, and then creates **RBE3** element to join the **RBE2** element to nodes on the target face.

Mid-Node Option

In this rollout, the **Connect Mid-Nodes** check box is available. This check box is used to create connections with the midnodes of parabolic elements.

Point to Point

Using this option, you can create connection between selected points. When you choose this option, the dialog box will be modified, as shown in Figure 8-16. You can use the rollouts, available for this option for creating connections in the following manner.

Source

The **Select Point (0)** option is highlighted in this rollout and you are prompted to select the source point for connection. Select the source point from the drawing area. You can select multiple points as source points. You can also create points using the **Point** dialog box by using the **Point Dialog** button available in this rollout.

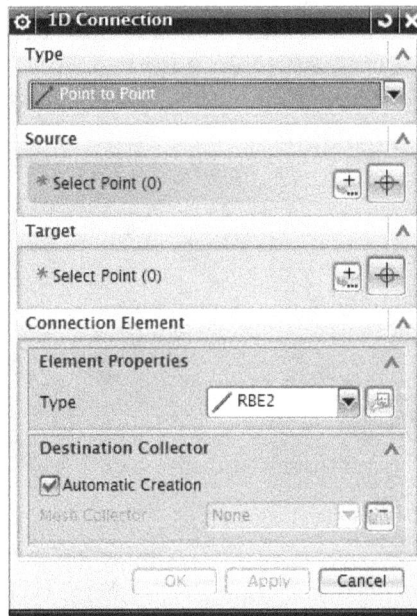

Figure 8-16 *The* **1D Connection** *dialog box for the* **Point to Point** *option*

Target
After selecting the source points, click in the **Select Point(0)** area of this rollout; you are prompted to select target point using which the connection will be created. Select the target point from the drawing area. You can select multiple points as target. You can also create target point using the **Point** dialog box by using the **Point Dialog** button available in the rollout.

Point to Edge
Using this option, you can create connection between the point and the edge of the geometry. When you choose this option, the dialog box will be modified, as shown in Figure 8-17. In this dialog box, the **Source** rollout is used to select point while the **Target** rollout is used to select edges of the given geometry. You can also create source point using the **Point** dialog box which is displayed on choosing the **Point Dialog** button available in the **Source** rollout. The options in other rollouts have already been discussed.

Note
If the model is not meshed, another rollout named **Mesh Density** *will also be available in the dialog box. When you click on the* **Mesh Control** *button in this rollout, the* **Mesh Control** *dialog box will be displayed. Using this dialog box, you can control the element distribution along the selected edge, as discussed in previous chapters.*

Point to Face
Using this option, you can create connection between the point and the faces of the geometry. When you choose this option, the dialog box will be modified, as shown in Figure 8-18. In this

dialog box, the **Source** rollout is used to a select point while the **Target** rollout is used to select a face of the geometry. You can also create source point using the **Point** dialog box by choosing the **Point Dialog** button available in the **Source** rollout. The **Point to Face** option also has the **Mesh Density** rollout as in the **Point to Edge** option by which you can control the mesh. The function of other rollouts and options are same as discussed earlier.

*Figure 8-17 The **1D Connection** dialog box for the **Point to Edge** option*

*Figure 8-18 The **1D Connection** dialog box for the **Point to Face** option*

Edge to Edge

Using this option, you can create connection between two edges of the geometry. When you choose this option, the dialog box will be modified, as shown in Figure 8-19. In this dialog box, the **Source** rollout is used to select source geometry edge for connection while the **Target** rollout is used to select target geometry edge. For creating edge to edge connection, three options are available in the **Method of Connection** drop-down list in the **Method of Connection** rollout; **Node to Node**, **RBE2 and RBE3 to Element Edge**, and **RBE2 and RBE3 to Element Face**. These options are discussed next.

Node to Node

For this option, software will create connection between the nodes of source and target edges. You can also select the 1D elements type from the **Type** drop-down list of the **Elements Properties** sub-rollout available in the **Connection Element** rollout. Note that if the edges have different node density, then the connection will follow the low density edge.

RBE2 and RBE3 to Element Edge

For this option, software will project nodes from the source edge to the target edge, and create connection between those nodes. Software also creates **RBE3** elements between each projected node on the target edge.

RBE2 and RBE3 to Element Face

For this option, software will project nodes from the source edge to the target edge, and create element along the projected nodes. Through projected nodes software create connection and also creates **RBE3** elements between each projected node on the target edge.

Similar to the previous option, this option also has the **Mesh Density** rollout which you can use for mesh controlling. The options in other rollouts have already been discussed.

Edge to Face

Using this option, you can create connection between edges to face of geometry. When you choose this option, the dialog box will be modified, as shown in Figure 8-20. In this dialog box, the **Source** rollout is used to select source geometry edge for connection while the **Target** rollout is used to select face. The options in other rollouts have already been discussed.

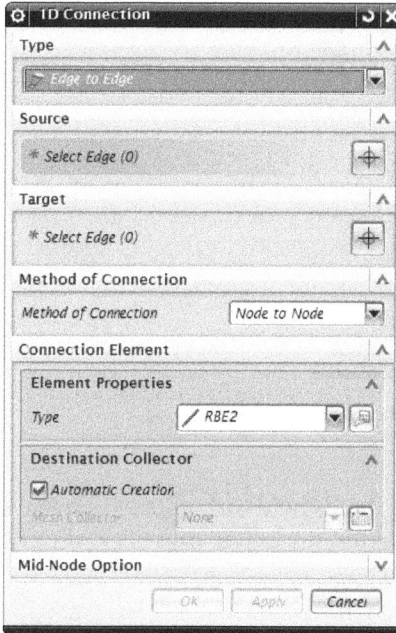

Figure 8-19 The *1D Connection* dialog box for the **Edge to Edge** option

Figure 8-20 The *1D Connection* dialog box for the **Edge to Face** option

Bolt Connection

Ribbon:	Home > Connections > More > Bolt Connection
Menu:	Insert > Mesh > Bolt Connection

The bolt connections are used to connect multiple sheets or bodies through bolted type connections in FEM environment. Note that to use this tool, you must be in Simulation FEM environment. To create connection, choose **Menu > Insert > Mesh > Bolt Connection** from the **Top Border Bar**; the **Bolt Connection** dialog box will be displayed, as shown in Figure 8-21. Three types of bolted connections can be created by using the options in the drop-down list of the **Type** rollout: **Bolt With Nut**, **Bolt in Tapped Hole**, and **Spider at Junction**. These options are discussed next.

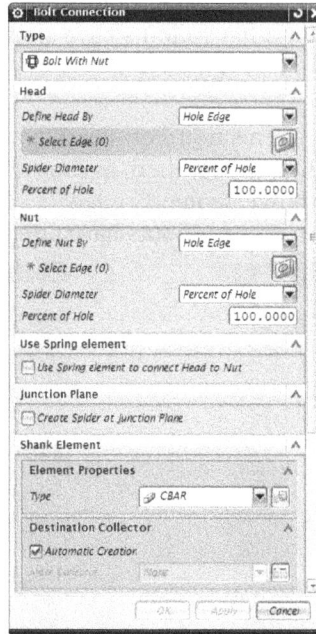

Figure 8-21 The **Bolt Connection** *dialog box*

Bolt with Nut

Using this option, you can create bolt type connection that is held with nuts. When you choose this option, the dialog box will be modified, refer to Figure 8-21. The rollouts and options for creating the **Bolt with Nut** connection are discussed next.

Head

This rollout is used to define the bolt head. Note that the head will be in a spider shaped diameter. You can define the bolt head position by using options in the **Define Head By** drop-down list: **Hole Edge** and **Center Point**. For the **Hole Edge** option, you need to select edge of hole along which you want to create hole. You can define head spider diameter using the **Percent of Hole** or **Diameter** option in the **Spider Diameter** drop-down list.

Using the **Center Point** option, you can define a point for bolt location by using the **Point** dialog box which will be available on clicking the **Point Dialog** button of the rollout. You also need to select face along which the head will be parallel. The **Diameter** edit box is used to define the head diameter.

Nut

This rollout is used to define the nut. Options in this rollout work same as the options in the **Head** rollout.

Use Spring element

In this rollout, the **Use Spring element to connect Head to Nut** check box is available. If you select this check box, the software will create a zero-length spring element instead of beam element.

Junction Plane

Using the **Create Spider at Junction Plane** check box in this rollout, you can create spider element between head and nut. When you select this check box, the options to create spider become available. Note that you cannot use this rollout along with the **Use Spring element** rollout.

Shank Element

In this rollout, there are two sub-rollouts: **Elements Properties** and **Destination Collector**. You can select the 1D element type from the **Type** drop-down list in the **Elements Properties** sub-rollout. The **Edit Mesh Associated Data** button is available aside the drop-down list. Using this, you can provide additional information related to the element. Note that if you have selected the **Use Spring element to connect Head to Nut** check box, only spring elements such as CBUSH, CELAS1, and CELAS2 will be available in the drop-down list.

Using the options available in the **Destination Collector** sub-rollout, you can define physical properties of the elements using the **New Collector** button. If you select the **Automatic Creation** check box, the software will create the element at default location with default properties. Note that you have to specify the properties of the elements, else error would be displayed while solving.

Spider Connection

This rollout is similar to the **Shank Element** rollout. Two types of elements are available in the **Type** drop-down list of the **Elements Properties** sub-rollout for spider connection: RBE2 and RBE3.

In this rollout, the **Connect Spider to Midnodes** check box is also available. When you select this check box, spider elements will get connected to the existing midnodes of the mesh.

Advanced Options

This rollout is used to set the tolerance value for bolt connection criteria. Three types of tolerance edit boxes are available in this rollout: **Alignment Tolerance**, **Junction Merge Tolerance**, and **Leg Node Tolerance**.

The **Alignment Tolerance** edit box is used to set tolerance for hole to nut position in a normal vector alignment. The **Junction Merge Tolerance** edit box is used for merging junction plane that you use for creating additional spider. The **Leg Node Tolerance** edit box is used to provide additional radial distance to the spider leg node for connection.

After selecting the top edge as head of the bolt and bottom as nut, choose the **Apply** button; the bolt connection will be created. Figure 8-22 shows the preview of the bolt with nut. In the preview, small spider shows the bolt head while large spider shows the nut. Figure 8-23 shows the bolt with nut when the **Use Spring element to connect Head to Nut** check box is selected and Figure 8-24 shows the bolt with nut when the **Create Spider at Junction Plane** check box is selected.

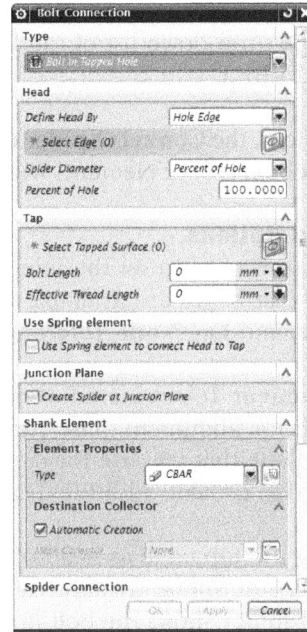

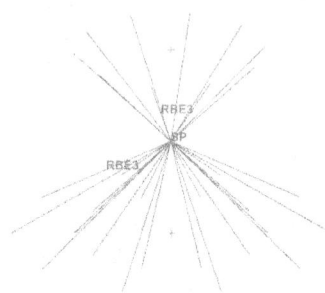

Figure 8-22 *Bolt with Nut*

Figure 8-23 *Bolt with Nut when the* **Use Spring element to connect Head to Nut** *check box selected*

Bolt in Tapped Hole

Using this option, you can create tapped (threaded) type bolt connection. When you choose this option, the dialog box will be modified, refer to Figure 8-25. All the rollouts of the dialog box for this option are same as that of the **Bolt With Nut** option except the **Tap** rollout. This rollout is discussed next.

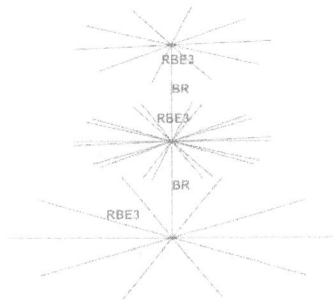

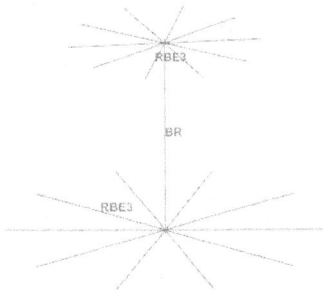

Figure 8-24 *Bolt with Nut when the* **Create Spider at Junction Plane** *check box selected*

Figure 8-25 *The* **Bolt Connection** *dialog box for the* **Bolt in Tapped Hole** *option*

Tap

This rollout is used to define the bolt length and thread length. After selecting the head position edge, click on the **Select Tapped Surface (0)** option in the rollout and choose the tapped (circular face) surface. Note that this tapped surface must pass (within tolerance) the head edge or head center point. Specify bolt and thread lengths in the respective **Bolt Length** and **Effective Thread Length** edit boxes of the rollout. Figure 8-26 shows the bolt length and effective thread length in a bolt.

Spider at Junction

Using the RBE2 or RBE3 element option in the **Spider at Junction** option, you can create spider type bolt connection. When you choose this option, the dialog box will be modified, refer to Figure 8-27. For connecting two mating parts, two rollouts are available in it: **Junction1** and **Junction2**. Both these rollouts work similar to the **Head** and **Nut** rollouts of the previous options. You can use **Junction1** rollout for defining edge or point of one mating part and **Junction2** for other mating part. Other rollouts for this option have already been discussed in the **Bolt With Nut** section.

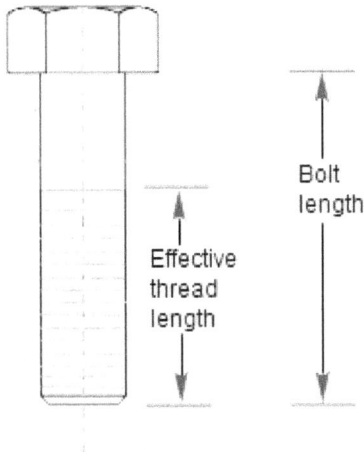

Figure 8-26 Threaded bolt

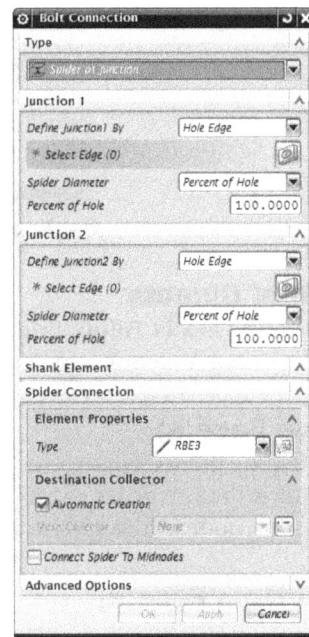

*Figure 8-27 The **Bolt Connection** dialog box for the **Spider at Junction** option*

Spot Weld

Ribbon:	Home > Connections > More > Spot Weld
Menu:	Insert > Mesh > Spot Weld

The **Spot Weld** tool is used to connect multiple sheets or bodies through 1D element. Note that to use this tool you must be in the Simulation FEM environment. To create spot weld, choose **Menu > Insert > Mesh > Spot Weld** from the **Top Border Bar**; the **Spot Weld** dialog box will be displayed, as shown in Figure 8-28. The options and rollouts of the dialog box are discussed next.

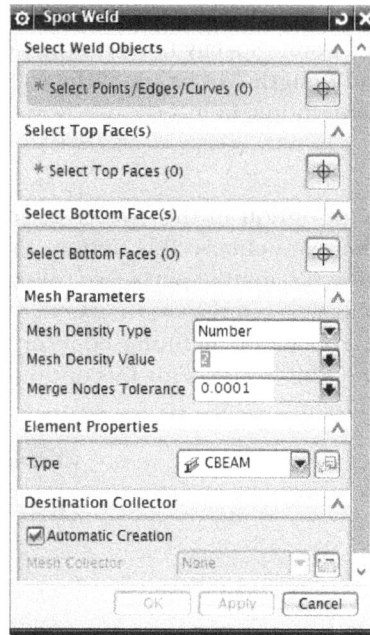

Figure 8-28 The **Spot Weld** *dialog box*

Select Weld Objects

In this rollout, the **Select Points/Edges/Curves (0)** option is highlighted and you need to select points, edges or curves along which the spot weld need to be created.

Select Top Face(s)

After selecting the points, edges or curves for spot weld position, choose the **Select Top Faces (0)** option in this rollout; you are prompted to select the face on which the selected weld objects will project start node of the spot weld.

Select Bottom Face(s)

After selecting the face for top node position of the spot weld, choose the **Select Bottom Faces (0)** option in the rollout. You need to select the face on which the selected weld objects will project bottom node of the spot weld.

Mesh Parameters

Options in this rollout are used to define the mesh parameters for spot welding. You can define mesh density using the options available in the **Mesh Density Type** drop-down list: **Size** and **Number**. You can specify the density value and nodes tolerance in the **Mesh Density Value** and **Merge Nodes Tolerance** edit boxes, respectively.

After specifying the above parameters, choose the element type from the **Element Properties** rollout and specify mesh collector from the **Destination Collector** rollout. Choose the **OK** button from the dialog box to create spot weld. Figure 8-29 shows the model for the selection of weld objects, top face, and bottom face to create spot weld. Figure 8-30 shows the model after creating spot weld.

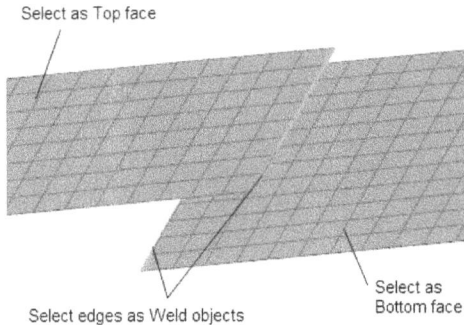

Figure 8-29 *Selection of weld objects, top face and bottom face*

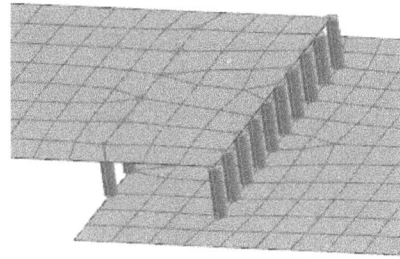

Figure 8-30 *Model after creating spot weld*

> **Note**
> *If you have meshed the model before creating the spot weld, then you need to update the mesh because spot weld creates nodes which need to be adjusted with the previous mesh. To update mesh, choose* **Menu > Edit > Update** *from the* **Top Border Bar;** *the mesh will be updated.*

Contact Mesh

Ribbon:	Home > Connections > More > Contact Mesh
Menu:	Insert > Mesh > Contact Mesh

The contact meshing is used to create point to point connection between two different edges of parts. To create connection, choose **Menu > Insert > Mesh > Contact Mesh** from the **Top Border Bar**; the **Contact Mesh** dialog box will be displayed, as shown in Figure 8-31. The method to create contact mesh using this dialog box is discussed next.

1. In the dialog box, the **Contact Edge** button is selected in the **Selection Steps** area. Select the required part.

2. Next, select the **Target Edge** button in this area and choose the edge of another part to which the edge is to be connected. Notice that after selecting the target edge, the **Apply** button becomes active in the dialog box. You can choose this button to create contact mesh with default settings.

3. You can use other buttons like **Start Point, End Point, Start Point on Contact Edge, Start Point on Target Edge, End Point on Contact Edge**, and **End Point on Target Edge** in the **Selection Steps** area to limit the contact region on the edges. Also, you can use the **Align Target Edge Nodes** and **Gap Tolerance** check boxes to specify the mesh node location. Note that the **Start Point on Target Edge** and **End Point on Target Edge** buttons will become active only when the **Align Target Edge Nodes** check box is cleared.

4. Specify the elements number in the **Number of Elements** edit box and choose the **Apply** button to create contact meshing with default settings.

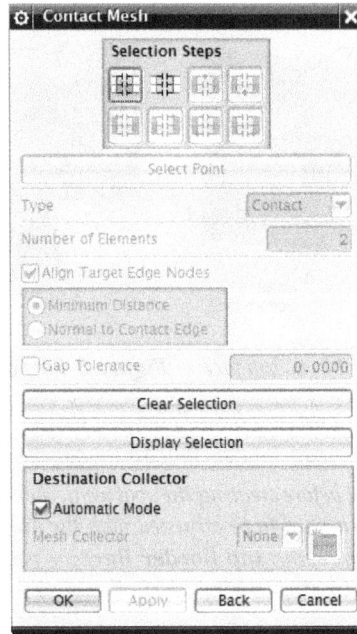

*Figure 8-31 The **Contact Mesh** dialog box*

Figure 8-32 shows the image for selecting contact and target edges. After choosing the **Apply** button, the contact mesh will be created, refer to Figure 8-33.

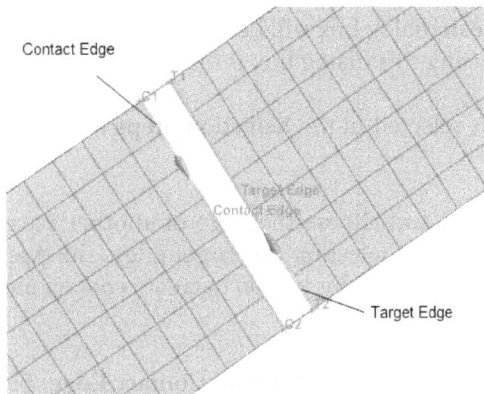

Figure 8-32 Selection of contact and target edges

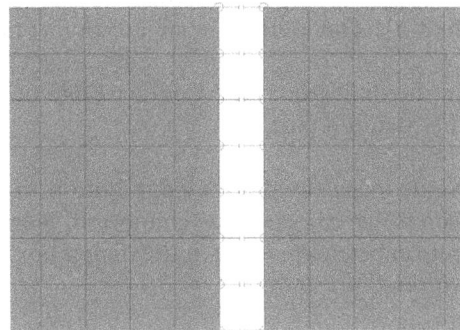

Figure 8-33 Created contact mesh

Note

*If you have already meshed the model before creating the contact mesh, then you need to update the mesh because the contact mesh creates nodes which need to be adjusted with the previous mesh. To update mesh, choose **Menu > Edit > Update** from the **Top Border Bar**; the mesh will be updated.*

Surface Contact Mesh

Ribbon: Home > Connections > More > Surface Contact
Menu: Insert > Mesh > Surface Contact Mesh

The **Surface Contact Mesh** tool is used to create a connection mesh between two contacted surfaces. To create surface contact mesh, choose **Menu > Insert > Mesh > Surface Contact Mesh** from the **Top Border Bar**; the **Surface Contact Mesh** dialog box will be displayed, as shown in Figure 8-34.

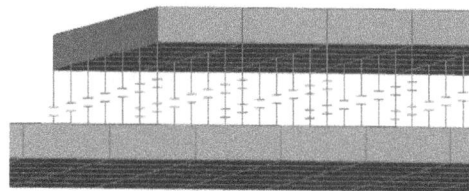

*Figure 8-34 The **Surface Contact Mesh** dialog box*

In the dialog box, by default the **Auto Create Contact Pairs** check box is cleared. For creating automatic mesh, select this check box; the **Capture Distance** edit box will become active. Specify the distance value in this edit box for two surfaces along which you want to create surface contact mesh. Next, choose the **Preview** button; the surfaces that will be in the specified range get highlighted. Now, choose the **OK** button; the surface contact mesh will be created.

You can also create surface contact mesh manually. To do so, invoke the **Surface Contact Mesh** dialog box and leave the **Auto Create Contact Pairs** check box cleared. In the **Selection Steps** area, the **Source** and **Target** buttons are highlighted and you need to select the surfaces. Select the source surface from the model. Next, click on the **Target** button from the **Selection Steps** area and select the target face from the model, refer to Figure 8-35. You can choose the **Flip Side** button for source and target faces replacement. After selecting both the surfaces, choose the **OK** button; the surface contact mesh will be created, refer to Figure 8-36.

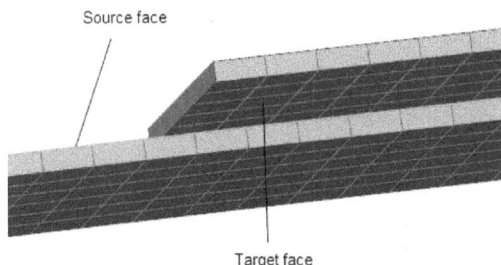

Figure 8-35 Selection of source and target faces   *Figure 8-36 Created surface contact mesh*

Mesh Mating Condition

Ribbon: Home > Connections > Mesh Mating
Menu: Insert > Model Preparation > Mesh Mating Condition

In Chapter 4, you have used the **Split Body** tool to create a mesh connection between two split bodies. You can also create this meshing connection between two separate bodies using the **Mesh Mating Condition** tool. To create mesh mating condition, choose **Menu > Insert > Model Preparation > Mesh Mating Condition** from the **Top Border Bar**; the **Mesh Mating Conditions** dialog box will be displayed, as shown in Figure 8-37. The rollouts in the dialog box are discussed next.

Type
The two options available in the drop-down list of the **Type** rollout are: **Automatic Creation** and **Manual Creation**. Using the **Automatic Creation** option, you can automatically create mesh mating condition between two bodies while the **Manual Creation** option is used to create mesh mating condition between two bodies by selecting their individual surface.

Selection
The options in this rollout will change according to the option selected from the drop-down list of the **Type** rollout. On selecting the **Automatic Creation** option from the drop-down list, the **Select faces or bodies (0)** option is highlighted and you need to select at least two different bodies between which you want to create mesh mating condition. On selecting the **Manual Creation** option, the **Select Source Face** and **Select Target Face** options will be highlighted. The **Select Source Face** option is used to select the source face for mating while the **Select Target Face** option is used to select the target face of the model.

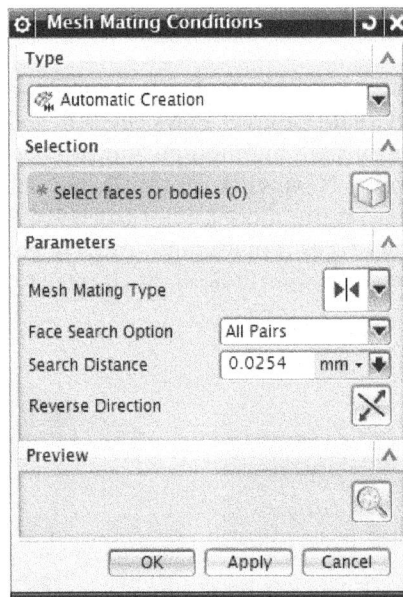

*Figure 8-37 The **Mesh Mating Conditions** dialog box*

Parameters

In this rollout, options are available for defining mesh mating parameters. The options in this rollout are discussed next.

Mesh Mating Type

Three options are available for mesh mating in the **Mesh Mating Type** drop-down list; **Glue-Coincident**, **Glue Non-Coincident**, and **Free-Coincident**. The functions of these are discussed next.

Glue-Coincident: This option is used for two identical geometry surfaces. For this option, software will merge the source and target surfaces to create a single surface having the meshing nodes of the source face only. Figure 8-38 shows the model of bolt and nut, where the bolt face is selected as the source face and nut inner face is selected as the target face. After choosing the **Apply** button the faces will be deformed to merge the source and target faces, refer to Figure 8-39.

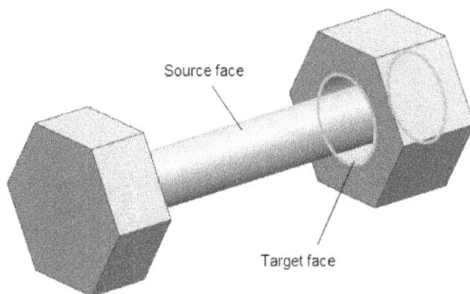

Figure 8-38 Selection of source and target faces

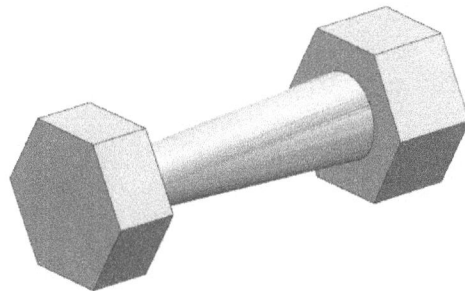

Figure 8-39 Mesh mating condition applied using **Glue-Coincident** option

Glue Non-Coincident: Using this option, software creates equations between the source and target faces. While meshing, these equations create node to node meshing connections using the RBE3 elements. Figure 8-40 shows the model of the bolt and nut after applying the glue non-coincident type mesh mating condition. After creating the mesh, the RBE3 elements create connection mesh between the faces, refer to Figure 8-41.

Free Coincident: For this option, software creates an alignment between the source and target faces. Using this alignment, a coincident mesh is created which has duplicate nodes between the faces. Figure 8-42 shows the model of bolt and nut after applying the free coincident type mesh mating condition. The break line along the bolt is shown for the node alignment.

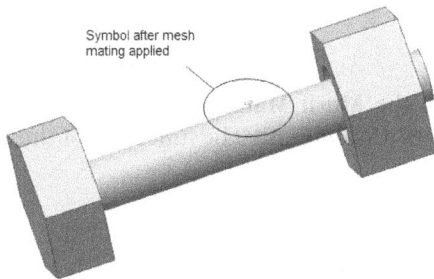

Figure 8-40 *Mesh mating condition apply using the* **Glue Non-Coincident** *option*

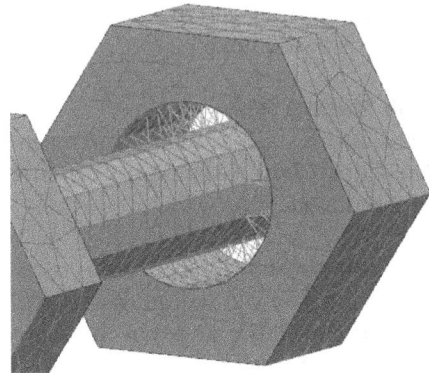

Figure 8-41 *Created mesh for* **Glue Non-Coincident** *option*

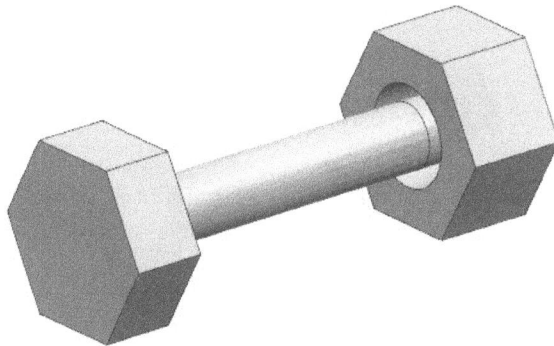

Figure 8-42 *Model after creating coincident mesh*

Face Search Option
Using the options in this drop-down list, you can control the search for face pair that takes part in the mesh mating. Options available in this drop-down list are discussed next.

All Pairs: Using this option, the software will search all the face pairs among the assembly either coincident or non-coincident within the specified tolerance distance.

Identical Pairs Only: Using this option, the software will limit the search the geometrically identical faces of the source and target faces within the specified tolerance distance.

Search Distance
Using this edit box, you can specify the minimum allowable search distance between which software will create the face pair.

Reverse Direction
Using this button, you can reverse the direction of dependency from source to target face for RBE3 element creation. It is mainly used in the **Glue Non-Coincident** type mating condition.

Figure 8-43 shows the meshed model of bolt and nut without mesh mating condition. As shown in figure, there is no connected node between bolt and nut. After applying the mesh mating condition between surfaces, a connection between all the nodes is established, refer to Figure 8-44.

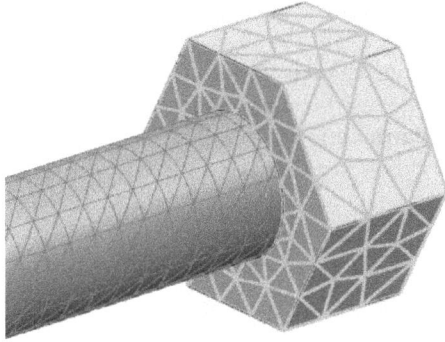

Figure 8-43 Model without applying mesh mating condition

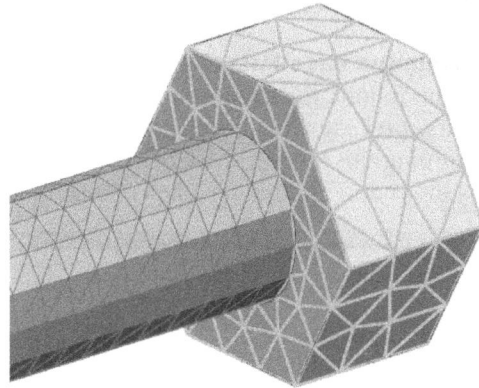

Figure 8-44 Created surface contact mesh

CONTACTS AND GLUING

In an assembly, while doing simulation operation, you need to transfer a load from one part to another part. These parts are not connected to each other with some bolt or any other type of connections as you have done earlier. In NX Nastran, you can use contacts and gluing conditions to connect two different parts. Note that these methods are available in the Simulation environment of NX Nastran so you need to be in the Simulation environment to use these methods. These contacts and gluing methods are discussed next.

Note
You will learn more about the Simulation environment and its dialog boxes in later chapter of this book.

Surface-to-Surface Contact

Ribbon:	Home > Loads and Conditions > Simulation Object Type > Surface-to-Surface Contact

The **Surface-to-Surface Contact** tool is used to create a contact between any two surfaces of an assembly. You can also create a frictional contact between the given surfaces. To create surface to surface contact, choose the **Surface-to-Surface Contact** tool from the **Simulation Object Type** drop-down list in the **Load and Conditions** group of the **Home** tab; the **Surface-to-Surface Contact** dialog box will be displayed, as shown in Figure 8-45. For creating surface to surface contact, two options are available in the drop-down list of the **Type** rollout: **Automatic Pairing** and **Manual**. These options are discussed next.

Automatic Pairing

The **Automatic Pairing** option is used to create automatic contact between two or more pairs. This option is selected by default. The rollouts and options in the **Surface-to-Surface Contact** dialog box for the **Automatic Pairing** option are discussed next.

Name

Using the **Name** rollout options, you can assign name and label to the contacts in their respective edit boxes. You can also provide the contact description in the area under the **Description** rollout of the **Name** rollout.

Automatic Face Pair Creation

In this rollout, the **Create Face Pairs** button is available which is used to generate face pairs, preview face pairs, and specify the properties of face pair generation. On choosing this button, the **Create Automatic Face Pairs** dialog box will be displayed, as shown in Figure 8-46. The rollouts in this dialog box are discussed next.

*Figure 8-45 The **Surface-to-Surface Contact** dialog box*

*Figure 8-46 The **Create Automatic Face Pairs** dialog box*

Face Pair Search Subset: Under this rollout, the **Face (0)** option is highlighted. You can select a subset of faces or parts of the given assembly along which you want to create surface to surface contact. If you do not select any part or faces, software will search for the surfaces in the whole assembly.

Properties: In this rollout, options are given for face pairing criteria. You can group the paired faces by using the options available in the **Grouping Option** drop-down list: **One**, **Per Body Pairs**, and **Per Face Pair**. The **One** option is used to group all face pairs into a single simulation object. The **Per Body Pairs** option groups the face pairs into separate bodies. The **Per Face Pair** is used to create a separate simulation object for each face pair. Below this drop-down list, the **Distance Tolerance** edit box is available which is used to specify the maximum search distance for the face pairs.

Preview: Using this rollout, you can view all the face pairs that qualify the distance tolerance criteria.

After specifying the face pairs, choose the **OK** button from the **Create Automatic Face Pairs** dialog box; the number of created face pairs will be updated in the **Automatic Face Pair Creation** rollout.

Contact Region Parameters (BCRPARA)

The options in this rollout are used to define the region parameters for surface to surface contacts. You can define the offset distances for the contact region in the **Source Offset** and **Target Offset** edit boxes. Using the **TOP** or **BOTTOM** option from the **Source Contact Side** and **Target Contact Side** drop-down lists, you can set the direction of the contact surface. You can also define the region type as flexible or rigid using the **FLEX** or **RIGID** option from the **Target Region Type** drop-down list.

Contact Set Properties (BCTSET)

In this rollout, the **Coefficient of Static Friction** edit box is available. Using this, you can specify the static friction value for the contact set. You can also specify the contacts minimum and maximum search distance in the **Min Search Distance** and **Max Search Distance** edit boxes, respectively.

Local Contact Pair Parameters

Using this rollout, you can define linear overrides and advanced nonlinear contact pair parameters which are discussed next.

Linear Overrides (BCTPARM): You can create linear overrides parameter using the **Create Modeling Object** button available next to the **Linear Overrides (BCTPARM)** drop-down list. By default, the **None** option is chosen in this drop-down list for which software uses the global parameter for linear overrides. When you choose the **Create Modeling Object** button, the **Contact Parameters - Linear Pair Overrides** dialog box will be displayed, as shown in Figure 8-47. In this dialog box, two rollouts are available: **Modeling Object** and **Properties**.

Using the options of the **Modeling Object** rollout, you can specify the name for the object and label in the **Name** and **Label** edit boxes respectively. Using the options of the **Properties** rollout, you can provide the description for the linear pair overrides. You can also specify the parameters for Penalty method of linear overrides. After specifying all the parameters in the dialog box, choose the **OK** button; the specified name of linear overrides will be displayed in the **Linear Overrides (BCTPARM)** drop-down list.

Advanced Nonlinear (BCTPARA): You can create advanced nonlinear parameter using the **Create Modeling Object** button available next to the **Advanced Nonlinear (BCTPARA)** drop-down list. By default, the **None** option is chosen in this drop-down list for which software uses the global parameter for advanced nonlinear parameter. When you choose the **Create Modeling Object** button; the **Contact Parameters - Advanced Nonlinear Pair** dialog box will be displayed, as shown in Figure 8-48. In this dialog box, two rollouts are available: **Modeling Object** and **Properties**.

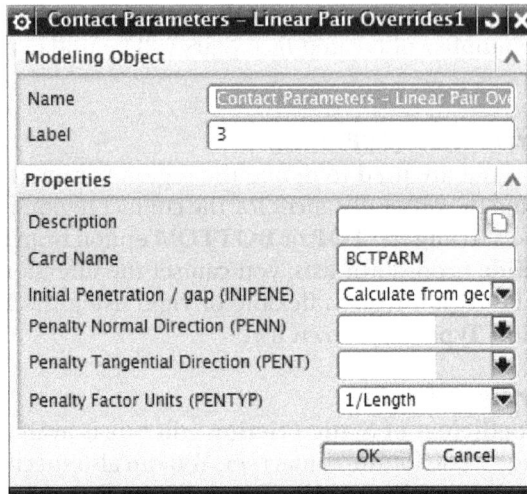

Figure 8-47 *The* **Contact Parameters - Linear Pair Overrides** *dialog box*

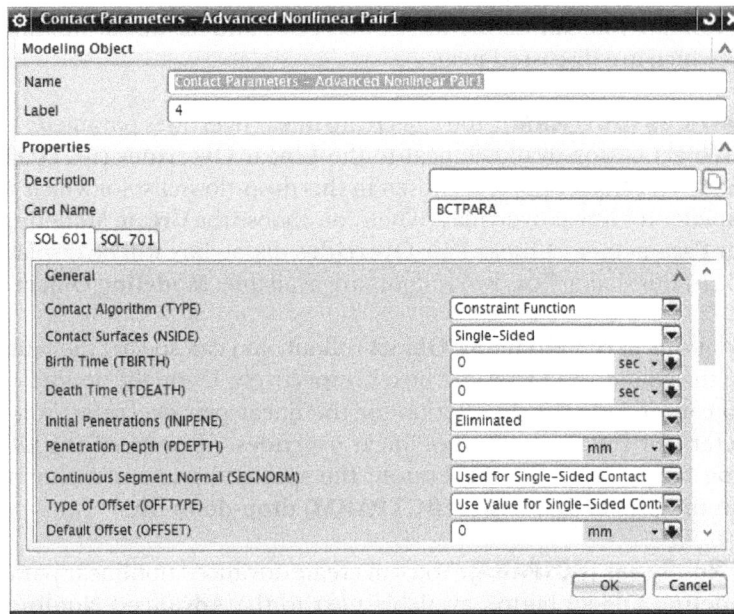

Figure 8-48 *The* **Contact Parameters - Advanced Nonlinear pair** *dialog box*

Using the options of the **Modeling Object** rollout, you can specify the name for the object and label in the **Name** and **Label** edit boxes respectively. Using the options of the **Properties** rollout, you can provide description for the advanced nonlinear pair. In this rollout, two tabs are available: **SOL 601** and **SOL 701**. You can specify specific values of the cards in their respective tabs. After specifying all the parameters in the dialog box, choose the **OK** button; the specified name of the advanced nonlinear pair will be displayed in the **Advanced Nonlinear (BCTPARA)** drop-down list.

After specifying all the parameters in their respective rollouts from the **Surface-to-Surface Contact** dialog box, the **OK** button will become active. After choosing the **OK** button, the surface to surface contact will be created.

Manual

The **Manual** option is used to manually create glued contact between the face pair. When you choose this option, the dialog box will be modified, as shown in Figure 8-49. The methods and functions of the rollouts of the dialog box are discussed next.

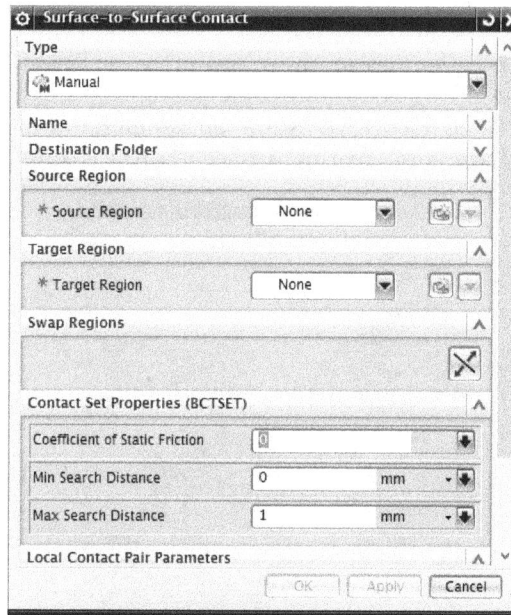

Figure 8-49 The Surface-to-Surface Contact dialog box for the Manual option

Source Region

In the **Source Region** rollout, the **Source Region** drop-down list is available which is used to select the source face of the contact pair. By default, the **None** option is selected in this drop-down list. To define a source region, choose the **Create Region** button of the rollout; the **Region** dialog box will be displayed, as shown in Figure 8-50.

In the dialog box, the **Name** rollout is used to specify the name and label of the region in their respective edit boxes. The **Region Objects** rollout is used to select the faces to create the respective region. If you select the **Group Reference** check box, you need to create a group of faces that you want to select as source region. In the **Common Contact Parameters (BCRPARA)** rollout you can define the contact face direction using the **TOP** and **BOTTOM** options in the **Surface** drop-down list. You can also define the offset contact using the **Offset** edit box for pre contact or interference contact. Using the **Nonlinear Contact Parameters (BCRPARA)** option, you can specify region as flexible or rigid for nonlinear cards such as SOL 601 and SOL 701. After specifying all the parameters, choose the **OK** button; the created region will be displayed in the **Source Region** drop-down list.

Target Region

In the **Target Region** rollout, the **Target Region** drop-down list is available in which the **None** option is chosen by default. The target region is used to select the target face of the contact pair. The target region is defined similar to the source region. After defining the region, the created region will be displayed in the **Target Region** drop-down list.

Swap Regions

In the **Swap Regions** rollout, the **Swap Regions** button is available which is used to reverse the source and target regions.

After specifying all the parameters in their respective rollouts, the **OK** button in the **Surface-to-Surface Contact** dialog box will become active. Choose the **OK** button, the surface to surface contact will be created. Figure 8-51 shows an assembly of two plate connected surfaces. Figure 8-52 shows, the surfaces after creating surface to surface contact.

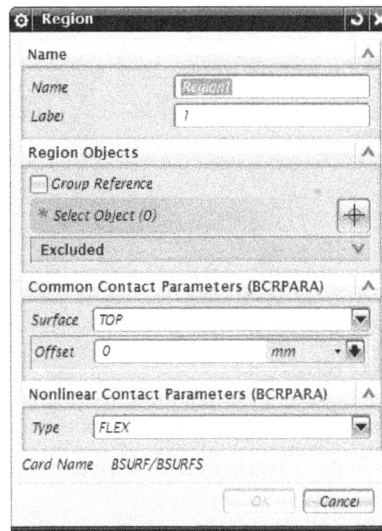

Figure 8-50 *The* ***Region*** *dialog box*

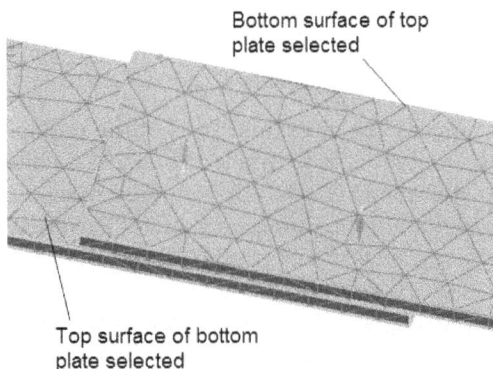

Figure 8-51 *Selection of source and target faces*   *Figure 8-52* *Created surface contact mesh*

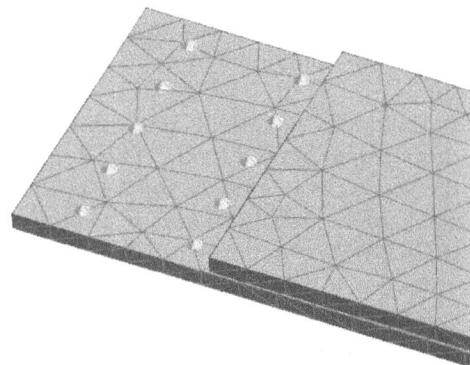

Surface-to-Surface Gluing

Ribbon: Home > Loads and Conditions > Simulation Object Type >
                Surface-to-Surface Gluing

The **Surface-to-Surface Gluing** tool is used to create a gluing contact between two
surfaces of an assembly. The gluing contact is used to prevent any directional movement
between the parts. To create surface to surface gluing contact, choose the **Surface-to-Surface
Gluing** tool from the **Simulation Object Type** drop-down list in the **Load and Conditions** group
of the **Home** tab; the **Surface-to-Surface Gluing** dialog box will be displayed, as shown in
Figure 8-53. For creating surface to surface gluing contact two options are available in the
drop-down list of the **Type** rollout: **Automatic Pairing** and **Manual**. The method of creating
gluing contacts using these options is discussed next.

Automatic Pairing

The **Automatic Pairing** option is used to create automatic contact between two or more pairs.
When you choose this option, the dialog box will be modified, refer to Figure 8-53. The **Name**
and **Automatic Face Pair Creation** rollouts for this option work similar to the options in the
Surface-to-Surface Contact tool. Other rollouts and functions available in the dialog box are
discussed next.

Linear Settings

The **Search Distance (BGSET)** edit box available in this rollout is used to define the distance
along which the software calculates glued region. You can also define glue parameters
by using the **Create Modeling Object** button available next to the **Override Parameters
(BGPARM)** drop-down list. By default, the **None** option is selected in this list. When you
choose the **Create Modeling Object** button, the **Glue Parameters - Linear Pair Overrides**
dialog box will be displayed, as shown in Figure 8-54.

In the **Glue Parameters - Linear Pair Overrides** dialog box, you can use the **Modeling Object**
rollout to define name and label of the glue parameter. Using the **Properties** rollout, you
can provide description, glue type, and penalty factor for the gluing contact. After choosing
the **OK** button from the dialog box, the names of the glue parameters will become available
in the **Override Parameters (BGPARM)** drop-down list. Choose the required option.

Advanced Nonlinear Settings

This rollout is used to specify the extension factor for nonlinear SOL 601 card. Default value
for extension factor is 0.01.

Boundary Condition Generation

In this rollout, the **Associate to Automatic Pair Recipe** check box is available. When you
select this check box while updating or changing the model, the face gluing objects will
recalculate and update accordingly.

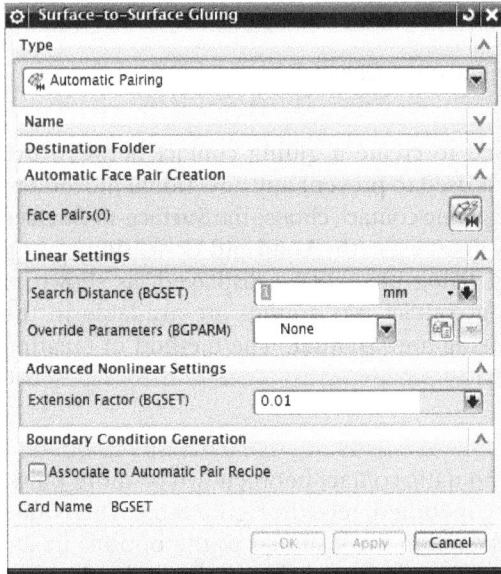

Figure 8-53   The Surface-to-Surface Gluing
dialog box

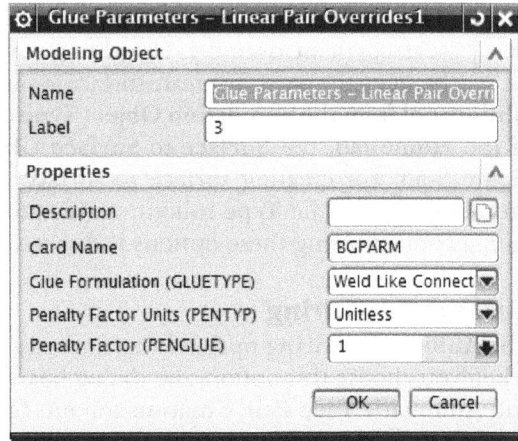

Figure 8-54   The Glue Parameters - Linear Pair
Overrides dialog box

Manual

The **Manual** option is used to manually create contact between face pairs. When you choose this option, the dialog box will be modified, as shown in Figure 8-55. The selection method of source and target region using the rollouts available in the dialog box is similar to the selection method in the **Surface-to-Surface Contact** dialog box. The functions of the other options of the dialog box for the **Manual** option are same as for the **Automatic Pairing** option.

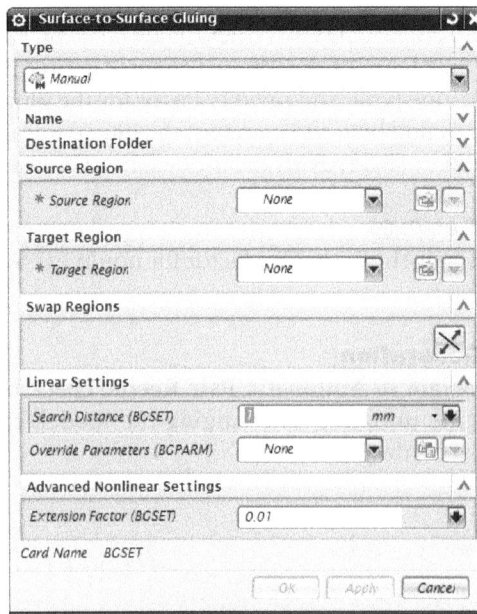

Figure 8-55   The Surface-to-Surface Gluing dialog box for the Manual option

After specifying all the parameters in their respective rollouts, the **OK** button in the **Surface-to-Surface Gluing** dialog box will become active. After choosing the **OK** button, the surface to surface gluing contact will be created.

Edge-to-Surface Gluing

Ribbon: Home > Loads and Conditions > Simulation Object Type >
                 Edge-to-Surface Gluing

The **Edge-to-Surface Gluing** tool is used to create a gluing contact between an edge and a surface of any given assembly for preventing any directional movement. To create edge to surface gluing contact, choose the **Edge-to-Surface Gluing** tool from the **Simulation Object Type** drop-down list in the **Load and Conditions** group of the **Home** tab; the **Edge-to-Surface Gluing** dialog box will be displayed, as shown in Figure 8-56. Method of creating edge to surface gluing contact using the dialog box is discussed next.

Name

Using the options in the **Name** rollout, you can assign name and label to the gluing contacts. You can also provide the contacts description in the area under the **Description** sub-rollout in the **Name** rollout.

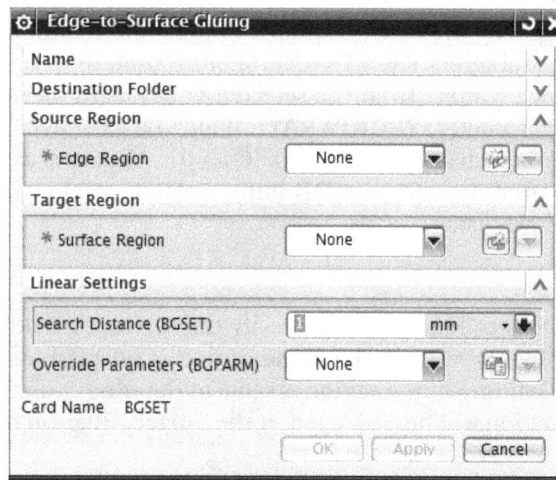

*Figure 8-56 The **Edge-to-Surface Gluing** dialog box*

Source Region

By using the options in this rollout, you can specify the region that defines the source edge of the gluing contact pair. In this rollout, the **None** option is selected by default in the **Edge Region** drop-down list. To define a source region, choose the **Create Region** button in the rollout; the **Region** dialog box will be displayed, as shown in Figure 8-57.

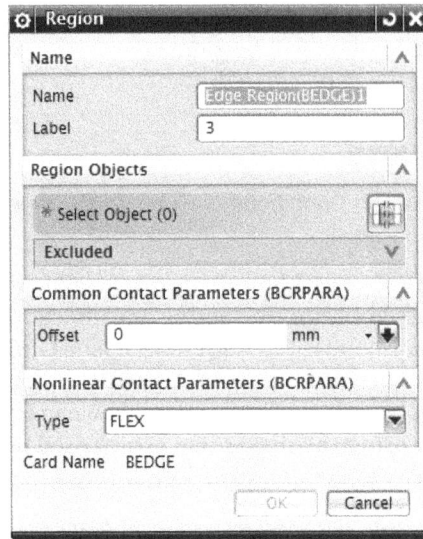

*Figure 8-57 The **Region** dialog box*

In the dialog box, the **Region Objects** rollout is used to select the edges for the region. The **Name** rollout is used to specify the name and label of the region in their respective edit boxes. The **Common Contact Parameters (BCRPARA)** rollout is used to define the offset contact for pre contact or interference contact from the selected edges using the **Offset** edit box. Using the **Nonlinear Contact Parameters (BCRPARA)** rollout, you can specify the region as flexible or rigid for nonlinear cards such as SOL 601 and SOL 701. After specifying all the parameters, choose the **OK** button. After choosing the **OK** button, the created region will be displayed in the **Edge Region** drop-down list.

Target Region

By using the options of this rollout, you can specify the region that defines the target face of the gluing contact pair. In this rollout, the **None** option is selected by default in the **Surface Region** drop-down list. Defining target region is same as defining source region. After defining the region, the created region will be displayed in the **Surface Region** drop-down list.

The options in the **Linear Settings** rollout work in the same way as in the **Surface-to-Surface Gluing** dialog box. After specifying all parameters in their respective rollouts, the **OK** button in the **Edge-to-Surface Gluing** dialog box will become active. After choosing the **OK** button, the edge to surface gluing contact will be created.

Edge-to-Edge Contact

Ribbon: Home > Loads and Conditions > Simulation Object Type >
                     Edge-to-Edge Contact

The **Edge-to-Edge Contact** tool is used to create a contact between edge to edge of an assembly. To create edge to edge gluing contact, choose the **Edge-to-Edge Contact** tool from the **Simulation Object Type** drop-down list in the **Load and Conditions** group of the **Home** tab; the **Edge-to-Edge Contact** dialog box will be displayed, as shown in Figure 8-58.

Method of creating edge to edge contact using the dialog box is similar to the method of creating the edge to surface gluing contact except for the **Target Region** rollout. For this rollout, you have to define the edges as the target region. You can also swap the edges using the **Swap Regions** button in the **Swap Regions** rollout.

After specifying all the parameters in their respective rollout, the **OK** button in the **Edge-to-Edge Contact** dialog box will become active. After choosing the **OK** button, the edge to edge contact will be created.

Edge-to-Edge Gluing

Ribbon: Home > Loads and Conditions > Simulation Object Type > Edge-to-Edge Gluing

The **Edge-to-Edge Gluing** tool is used to create a gluing contact between edge to edge of an assembly for preventing any directional movement. To create edge to edge gluing contact, choose the **Edge-to-Edge gluing** tool from the **Simulation Object Type** drop-down list of the **Load and Conditions** group of the **Home** tab; the **Edge-to-Edge Gluing** dialog box will be displayed, as shown in Figure 8-59. Method of creating edge to edge gluing contact using the dialog box is similar to the method of creating edge to edge contact.

Figure 8-58 The **Edge-to-Edge Contact** dialog box

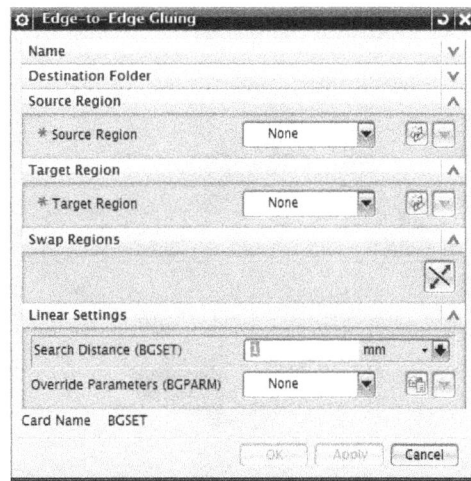

Figure 8-59 The **Edge-to-Edge Gluing** dialog box

After specifying all the parameters in their respective rollout, the **OK** button in the **Edge-to-Edge Gluing** dialog box will become active. After choosing the **OK** button, the edge to edge gluing contact will be created.

TUTORIAL

To proceed the tutorials, you need to download the zipped file named as *c08_NX_Nastran_input* from the Input Files section of the CADCIM website. The complete path for downloading the file is:

Textbooks > CAE Simulation > NX Nastran > NX Nastran 9.0 > Input Files

After the file is downloaded, extract the folder to the location *C:\NX Nastran* and rename it as *c08*.

Tutorial 1

In this tutorial, you will open the Flange Coupling assembly, shown in Figure 8-60, from the extracted folder and delete bolts and nuts. Import the file into NX Nastran and apply bolt connection and surface to surface contact between the contacted surface. Model after applying bolt connection and contact is shown in Figure 8-61.   **(Expected time: 45 min)**

The following steps are required to complete this tutorial:

a. Open the assembly file into the NX and delete all the bolts and nuts.
b. Invoke Advanced Simulation environment and create new assembly FEM file.
c. Map both flanges with NX Nastran.
d. Create mesh on the parts.
e. Use Assembly Label Manager for label correction.
f. Create mesh collector for 1D connection of bolt.
g. Create bolt and nut connection.
h. Create Surface to Surface contact.
i. Save and close the model.

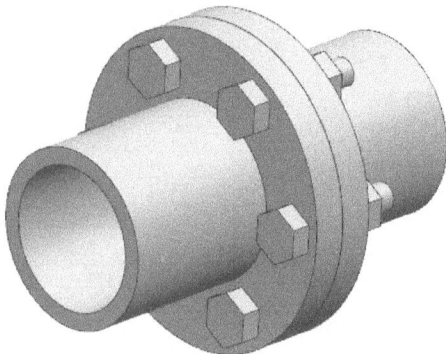

Figure 8-60 Flange assembly for creating bolt connection and contact

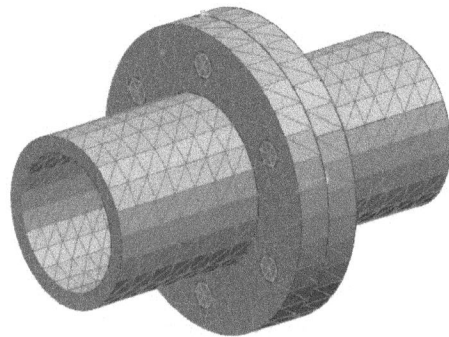

Figure 8-61 Created surface contact mesh

Open the Assembly File

1. Start NX 9.0 by double-clicking on the shortcut icon of **NX 9.0** on the desktop of your computer, if you are initialing from this tutorial only.

2. Download the **c08_tut01** zipped folder file from *www.cadcim.com*. Extract the folder to the location *C:\NX Nastran\c08\Tut01*.

![Note icon] **Note**
You must create the Tut01 folder in c08 folder at the location C:\NX Nastran as created in the previous chapter.

3. To open the flange assembly file, choose **File > Open** from the **Ribbon**; the **Open** dialog box is displayed. Select the **Flange_Coupling.prt** file from the location you extracted and choose the **OK** button; the file is opened in the Modeling environment, as shown in Figure 8-62.

4. Delete the bolts and nuts from the assembly. The assembly after deleting bolts and nuts is shown in Figure 8-63.

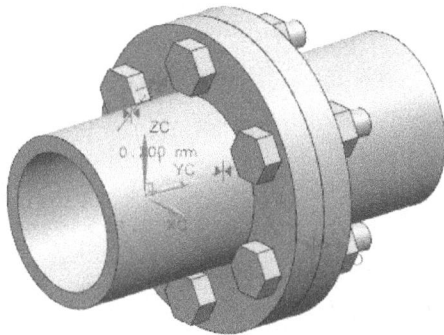

Figure 8-62 Flange assembly for creating bolt connection and contact

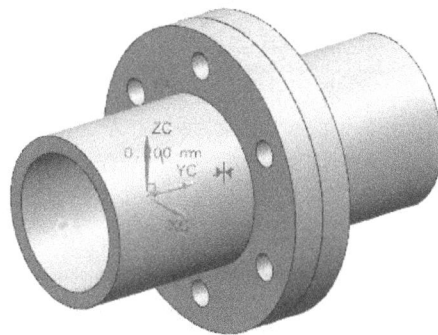

Figure 8-63 Assembly after deleting nuts and bolts

Creating Assembly FEM

1. For creating Assembly FEM, you need to invoke the Advanced Simulation environment. To do so, choose **Advanced Simulation** from the **Application** area of the **File** tab; you enter the Advanced Simulation environment.

2. Choose the **New Assembly FEM** tool from the **Change Window Drop-Down** list in the **Context** group from the **Home** tab; the **New Part File** dialog box is displayed, as shown in Figure 8-64. Choose the **OK** button from the dialog box. The **New Assembly FEM** dialog box is displayed, as shown in Figure 8-65. Leave the default settings and choose the **OK** button from the dialog box; the Assembly FEM environment is opened.

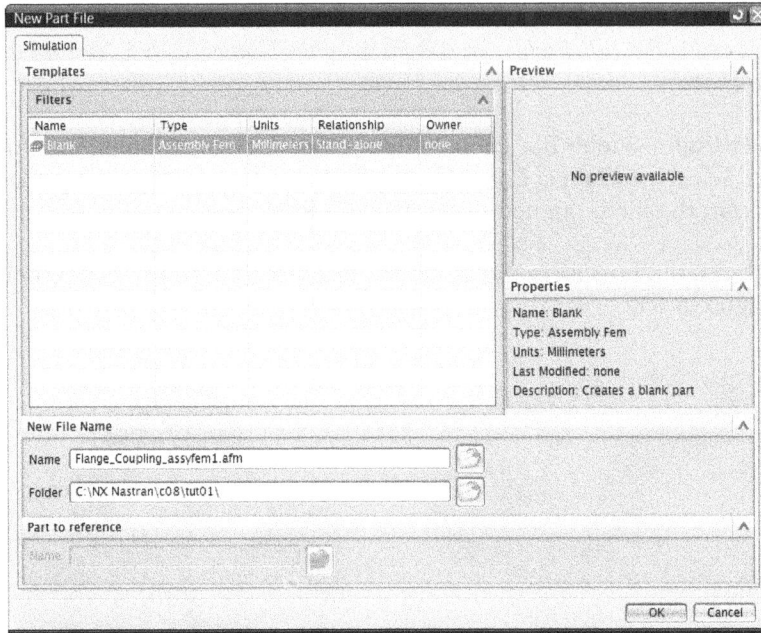

Figure 8-64 The **New Part File** dialog box

Figure 8-65 The **New Assembly FEM** dialog box

Mapping the Part Files

Expand the **Flange_Coupling.prt** node from the **Simulation Navigator**. You will see both the flanges are in ignored status. You need to map both parts with Nastran FEM type.

1. For mapping the flange, right-click on **Flange.prt** under the **Flange_Coupling.prt** node in the **Simulation Navigator**; a shortcut menu is displayed, as shown in Figure 8-66. Choose the

Map New option from the shortcut menu; the **New Part File** dialog box is displayed. Select the **NX Nastran** from the **Name** column of the **FEM** type from the dialog box and choose the **OK** button; the **New FEM** dialog box is displayed with the preview of the flange in the **CAD Part** window. Choose the **OK** button from the dialog box; the part is mapped and displayed in the environment. Also, the FEM file of the flange is placed in the **Simulation Navigator**.

2. Similarly, map another flange of the assembly.

Creating Mesh on the Parts

To create meshing on the parts, you must make them worked part. After that you will create 3D tetrahedral mesh on the part.

1. For making the part workable, right-click on **FLANGE_fem1.fem** in the **Simulation Navigator**; a shortcut menu is displayed, as shown in Figure 8-67. Choose the **Make Work Part** option; other parts will become faded and meshing tools will become available in the environment.

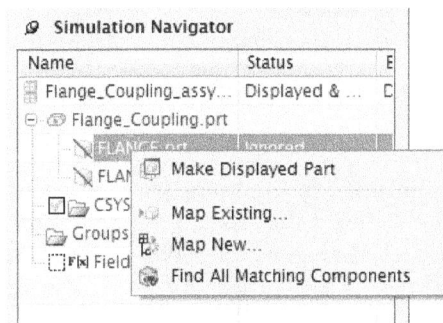

Figure 8-66 Shortcut menu for mapping part

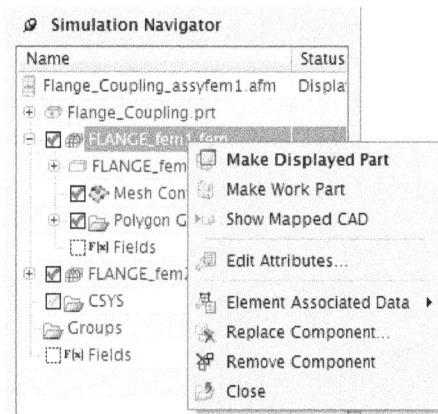

Figure 8-67 Shortcut menu for making part workable

2. To create mesh, choose the **3D Tetrahedral Mesh** tool from **Menu > Insert > Mesh** from the **Top Border Bar**; the **3D Tetrahedral Mesh** dialog box is displayed. Select the body and choose CTETRA(4) from the **Type** drop-down list in the **Element Properties** rollout. Specify **20** in the **Element Size** edit box in the **Mesh Parameters** rollout, refer to Figure 8-68. Leave the default settings and choose the **OK** button; the mesh is created on the part.

3. Similarly, create mesh on the other part also. After creating mesh, double-click on the main assembly from the **Simulation Navigator** to make the complete assembly visible. Figure 8-69 shows the assembly after meshing.

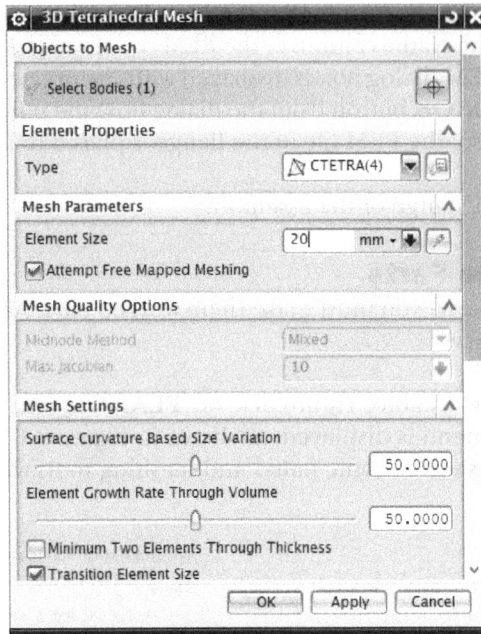

*Figure 8-68 The **3D Tetrahedral Mesh** dialog box*

The label of the elements of both the parts conflict with each other. You need to check and solve this.

4. Right-click on the **Flange_Coupling_assyfem1.afm** node in the **Simulation Navigator**; a shortcut menu is displayed, as shown in Figure 8-70. Select the **Assembly Label Manager** option from the shortcut menu; the **Assembly Label Manager** dialog box is displayed, as shown in Figure 8-71.

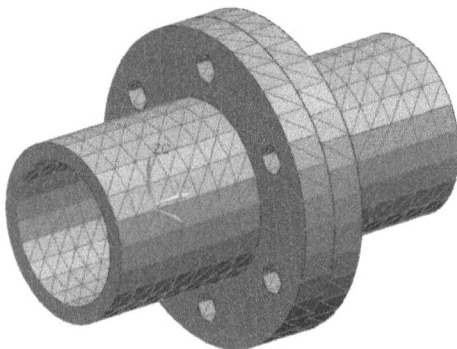

Figure 8-69 Meshed assembly

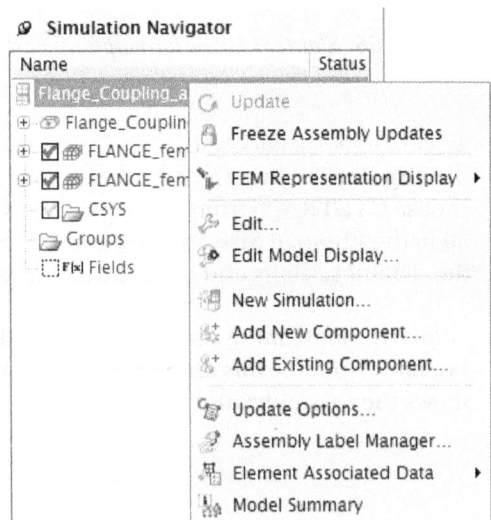

Figure 8-70 Shortcut menu for label manager

You can see in this dialog box that status of the nodes and elements under the **Labels** rollout is in conflict. You need to solve this.

5. Choose the **Automatically Resolve** button in the **Automatic Label Resolution** rollout; all the labels become valid. Choose the **OK** button to accept and exit the dialog box.

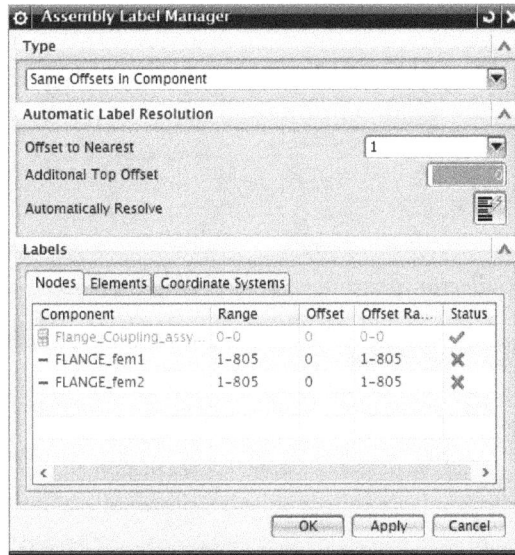

Figure 8-71 The Assembly Label Manager dialog box

Creating Mesh Collector for Bolt Connection

Before creating the bolt connection, you need to create a mesh collector. It will include the bolt cross section.

1. Choose **Menu > Insert > Mesh Collector** from the **Top Border Bar**; the **Mesh Collector** dialog box is displayed.

2. Choose the **1D** option from the **Element Family** drop-down list and the **Bar Collector** option from the **Collector Type** drop-down list in the **Element Topology** rollout, refer to Figure 8-72. Next, choose the **Create Physical** button available next to the **Bar Property** drop-down list in the **Physical Property** sub-rollout under the **Properties** rollout; the **PBAR** dialog box is displayed, as shown in Figure 8-73.

3. Choose the **Show Section Manager** button available next to the **Fore Section** drop-down list in the **Properties** rollout; the **Beam Section Manager** dialog box is displayed, as shown in Figure 8-74.

4. Choose the **Create Section** button from the dialog box; the **Beam Section** dialog box is displayed, as shown in Figure 8-75.

5. Make sure that the **ROD** type is selected from the **Type** rollout and specify **8** mm diameter in the **DIM1** edit box of the **Dimensions** sub-rollout in the **Properties** rollout, refer to Figure 8-75.

Figure 8-72 The Mesh Collector dialog box

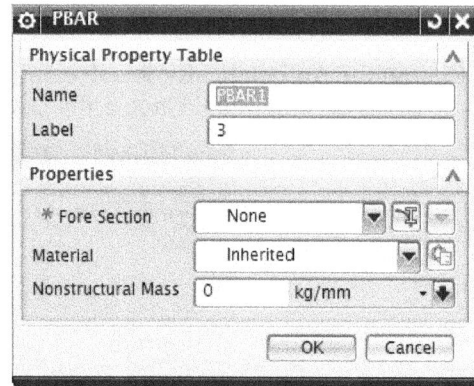

Figure 8-73 The PBAR dialog box

Figure 8-74 The Beam Section Manager
dialog box

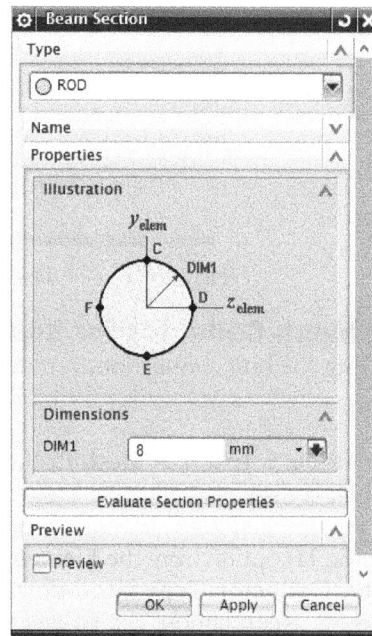

Figure 8-75 The Beam Section dialog box

6.   Choose the **OK** button to accept section. Close other dialog boxes by accepting the default values.

Creating Bolt Connection

Now you will create and apply a bolt with nut connection to this Assembly FEM model.

1.   To create bolt connection, choose **Menu > Insert > Mesh > Bolt Connection** from the **Top Border Bar**; the **Bolt Connection** dialog box is displayed, refer to Figure 8-76. Select the **Bolt With Nut** option from the **Type** rollout if it is not selected by default.

2. Click in the **Select Edge(0)** area from the **Head** rollout and select all the edges of hole of the first flange, refer to Figure 8-77. Also, enter **150** in the **Percent of Hole** edit box.

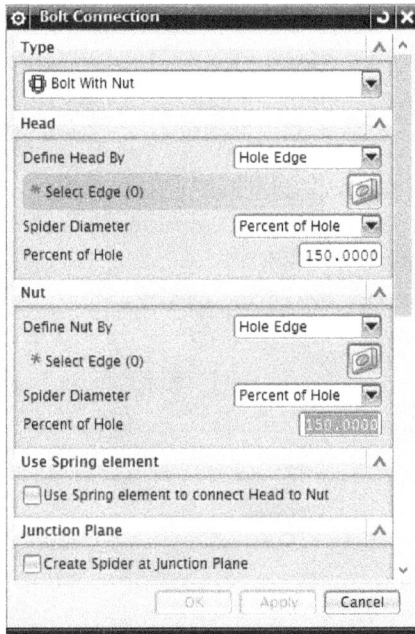

*Figure 8-76 The **Bolt Connection** dialog box*

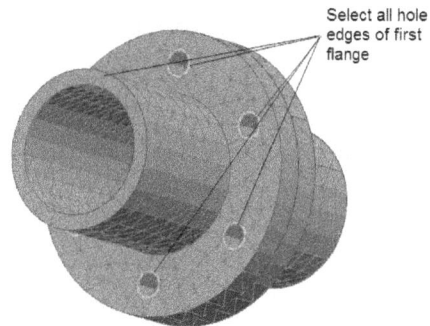

Figure 8-77 Edges of the holes selected for bolt head

3. Similarly, choose the **Select Edge(0)** area from the **Nut** rollout and select all the edges of hole of the second flange. Enter **150** in the **Percent of Hole** edit box.

4. Clear the **Automatic Creation** check box from the **Destination Collector** sub-rollout of the **Shank Element** rollout; the created mesh collector name is displayed and applied to the bolt.

5. Leave all the default settings and choose the **OK** button to create bolt with nut connection; the bolt connection is applied to the model, refer to Figure 8-78. You can also preview this connection by hiding both the flanges FEM, refer to Figure 8-79.

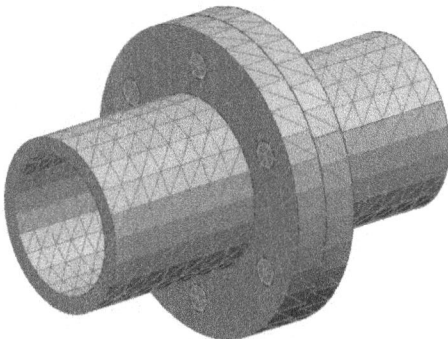

Figure 8-78 Assembly after creating bolt connection

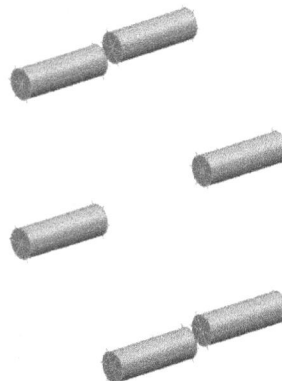

Figure 8-79 Bolt connection after hiding the both meshed flanges

Creating Surface to Surface Contact

For creating surface to surface contact between the surfaces in contact of both the flanges, you have to enter in the Simulation environment.

1. To enter in Simulation environment, choose the **New Simulation** tool from the **Change Window Drop-Down** list of the **Context** group from the **Home** tab; the **New Part File** dialog box is displayed, as shown in Figure 8-80. Choose the **OK** button from the dialog box; the **New Simulation** dialog box is displayed, as shown in Figure 8-81. Retain the default settings and choose the **OK** button from the dialog box; the **Solution** dialog box is displayed, as shown in Figure 8-82. Again, retain the default settings in the **Solution** dialog box and choose the **OK** button; you enter in the Simulation environment.

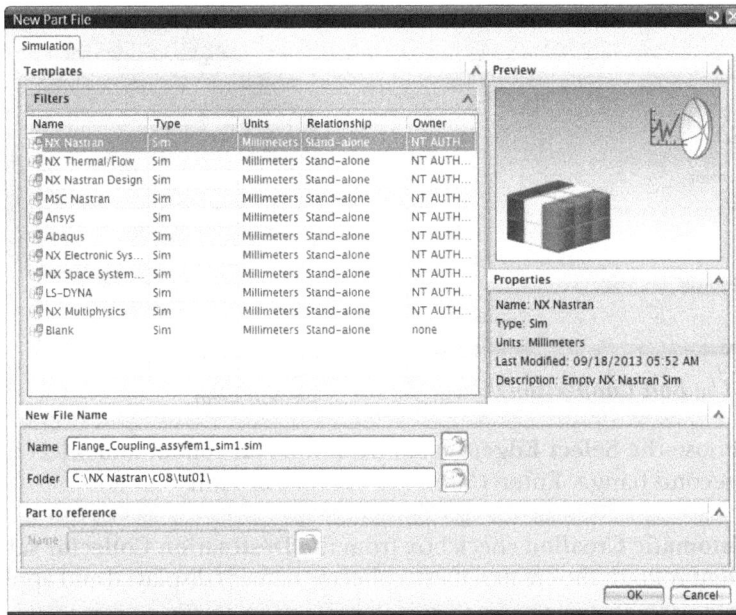

*Figure 8-80 The **New Part File** dialog box*

Note
You will learn more about the Simulation environment and its dialog boxes in later chapters of this book.

Next, you will create surface to surface contact between the surfaces in contact of both the flanges.

2. Choose the **Surface-to-Surface Contact** tool from the **Simulation Object Type** drop-down list in the **Load and Conditions** group from the **Home** tab; the **Surface-to-Surface Contact** dialog box is displayed, as shown in Figure 8-83.

3. Make sure that the **Automatic Pairing** option from the **Type** rollout is selected and choose the **Create Face Pairs** button from the **Automatic Face Pair Creation** rollout; the **Create Automatic Face Pairs** dialog box is displayed.

4. Enter **1** mm in the **Distance Tolerance** edit box in the **Properties** rollout, refer to Figure 8-84,

and select all the surfaces of the model by dragging a window around it. Choose the **Preview** button available in the **Preview** rollout; contact surfaces are highlighted and a message is displayed as **1 face pair found**.

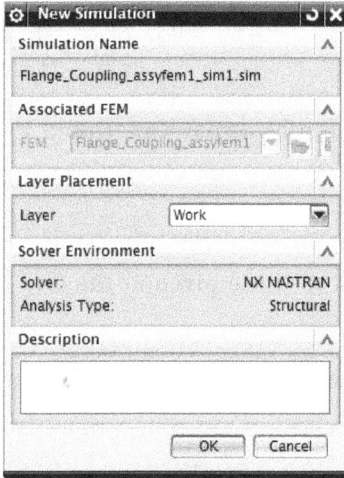

Figure 8-81 The *New Simulation* dialog box

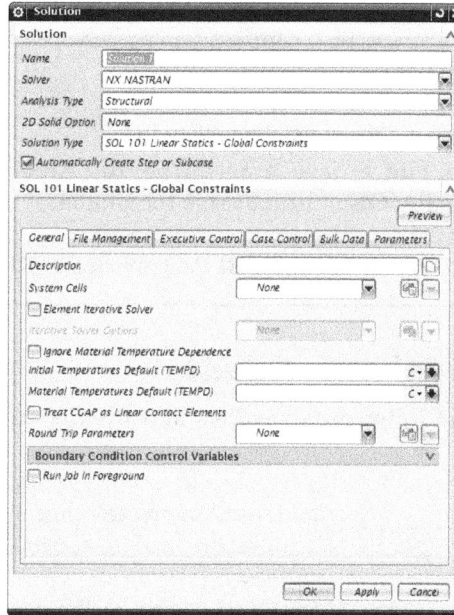

Figure 8-82 The *Solution* dialog box

Figure 8-83 The *Surface-to-Surface Contact* dialog box

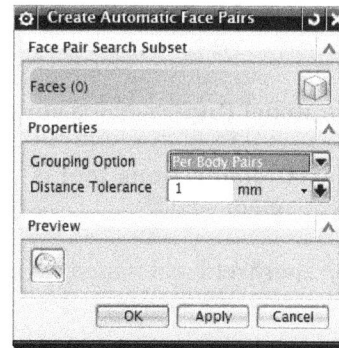

Figure 8-84 The *Create Automatic Face Pairs* dialog box

5. Choose the **OK** button from the **Create Automatic Face Pairs** dialog box; the face pair is added under the **Automatic Face Pair Creation** rollout.

6. Leave the default settings in the dialog box and choose the **OK** button from the **Surface-to-Surface Contact** dialog box; the surface to surface contact is created between the surfaces.

Saving and Closing the Model

1. Choose **File > Save > Save** from the **Ribbon**; the **Name Parts** dialog box is displayed. Choose the **OK** button to save the model.

2. Choose **File > Close > All Parts** from the **Ribbon**; all displayed parts are closed.

Self-Evaluation Test

Answer the following questions and then compare them to those given at the end of this chapter:

1. The _____ tool is used to create point to point connection between two different part edges.

2. You can merge the source and target surfaces to create a single surface by using the _____ option of the **Mesh Mating Condition** tool.

3. The _____ tool is used to create a contact between two surfaces of an assembly.

4. The _____ tool is used to create a gluing contact between two surfaces of an assembly.

5. The _____ tool is used to create a gluing contact between an edge and a surface of an assembly for preventing any directional movement.

6. You can also create Assembly FEM in NX Nastran. (T/F)

7. The **Assembly Label Manager** option is used to resolve the label conflicts of nodes and elements. (T/F)

8. The **Model Display** dialog box helps you to differentiate parts in a meshed assembly. (T/F)

9. The **1D connections** tool is used to connect multiple sheets or bodies in the FEM environment. (T/F)

10. The **Spot Weld** tool is used to connect multiple sheets or bodies through 1D element. (T/F)

Review Questions

Answer the following questions:

1. Which of the following file types is created by NX Nastran for Assembly FEM?

    (a) *afm*   (b) *prt*
    (c) *fem*   (d) None of these

2. Which of the following tools is used to connect multiple sheets or bodies through bolted type connections?

    (a) **1D Connection**   (b) **Bolt Connection**
    (c) **Surface Contact Mesh**   (d) None of these

3. Which of the following tools is used to create a connection mesh between two contacted surfaces?

    (a) **Mesh Mating Condition**   (b) **Surface Contact Mesh**
    (c) **Bolt Connection**   (d) None of these

4. Which of the following tools is used to create the meshing connection between two separate bodies?

    (a) **Bolt Connection**   (b) **Mesh Mating Condition**
    (c) **1D Connection**   (d) None of these

5. Which of the following options of the **Mesh Mating Condition** tool is used to create an alignment between source and target face for mesh mating condition?

    (a) **Glue-Coincident**   (b) **Free Coincident**
    (c) **Glue Non-Coincident**   (d) None of these

6. The _____ tool is used to update mesh after creating spot weld.

7. The _____ button is used to create face pairs between which you can apply surface contact.

8. The _____ tool is used to create a gluing contact between edge and surface of an assembly for preventing any directional movement.

9. The _____ tool is used to create a contact between edge to edge of an assembly.

10. The _____ tool is used to create a gluing contact between edge to edge of an assembly for preventing any directional movement.

EXERCISES

To proceed the exercises, you need to extract the *c08_NX_Nastran_input.zip* file downloaded at the beginning of the tutorials.

Exercise 1

In this exercise, you will import the *c08_exr01.prt* file, as shown in Figure 8-85, from the extracted folder and delete bolts from the assembly. Open Advanced Simulation environment and apply bolt connection. Also, create spot weld connection on the side edges of the sheet part then create surface to surface contact. Model after applying bolt connection and spot weld is shown in Figure 8-86.   **(Expected time: 45 min)**

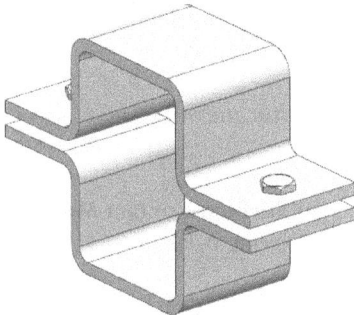

Figure 8-85 Model for Exercise 1

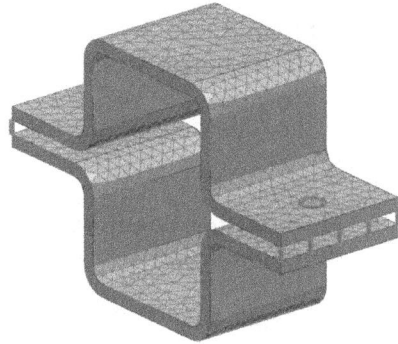

Figure 8-86 Final model after applying bolt connection, spot weld, and contacts

Hint

1. Import the assembly and delete the bolt.
2. Enter in the FEM environment.
3. Mesh the model with CTETRA(4) element with 8 mm size.
4. Choose the **Mesh Collector** tool and create a 1D type **Beam Collector** with a cross section of **1 mm** and name **Spot Weld**.
5. Invoke the **Spot Weld** tool and select one of the side edges as weld objects and also select adjacent faces as top and bottom faces, refer to Figure 8-87. Enter **5** in the **Mesh Density Value** edit box and select CBEAM from the **Type** drop-down list available in the **Element Properties** rollout. Also, select created mesh collector from the **Destination Collector** rollout. Choose the **Apply** button to create the spot weld.
6. Similarly, create spot weld on the other side of the edges and choose the **Update** tool from the **Context** group of the **Home** tab to update the mesh.
7. Create another mesh collector by 1D type **Bar Collector** with a cross section of **3 mm** and name **Bolt Connection**.
8. Choose the **Bolt Connection** tool and select the **Bolt With Nut** option.
9. Specify **150** in the **Percent of Hole** edit box under the **Head** rollout and **150** in the **Percent of Hole** edit box under the **Nut** rollout. Select the **Bolt Connection** mesh collector created from the **Destination Collector** sub-rollout under the **Shank Element** rollout.
10. Select the upper edge of the top plate hole and lower edge of the bottom plate hole as head and nut to define bolt connection. Similarly, create bolt connection for the other side bolt.

11. Enter in the Simulation Environment.
12. Create surface to surface contact by using the **Surface-to-Surface Contact** tool. Note that for selecting the face pairs of contact, enter **6** in the **Distance Tolerance** edit box under the **Properties** rollout of the **Create Automatic Face Pairs** dialog box. Model after creating contact is shown in Figure 8-88.

Note

*In this exercise, you will not create Assembly FEM, as the **Spot Weld** tool is available in FEM environment only.*

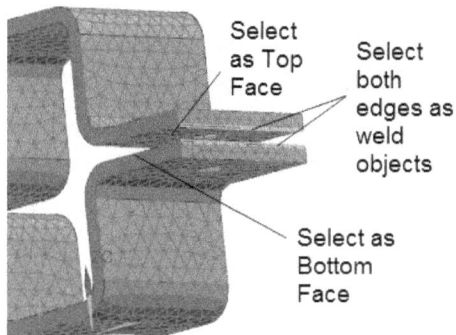

Figure 8-87 Edges and faces selected for creating spot weld

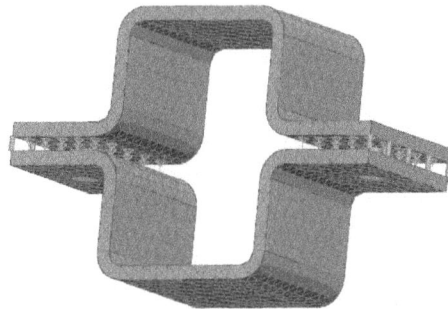

Figure 8-88 Model after creating contact between surfaces

Answers to Self-Evaluation Test

1. Contact Mesh, 2. Glue-Coincident, 3. Surface-to-Surface Contact, 4. Surface-to-Surface Gluing, 5. Edge-to-Surface Gluing, 6. T, 7. T, 8. T, 9. T, 10. T

Chapter 9

Defining Material and Boundary Conditions

Learning Objectives

After completing this chapter, you will be able to:

- *Assign material using the Assign Material tool*
- *Define material using Mesh Collector*
- *Manage material*
- *Create different type of constraints*
- *Apply different type of loads*
- *Use fields and expressions to apply boundary conditions*

INTRODUCTION

To perform analysis in a model, you need to define material as well as boundary conditions. The material of the part defines the properties and strength of the model while boundary conditions represent the effect of the surrounding environment such as forces, pressures, displacement, and so on. Methods to define material and boundary conditions are discussed next.

DEFINING MATERIAL

After different operations like meshing and connection, you need to define material to the body to compare them to the real life models. In NX Nastran, you can select existing material from material library or create your own customized material according to the need and specified data. Different types of materials are used in real life, such as Isotropic, Orthotropic, Anisotropic, and so on. You can also manage the existing library of materials. The tools and methods for defining materials to the model are discussed next. Note that the tools to assign material are available in the FEM environment.

> **Note**
> *You can also define material to the body before meshing and other connection operations.*

Assign Material

Ribbon: Home > Properties > More > Materials > Assign Materials
Menu: Tools > Materials > Assign Materials

The **Assign Materials** tool is used to define material to a body. Using this, you can also import your user defined material library. To do so, choose **Menu > Tools > Materials > Assign Materials** from the **Top Border Bar**; the **Assign Material** dialog box will be displayed, as shown in Figure 9-1. The rollouts and options in this dialog box are discussed next.

Type

In the drop-down list of the **Type** rollout, three options are there: **Select Body(s)**, **Body(s) Without Assigned Material**, and **All Bodies in Work Part**. These options are discussed next.

Select Body(s)

By using the **Select Body(s)** option, you can select an individual body from the model to assign the material. When you select this option from the drop-down list of the **Type** rollout; the **Select Body** rollout will be displayed. In this rollout, the **Select Body (0)** option is highlighted and you will be prompted to select the object to assign material. You can also select multiple bodies at a time.

Body(s) Without Assigned Material

By using the **Body(s) Without Assigned Material** option

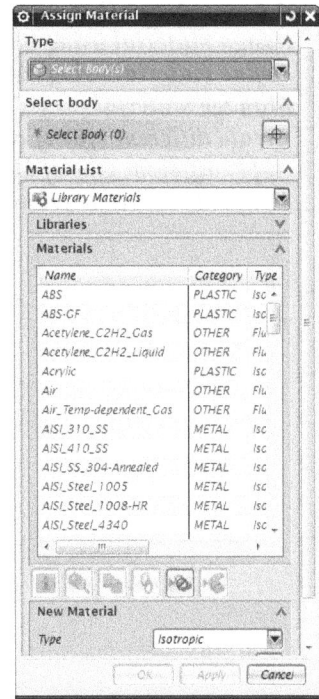

Figure 9-1 *The Assign Material dialog box*

from the drop-down list of the **Type** rollout, you can apply a material to all the bodies at a time on which the material is not assigned.

All Bodies in Work Part

By using the **All Bodies in Work Part** option from the drop-down list of the **Type** rollout, you can apply a material to all the bodies displayed in the window.

Note
If you apply a material on a body to which a different material is already assigned, then the latest assigned material will replace the previous one.

Material List

In this rollout, you can apply material to the model from existing NX material library or by importing local library. You can also create a user defined material. For selecting materials, two options are available in drop-down list of this rollout: **Library Materials** and **Local Materials**. For the **Library Materials** option, three check boxes will be available in the **Libraries** sub-rollout: **NX Material Library**, **Site MatML Library**, and **User MatML Library**. Materials displayed in the **Materials** sub-rollout will depend on the check boxes selected from the **Libraries** sub-rollout. Note that using the **Site MatML Library** and **User MatML Library** options, you can import a *.xml type material format in environment.

The **Local Materials** option is used to display the edited materials in the **Materials** sub-rollout.

To assign the material to the body, select the body and choose the material from the **Materials** sub-rollout, and then choose the **OK** button from the dialog box; the material will be assigned to the body.

Note
Remaining options of the dialog box will be discussed later. Also, you will learn about managing the libraries and local materials later in this chapter.

Defining Material using Mesh Collector

Ribbon: Home > Properties > Mesh Collector
Menu: Insert > Mesh Collector

You can also define the material using the **Mesh Collector** dialog box. To define material using the **Mesh Collector** dialog box, choose **Menu > Insert > Mesh Collector** from the **Top Border Bar**; the **Mesh Collector** dialog box will be displayed, refer to Figure 9-2. To define a material, choose the **Create Physical** button; the respective dialog box will be displayed with physical properties. Figure 9-3 shows the **PSOLID** dialog box for **3D** element type.

To assign the material, select the **Choose Material** button available next to the **Material** drop-down list in the **Properties** rollout; the **Material List** dialog box will be displayed, as shown in Figure 9-4. Select the material from the **Materials** sub-rollout of the **Material List** rollout and choose the **OK** button; the selected material will be displayed in the **Material** drop-down list of the **PSOLID** dialog box. Now, choose the **OK** button from all the dialog boxes to close them.

Figure 9-2 *The* **Mesh Collector** *dialog box*

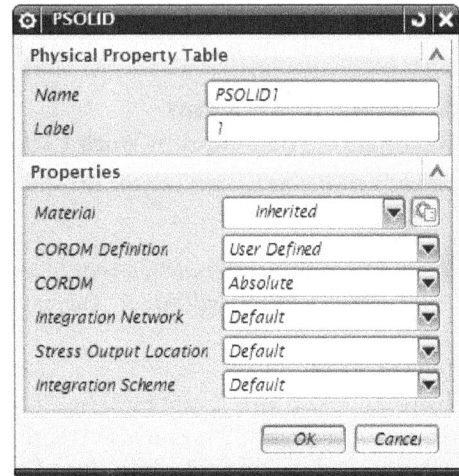

Figure 9-3 *The* **PSOLID** *dialog box*

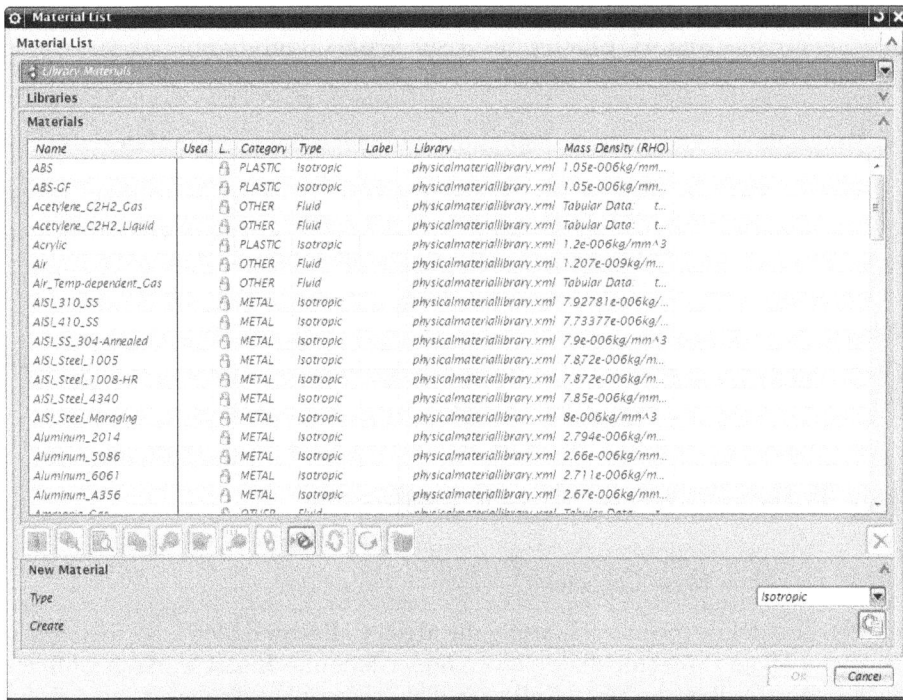

Figure 9-4 *The* **Material List** *dialog box*

Next, while creating a mesh on a body, choose the created mesh collector from the **Destination Collector** rollout of the respective mesh dialog box. Now, select the body and specify the mesh settings as discussed in earlier chapters and choose the **OK** button; the body will be meshed and material will be assigned to it. Similarly, you can assign material to 0D, 1D, or 2D element/mesh.

Note

The options in the rollout of the **Material List** *dialog box work similar to the options in the* **Assign Materials** *dialog box.*

MANAGING MATERIAL

As discussed earlier, besides using the existing NX material library, you can import user defined XML type material library. You can also edit the existing material from the library. You will learn more about managing the materials next.

Manage Materials

Ribbon:	Home > Properties > Manage Materials
Menu:	Tools > Materials > Manage Materials

For using regular materials or some specific materials, you can use the **Manage Materials** tool. Using this tool, you can customize and manage the material to be displayed. To do so, choose **Menu > Tools > Materials > Manage Materials** from the **Top Border Bar**; the **Manage Materials** dialog box will be displayed, as shown in Figure 9-5. In this dialog box, the **Material List** rollout is available and the options of this rollout are used to manage the materials. The options available in the drop-down list of the **Material List** rollout are: **Library Materials** and **Local Materials**. The rollouts and options of the dialog box depend upon the option chosen from the drop-down list of the **Material List** rollout and are discussed next.

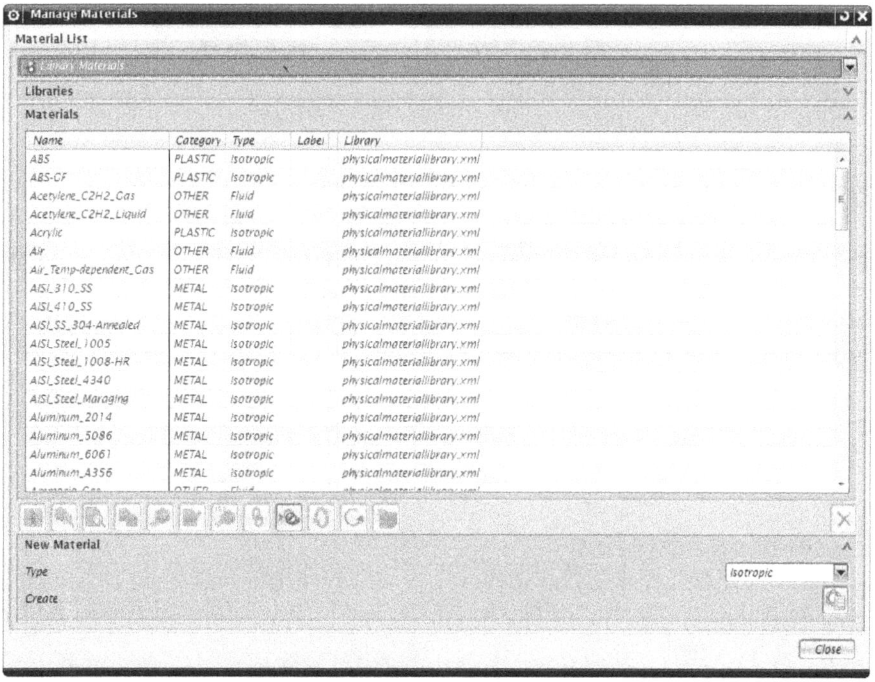

Figure 9-5 The **Manage Materials** *dialog box*

Libraries Sub-rollout

This sub-rollout will be available, when you select the **Library Materials** option from drop-down list of the **Material List** rollout. In this sub-rollout, three check boxes are available: **NX Material Library**, **Site MatML Library**, and **User MatML Library**. If you select the **NX Material Library** check box, the materials from NX library will be displayed under the **Materials** sub-rollout. Using the **Site MatML Library** check box, you can export a XML type material file from a site. It can be a combination of material files. Using the **User MatML Library** option, you can import a user defined XML type material file.

Materials Sub-rollout

In this sub-rollout, materials displayed depend upon the check boxes selected from the **Libraries** sub-rollout. On selecting the **NX Material Library** check box, different types of materials will be displayed. You can sort the materials to be displayed by using the options of the shortcut menu available on right-clicking on any heading of the area, as shown in Figure 9-6. You can select the material name, type, or category from this shortcut menu to display the materials.

Below the **Materials** sub-rollout, some command buttons are available. These buttons are discussed next.

Display material properties for the selected material(s)

When you choose this button, the **Information** window will be displayed for the selected material with all the material properties.

Inspect Material

Using this button, you can view the material properties in the respective material dialog box. Figure 9-7 shows the **Isotropic Material** dialog box. In the dialog box, the material properties are in read-only mode.

Preview Solver Syntax

Using this button, you can view the specific properties of material such as density, young's modulus, poisson's ratio and so on that play a major role in analyzing a part. When you choose this button, the **Information** window will be displayed with all the properties.

Copy the selected material

Using this button, you can create a copy of the selected material. When you choose this button, a corresponding material dialog box will be displayed. You can change the default values of the material. Choose the **OK** button from dialog box after specifying all the values; the created copy of the material will be located in the **Material** sub-rollout and displayed when you select the **Local Materials** option from the drop-down list.

Edit the selected material

This button will be activated only when you select a copied material created under local materials. You can edit the properties of the local material that you created by copying from the library material.

Rename the selected material

This button will be activated only when you select a copied material created under local materials. Using this button, you can rename the local material that you created by copying from the library material.

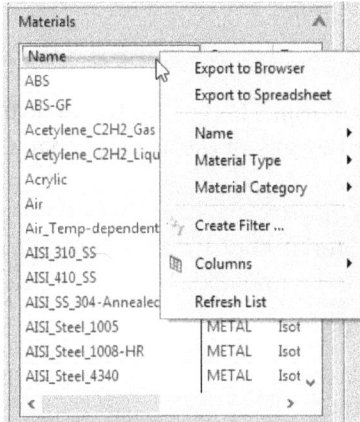

Figure 9-6 Shortcut menu for material sorting

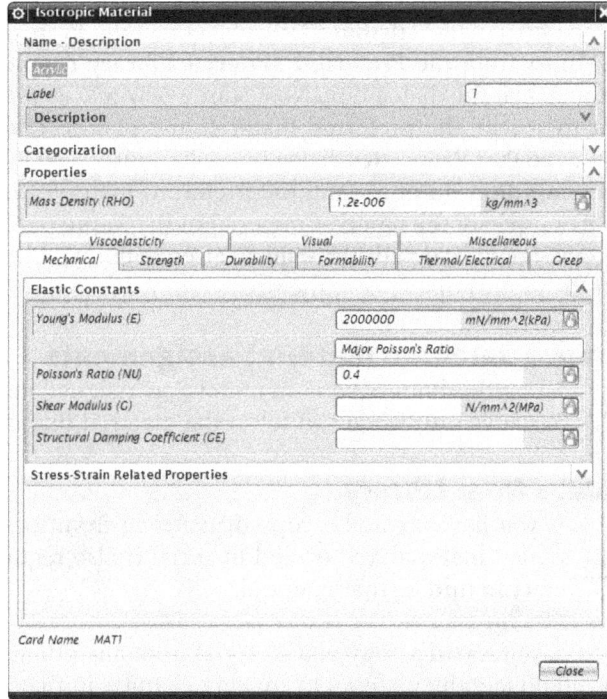

Figure 9-7 The **Isotropic Material** dialog box

Relabel the selected material

This button will be activated only when you select a copied material created under local materials. Using this button, you can relabel the local material that you created by copying from the library material.

Show where the selected material(s) are used

Using this button, you can highlight the bodies where a selected material is used. When you choose this button, the **Information** window will be displayed informing you about the selected material and the bodies on which that material is used. This button will only be available when you select a material that is already assigned to a part.

Highlight bodies without a material assigned

Using this button, you can highlight the body having no material assigned.

Highlight bodies that don't have the preferred material assigned

Using this button, you can highlight the body having no preferred material assigned. Note that this button will be available only when you have created the preferred material. You can create preferred material list for a specific analysis project. To create the preferred material, choose **File > Utilities > Customer Defaults** from the **Ribbon**; the **Customer Defaults** dialog box will be displayed. Next, choose **Gateway > Material/Mass** from the tree located at the left in the dialog box. Next, choose the **Attributes** tab and specify the name of the preferred part material in the **Attribute Title** text box of the **Preferred Part Material** area and then choose the library from the **Preferred Material Library** drop-down list of this area. Choose the **OK** button to exit the dialog box. Now, start a new NX session and open the FEM file for which you want to specify the preferred material and choose **File > Properties** from the **Ribbon**; the **Displayed Part Properties** dialog box will be displayed. Now, select **Materials** from the **Category (optional)** drop-down list and the preferred material created from the **Title/Alias** drop-down list. Also, specify the name of material in the **Value** text box. Note that the name in the **Value** text box should be exactly the same as given in library. Choose the **OK** button from the **Displayed Part Properties** dialog box; the preferred material will be created.

Remove selected material assignments

Using this button, you can remove an assigned material from a body. This button will be available only when you select the material that is used on any of the assembly parts.

Update from Library

If you have created a copy of material from local material library and if you edit the parent material, the copied material is also required to be updated. Using this button, you can update that material.

If you have created a copy of a material from any other XML file material loaded earlier then you can update the material by making changes in that XML file.

Load library material into the file

Using this button, you can load any regular material from library to local material.

Delete selected materials

Using this button, you can delete a local material from the list.

Besides importing a material, you can also create a user defined material using the **New Material** sub-rollout available in the **Material List** rollout. The **New Material** sub-rollout is discussed next.

New Material

Using this sub-rollout, you can create a user defined material. To do so, select the material type from the **Type** drop-down list and choose the **Create material** button; the respective material dialog box will be displayed. Specify all specifications related to the material and choose the **OK** button from the dialog box. The created material will be displayed under the **Material** sub-rollout, when the **Local Materials** option is selected from the drop-down list in the **Material List** rollout. The material types that you can create are available in the **Type** drop-down list and are shown in Figure 9-8.

Isotropic
Orthotropic
Anisotropic
Fluid
Hyperelastic - general
Mooney-Rivlin
Ogden
Foam
Arruda-Boyce
Gasket - Strain
Shape Memory Alloy
Sussman-Bathe

Figure 9-8 The material type list

BOUNDARY CONDITION

Boundary conditions are the loads/constraints that are applied on the boundary of an FEA model to represent the effect of the surrounding environment. It includes forces, pressures, displacement, and so on. You can apply the boundary condition to the model in Simulation environment of NX Nastran. The boundary conditions are divided into two categories, Constraints and Loads, and are discussed next.

Constraints

Constraints are defined as the boundary conditions that prevent the model from rotational or translational movements. In other words, the constraints are used to restrict the degrees of freedom of a model. The applied constraints act as a rigid support between the model and the ground. You can select nodes, edges, or surfaces of a model to apply constraints. The tools used for applying constraints are discussed next.

User Defined Constraint

Ribbon: Home > Loads and Conditions > Constraint Type > User Defined Constraint

Using the **User Defined Constraint** tool, you can selectively restrict the degrees of freedom of the selected part of the model. To use this tool, choose the **User Defined Constraint** tool from the **Constraint Type** drop-down list of the **Loads and Conditions** group in the **Home** tab; the **User Defined Constraint** dialog box will be displayed, as shown in Figure 9-9. The options and rollouts available in this dialog box are discussed next.

Type

In the drop-down list of the **Type** rollout, two options are available: **SPC** and **SPC1**. The **SPC** option defines the single point constraint using SPC entry which is the most common way to specify single point constraint while the **SPC1** defines single point constraint using the SPC1 entry which is a less common entry.

Name

Under the **Name** rollout, you can specify the name of constraint. You can also provide the description of the constraint under the **Description** sub-rollout.

Destination Folder

In the **Destination Folder** rollout, you can specify the folder in which the created constraint will be stored. To create a folder, click on the **Folder Manager** button; the **Folder Manager** dialog box will be displayed. To create a new folder for constraint destination, right-click on an available root folder; the **New Folder** option will be displayed, as shown in Figure 9-10. Choose the **New Folder** option, a new folder node will be created below the root folder as well as in the **Simulation Navigator**. You can change its name also. After creating the folder, choose the **OK** button from the dialog box; all the created folders will be displayed under the **Constraint Container** drop-down list.

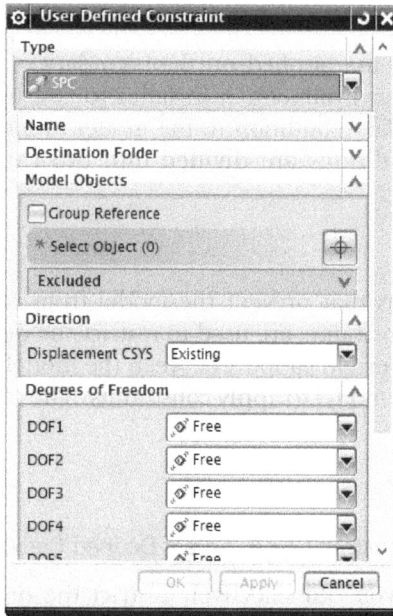

Figure 9-9 The *User Defined Constraint* dialog box

Figure 9-10 The *Folder Manager* dialog box

Model Objects

In the **Model Objects** rollout, you can specify the part of the model to be constrained. The part may be any geometry of the model like edge, point, face or meshed part like nodes and elements. The options in this rollout are discussed next.

Group Reference: By using this option, you can create a group for constraining. When you select the **Group Reference** check box, the options for creating a group of constraints will be displayed in the rollout. You can create a group by using the **New Group** button, available while selecting the check box. When you click on the **New Group** button, the **New Group** dialog box will be displayed, as shown in Figure 9-11. In this dialog box, the **Select Object(0)** option is highlighted and you are prompted to select the part of model to make a group. You can also select the elements and nodes of the model by specifying their label in their respective label edit boxes. Under the **Name** rollout, you can specify the name of the group. After specifying all settings, choose the **OK** button from the dialog box, the name of the new group will be displayed next to the **Group Reference** check box.

Figure 9-11 The *New Group* dialog box

Filter Type: Using this drop-down list, you can filter the type of object on which you want to apply the constraint from created group. Four type of objects are displayed in this drop-down list: **Polygon Face**, **Curves**, **Points**, and **Nodes**.

Excluded: Using this rollout, you can exclude the individual object or element from the selection.

Direction
In this rollout, you can specify the displacement CSYS along with the constraint applied. To specify coordinate system, four options are available in the **Displacement CSYS** drop-down list: **Existing**, **Cartesian**, **Cylindrical**, and **Spherical**. To use the existing coordinate system, choose the **Existing** option from the **Displacement CSYS** drop-down list. When you choose the option other than the **Existing** option, then you need to create a user defined coordinate system.

Degrees of freedom
In this rollout, you can define the individual degree of freedom for the selected part of the model. In the rollout, six drop-down lists are available: the **DOF1**, **DOF2**, and **DOF3** options are used for displacement constraint, while the **DOF4**, **DOF5**, and **DOF6** options are used for rotational constraint. You can apply Free, Fixed, or Displacement/Rotational constraint to DOFs from their respective drop-down lists.

After specifying all values for the part of the model, the **OK** and **Apply** buttons in the dialog box will become active. Choose **OK** to apply the constraint and exit the dialog box. Figure 9-12 shows a model for defining constraint on one side of the face and Figure 9-13 shows model after defining the constraint.

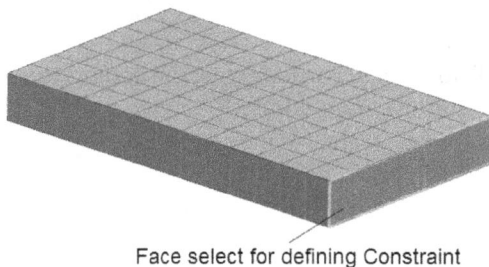

Face select for defining Constraint

Figure 9-12 Model for defining constraint

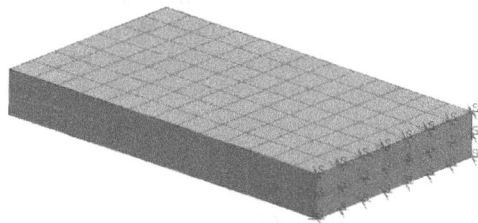

Figure 9-13 Model after defining user defined constraint

Enforced Displacement Constraint

Ribbon: Home > Loads and Conditions > Constraint Type > Enforced
                Displacement Constraint

The **Enforced Displacement Constraint** tool is used to define a displacement force on the selected geometry, along an individual degree of freedom or in a specified direction. To use this tool, choose the **Enforced Displacement Constraint** tool from the **Constraint Type** drop-down list of the **Loads and Conditions** group of the **Home** tab; the **Enforced Displacement Constraint** dialog box will be displayed, as shown in Figure 9-14. To apply enforced displacement on the model, six options are available in the drop-down list of the **Type** rollout: **Components**, **Magnitude and Direction**, **Normal**, **Magnitude and Direction - Spatial**, **Normal - Spatial**, and **Translational Components - Field**. Methods to apply enforced displacement using these options are discussed next.

Components

Using this option, you can enforce the selected geometry along the different degrees of freedom. When you choose this option, six rollouts will be displayed in the dialog box, besides **Type** rollout, **Name**, **Destination Folder**, **Model Objects**, **Direction**, **Degrees of Freedom**, and **Distribution**. The **Name** and **Destination Folder** rollouts work same as discussed earlier. While using the **Model Objects** rollout options, you can select the geometry on which you want to apply enforced displacement. The other rollouts and options are discussed next.

Direction: Using the options of this rollout, you can define the direction using the different displacement coordinate system.

Degrees of freedom: Using the options of this rollout, you can define the enforced displacement for different degrees of freedom. You can specify the displacement using the **Expression** or **Field** option.

Distribution: Using the options of this rollout, you can define the distribution of the enforced load. Two options are available in the **Method** drop-down list: **Constant** and **Spatial**. The **Constant** option will apply load constantly on the geometry while the **Spatial** option will distribute the load in a unitless mapped field.

Figure 9-14 The Enforced Displacement Constraint dialog box

Note

*You will learn more about using the **Field** and **Spatial** options for specifying a boundary condition, later in this chapter.*

Magnitude and Direction

Using this option, you can enforce the selected geometry along a specified direction. When you choose this option, apart from the **Type** rollout, six other rollouts will be displayed in the dialog box, besides the **Type** rollout, namely **Name**, **Destination Folder**, **Model Objects**, **Direction**, **Magnitude**, and **Distribution**. Using the **Direction** rollout, you can specify the direction for enforced displacement. The options of the **Magnitude** rollouts are used to specify the displacement value along the specified direction. You can specify the displacement using the **Expression** or **Field** option. The functions of other options in this rollout are same as discussed earlier.

Normal

Using this option, you can enforce the selected geometry along its perpendicular direction. When you choose this option, apart from the Type rollout, the **Name**, **Destination Folder**, **Model Objects**, **Magnitude**, and **Distribution** rollouts will be displayed. The displacement value specified using the options of the **Magnitude** rollouts will be applied perpendicularly on the geometry. You can specify the displacement using the **Expression** or **Field** option. The functions of the options in other rollouts are same as discussed earlier.

Note

The function of the **Magnitude and Direction - Spatial** *and* **Normal - Spatial** *options are same as that of* **Magnitude and Direction** *and* **Normal** *options respectively. But in the* **Magnitude and Direction - Spatial** *and* **Normal - Spatial** *options, you can vary the magnitude of displacement along the selected geometry.*

Translational Components - Field

Using this option, you can enforce the selected geometry along the translational degree of freedom. When you choose this option, apart from the **Type** rollout, **Name**, **Destination Folder**, **Model Objects**, **Direction**, and **Degrees of Freedom** rollouts are displayed. The options of the **Scale Factors** sub-rollout of the **Degrees of Freedom** rollout are used to define a unitless field displacement that can be scalable along the translational degree of freedom. The functions of the options of other rollouts are same as discussed earlier.

After specifying all the value for the part of the model, the **OK** and **Apply** buttons in the dialog box will become active. Choose the **OK** button to apply the constraint and exit the dialog box. Figure 9-15 shows a model for defining Enforced Displacement Constraint on one side of the face and Figure 9-16 shows model after defining constraint in a downward direction.

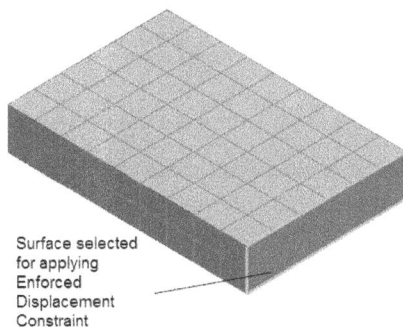

Figure 9-15 Model for defining Enforced Displacement Constraint

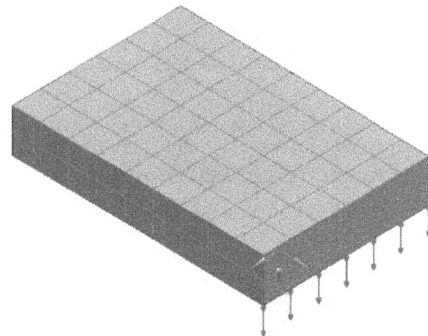

Figure 9-16 Model after defining Enforced Displacement Constraint at downward direction

Fixed Constraint

Ribbon: Home > Loads and Conditions > Constraint Type > Fixed Constraint

Using the **Fixed Constraint** tool, you can restrict all the degrees of freedom like translation and rotational for the selected part of the model. To use this tool, choose the **Fixed Constraint** tool from the **Constraint Type** drop-down list of the **Loads and Conditions** group of the **Home** tab; the **Fixed Constraint** dialog box will be displayed, as shown in Figure 9-17. In the dialog box, three rollouts are available, **Name**, **Destination Folder**, and **Model Objects**. The **Name** rollout is used to define the name of constraint, and the **Destination Folder** is used to define the constraint location in the **Simulation Navigator**. Using the **Model Objects** rollout, you can specify the object of the model on which you want to apply fixed constraint. All these rollouts have already been discussed earlier.

After specifying all the value for the part of the model, the **OK** and **Apply** buttons in the dialog box will become active. Choose the **OK** button to apply the constraint and exit the dialog box.

Fixed Translation Constraint

Ribbon: Home > Loads and Conditions > Constraint Type > Fixed Translation Constraint

Using the **Fixed Translation Constraint** tool, you can restrict the translational degree of freedom of the selected part of the model. To use this tool, choose the **Fixed Translation Constraint** tool from the **Constraint Type** drop-down list of the **Loads and Conditions** group of the **Home** tab; the **Fixed Translation Constraint** dialog box will be displayed, as shown in Figure 9-18. In the dialog box, three rollouts are available: **Name**, **Destination Folder**, and **Model Objects**. The functions of these rollouts are same as discussed earlier.

After specifying all the values for the part of the model, the **OK** and **Apply** button in the dialog box will become active. Choose the **OK** button to apply the constraint and exit the dialog box.

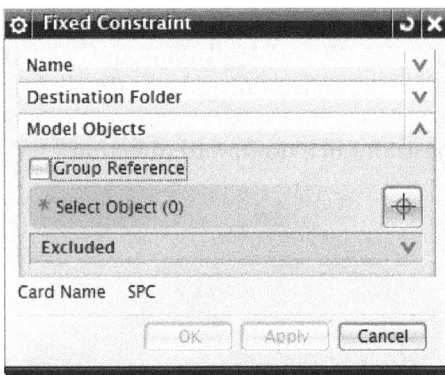

Figure 9-17 The *Fixed Constraint* dialog box

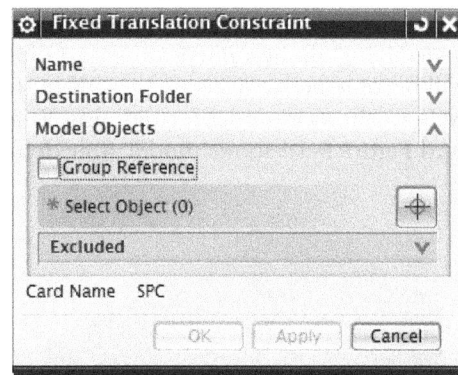

Figure 9-18 The *Fixed Translation Constraint* dialog box

Fixed Rotation Constraint

Ribbon: Home > Loads and Conditions > Constraint Type > Fixed Rotation Constraint

Using the **Fixed Rotation Constraint** tool, you can restrict the rotational degree of freedom of the selected part of the model. To use this tool, choose the **Fixed Rotation Constraint** tool from the **Constraint Type** drop-down list of the **Loads and Conditions** group of the **Home** tab; the **Fixed Rotation Constraint** dialog box will be displayed, as shown in Figure 9-19. In the dialog box, three rollouts are available, **Name**, **Destination Folder**, and **Model Objects**. The functions of these rollouts are same as discussed earlier.

After specifying all the values for the part of the model, the **OK** and **Apply** buttons in the dialog box will become active. Choose the **OK** button to apply the constraint and exit the dialog box.

Simply Supported Constraint

Ribbon: Home > Loads and Conditions > Constraint Type > Simply Supported Constraint

Using the **Simply Supported Constraint** tool, you can restrict the Z axis (displacement) degree of freedom for the selected part of the model whereas the rest five degrees of freedom will be free. You can define the Z axis in the required direction or make it normal to the selected object. To use this tool, choose the **Simply Supported Constraint** tool from the **Constraint Type** drop-down list of the **Loads and Conditions** group of the **Home** tab; the **Simply Supported Constraint** dialog box will be displayed, as shown in Figure 9-20. To define the simply supported constraint, two options are available in the drop-down list of the **Type** rollout: **Directed** and **Normal Simply Supported**. These options are discussed next.

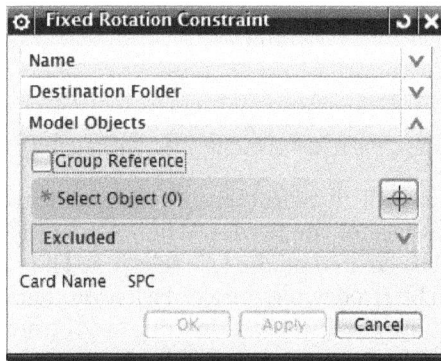

*Figure 9-19   The **Fixed Rotation Constraint** dialog box*

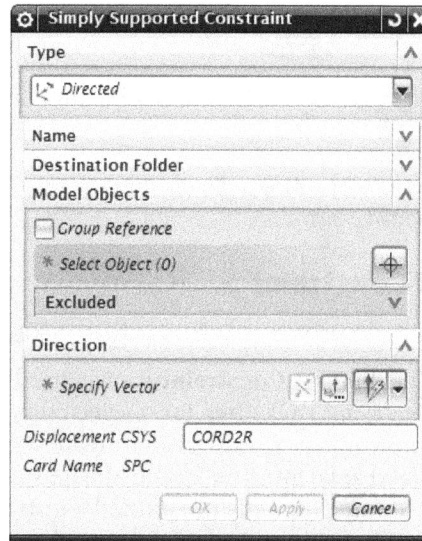

*Figure 9-20 The **Simply Supported Constraint** dialog box*

Directed
Using the **Directed** option, you can define the Z axis direction for simply supported constraint on the selected object. When you select this option, four rollouts will be displayed, **Name**, **Destination Folder**, **Model Objects**, and **Direction**. The **Name**, **Destination Folder**, and **Model Objects** rollouts have been discussed earlier. Using the **Direction** rollout, you can define the direction of Z axis for the constraint. You can use the default option for Z axis from the drop-down list available in the rollout. You can also create the axis by using the **Vector** dialog box which will be available on choosing the **Vector Dialog** button. After defining the Z axis, you can also reverse the direction by using the **Reverse Direction** button.

Normal simply supported
When the **Normal simply supported** option is chosen, software uses the Z axis of the default coordinate system of the environment to apply constraint.

After specifying all the values for the part of the model, the **OK** and **Apply** buttons of the dialog box will become active. Choose the **OK** button to apply the simply supported constraint and exit the dialog box. Figure 9-21 shows the model for defining simply supported constraint on one side of the face and the direction arrow displayed for fixing the Z axis. Figure 9-22 shows the model after defining constraint.

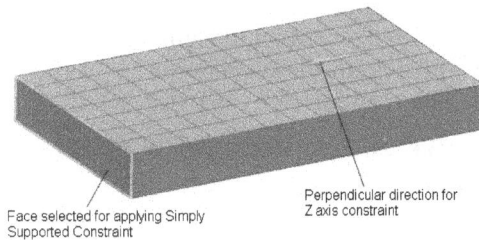

Figure 9-21 Model for defining simply supported constraint

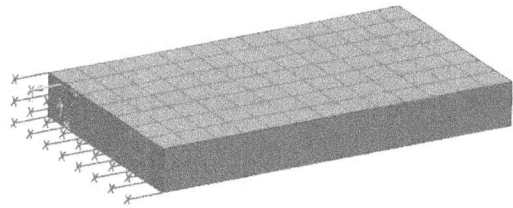

Figure 9-22 Model after defining simply supported constraint

Pinned Constraint

Ribbon: Home > Loads and Conditions > Constraint Type > Pinned Constraint

The **Pinned Constraint** tool is used to restrict all displacement DOFs and leave all rotational DOFs free for a selected cylindrical face of the model. When you apply this constraint, a cylindrical coordinate system will be created on the face selected. The radial growth R and axial growth Z of coordinate system will be fixed while the axial rotation theta will be free to rotate. To apply the pinned constraint, choose the **Pinned Constraint** tool from the **Constraint Type** drop-down list of the **Loads and Conditions** group of the **Home** tab; the **Pinned Constraint** dialog box will be displayed, as shown in Figure 9-23. In the dialog box, three rollouts are available: **Name**, **Destination Folder**, and **Cylindrical Object**. Using the **Name** rollout options, you can specify the name of the constraint and the **Destination Folder** is used to define the constraint location in the **Simulation Navigator**. The options of the **Cylindrical Object** rollout work similar to **Model Objects** rollout, as discussed earlier. However, here you can select the cylindrical face of the model to apply pinned constraint.

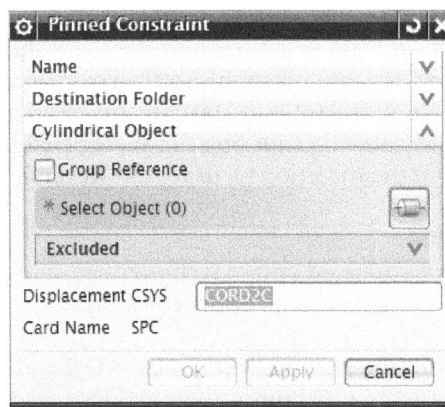

Figure 9-23 The **Pinned Constraint** dialog box

After specifying all the values for the part of the model, the **OK** and **Apply** buttons in the dialog box will become active. Choose the **OK** button to apply the pinned constraint and exit the dialog box. Figure 9-24 shows a model for defining pinned constraint on cylindrical face and Figure 9-25 shows a model after defining pinned constraint.

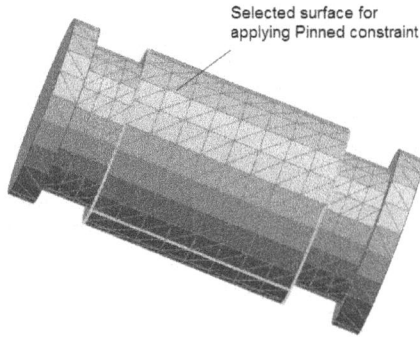

Selected surface for
applying Pinned constraint

Figure 9-24 *Model for defining pinned constraint*

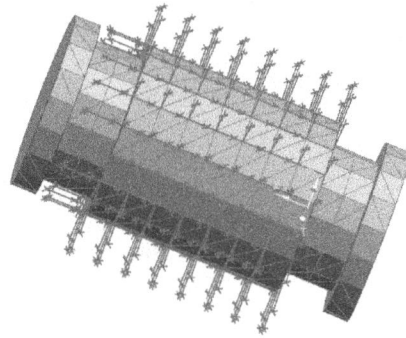

Figure 9-25 *Model after defining pinned constraint*

Cylindrical Constraint

Ribbon: Home > Loads and Conditions > Constraint Type > Cylindrical Constraint

Using the **Cylindrical Constraint** tool, you can restrict the degrees of freedom for the cylindrical face of the model with respect to cylindrical coordinate system. You can assign fix, free or displacement value to R, theta or Z coordinate system. To create the constraint, choose the **Cylindrical Constraint** tool from the **Constraint Type** drop-down list of the **Loads and Conditions** group of the **Home** tab; the **Cylindrical Constraint** dialog box will be displayed, as shown in Figure 9-26. In the dialog box, four rollouts are available, **Name**, **Destination Folder**, **Cylindrical Object**, and **Components**. The options of the **Name**, **Destination Folder**, and **Cylindrical Object** rollouts have been discussed earlier under different rollouts. The functions of the **Components** rollout is discussed next.

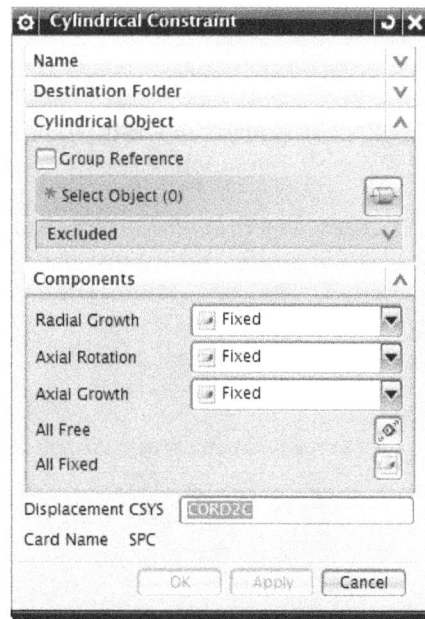

Figure 9-26 The Cylindrical Constraint dialog box

Components

Using the options of this rollout, you can apply constraint to cylindrical face along R, theta, or Z axis. Under this rollout, three drop-down lists are available, **Radial Growth**, **Axial Rotation**, and **Axial Growth**. In each of these drop-down lists, the **Free**, **Fixed**, and **Displacement** options are available. Using these options, you can make the axis free or fixed or assign a displacement value to each axis individually. You can also make the entire cylindrical axis free or fixed using the **All Free** or **All Fixed** button, respectively.

After specifying all the values for the part of the model, the **OK** and **Apply** buttons in the dialog box will become active. Choose the **OK** button to apply the cylindrical constraint and exit the dialog box.

Slider Constraint

Ribbon: Home > Loads and Conditions > Constraint Type > Slider Constraint

The **Slider Constraint** tool is used to restrict all the DOFs except the displacement along X axis. You can apply this constraint on a planar face of the model. To apply the constraint, choose the **Slider Constraint** tool from the **Constraint Type** drop-down list of the **Loads and Conditions** group of the **Home** tab; the **Slider Constraint** dialog box will be displayed, as shown in Figure 9-27. In the dialog box, four rollouts are available: **Name**, **Destination Folder**, **Planar Object**, and **Sliding Direction**. The options of the **Name** and **Destination Folder** rollouts have already been discussed under different rollouts. Using the **Planar Object** rollout, you can select the planar face of the model. The options in this rollout work similar to the options in the **Model Objects** rollout and have been discussed earlier. Using the options of the **Sliding Direction** rollout, you can define the X axis direction along which you want to slide the model.

After specifying all the values for the part of the model, the **OK** and **Apply** buttons in the dialog box will become active. Choose the **OK** button to apply the slider constraint and exit the dialog box.

Roller Constraint

Ribbon: Home > Loads and Conditions > Constraint Type > Roller Constraint

The **Roller Constraint** tool is used to restrict two directional DOFs for displacement as well as rotation. It is used for a planar face which can be translated and rotated along a selected direction. To apply the roller constraint, choose the **Roller Constraint** tool from the **Constraint Type** drop-down list of the **Loads and Conditions** group of the **Home** tab; the **Roller Constraint** dialog box will be displayed, as shown in Figure 9-28. In the dialog box, four rollouts are available, **Name**, **Destination Folder**, **Planar Object**, and **Rolling Direction**. The options in the **Name**, **Destination Folder**, and **Planar Object** rollouts have the same functions as discussed earlier. Using the options of the **Rolling Direction** rollout, you can select an axial direction along which you want the model to translate and rotate.

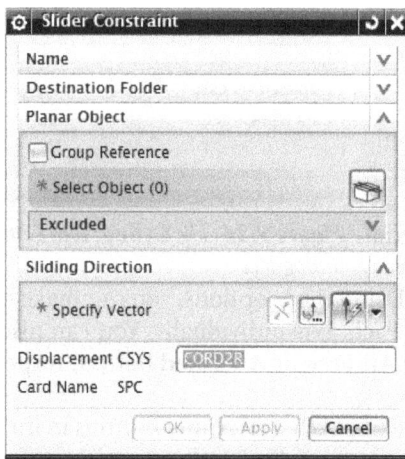

Figure 9-27 The Slider Constraint dialog box

Figure 9-28 The Roller Constraint dialog box

After specifying all the values for the part of the model, the **OK** and **Apply** buttons in the dialog box will become active. Choose the **OK** button to apply the roller constraint and exit the dialog box.

Symmetric Constraint

Ribbon: Home > Loads and Conditions > Constraint Type > Symmetric Constraint

Using the **Symmetric Constraint** tool, you can apply symmetric loading conditions to a model. When a model is symmetric about an axis or a plane and the boundary conditions are also symmetric, then you can split the model and apply the boundary conditions to one half. To use this tool, choose the **Symmetric Constraint** tool from the **Constraint Type** drop-down list of the **Loads and Conditions** group of the **Home** tab; the **Symmetric Constraint** dialog box will be displayed, as shown in Figure 9-29. In the dialog box, three rollouts are available, **Name, Destination Folder**, and **Planar Object**. The options in the **Name** and **Destination Folder** rollouts have the same functions as discussed earlier. Using options of the **Planar Object** rollout, you can select the planar face of the model. The options in this rollout work similar to those in the **Model Objects** rollout. You can use the following steps to apply the symmetric constraint.

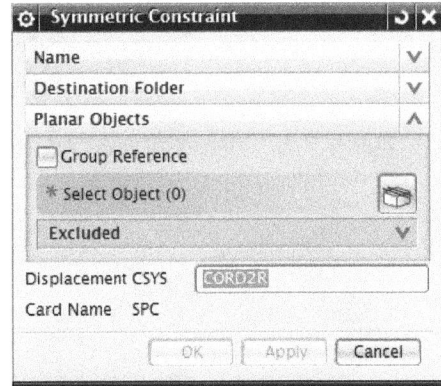

Figure 9-29 The Symmetric Constraint dialog box

1. Choose for the idealize file while creating the FEM file, as discussed in earlier chapters.
2. Split the part using the idealize file, along the symmetric axis by using the **Split** tool, as discussed in earlier chapters.
3. Hide the half of the split part using the **Simulation Navigator** from the **Polygon Geometry** node, as discussed in previous chapters.
4. Mesh the model and open the Simulation Environment.
5. Apply the symmetric constraint using the **Symmetric Constraint** tool on the splitted face of the part.
6. After specifying all the values for the part of the model, the **OK** and **Apply** buttons in the dialog box will become active. Choose the **OK** button to apply the symmetric constraint and exit the dialog box.

Figure 9-30 shows a model for defining symmetric constraint. Figure 9-31 shows the model after splitting and hiding the half of the portion. In Figure 9-32, remaining model is meshed and the two split faces are selected for defining symmetric constraint. Figure 9-33 shows the model after defining symmetric constraint.

Note that after applying the symmetric constraint, the translation along X and Y axes and the rotation along Z axis will become free. Similarly, the rotation along X and Y axes and the translation along Z axis will become fixed.

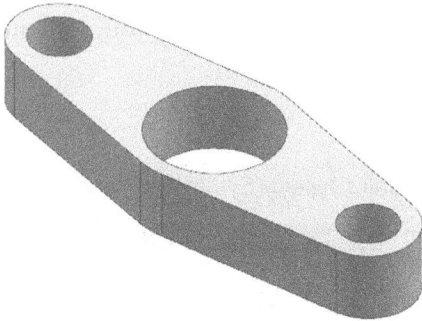

Figure 9-30 Symmetric model for defining constraint

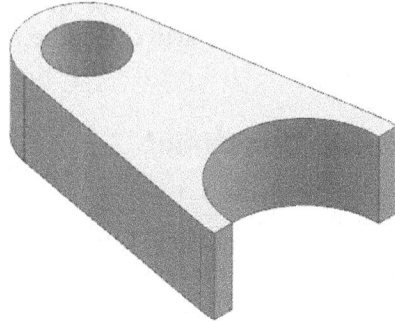

Figure 9-31   Model after splitting and hiding half of the portion

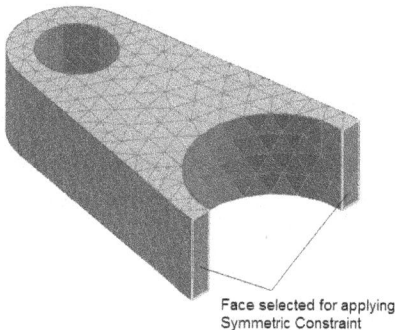

Face selected for applying
Symmetric Constraint

Figure 9-32   Face selected of the meshed model for defining symmetric constraint

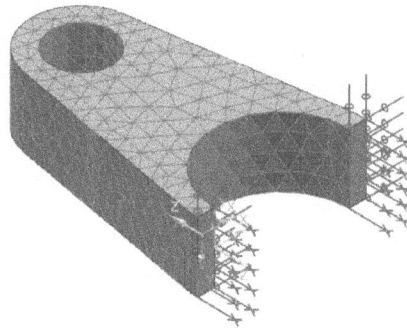

Figure 9-33   Model after defining the symmetric constraint

Anti-Symmetric Constraint

Ribbon: Home > Loads and Conditions > Constraint Type > Anti-Symmetric Constraint

The **Anti-Symmetric Constraint** tool works similar to the **Symmetric Constraint** tool but it constrains the symmetric coplanar face opposite to the **Symmetric Constraint** tool. Using the anti-symmetric constraint, the rotation along X and Y axes and translation along Z axis will become free. Similarly, the translation along X and Y axes and the rotation along Z axis will become fixed.

To use this tool, choose the **Anti-Symmetric Constraint** tool from the **Constraint Type** drop-down list of the **Loads and Conditions** group of the **Home** tab; the **Anti-Symmetric constraint** dialog box will be displayed, as shown in Figure 9-34. In the dialog box, three rollouts are available, **Name**, **Destination Folder**, and **Planar Object**. The working of the options of the **Name** and **Destination Folder** rollout are similar, as discussed earlier. Using the **Planar Object** rollout, you can select the planar face of the model. This rollout works similar to the **Model Objects** rollout as discussed earlier. You can use following steps to apply the anti-symmetric constraint.

1. Choose the idealize file while creating the FEM file as discussed in earlier chapters.
2. Split the part using the idealize file along the symmetric axis by using the **Split** tool as discussed in previous chapters.

3. Hide the half of the splitted part using the **Simulation Navigator** from the **Polygon Geometry** node, as discussed in previous chapters.
4. Mesh the model and open the Simulation Environment.
5. Apply the anti-symmetric constraint using the **Anti-Symmetric Constraint** tool, on the splitted face of the part.
6. After specifying all the values for the part of the model, the **OK** and **Apply** buttons in the dialog box will become active. Choose the **OK** button to apply the anti-symmetric constraint and exit the dialog box.

Figure 9-35 shows the model used for symmetric constraint after defining anti-symmetric constraint.

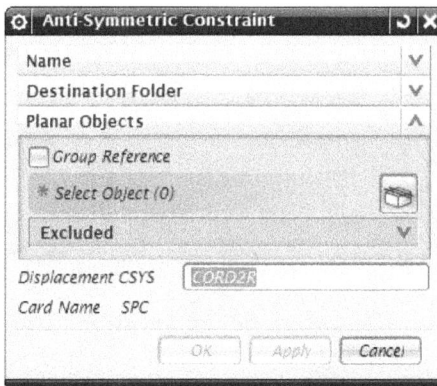

Figure 9-34 The **Anti-Symmetric Constraint** dialog box

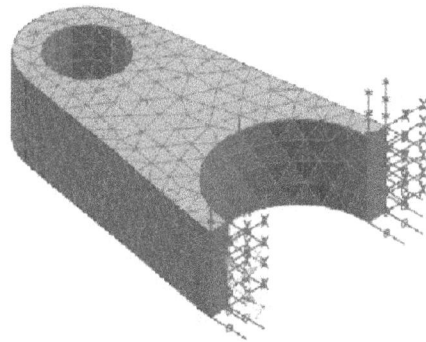

Figure 9-35 Model after defining anti-symmetric constraint

Loads

Loads are defined as external forces acting on a body. They include force, pressure, moment, and so on. The tools that are used for applying different loading conditions are discussed next.

Acceleration

Ribbon: Home > Loads and Conditions > Load Type > Acceleration

Using the **Acceleration** tool, you can apply the acceleration on the selected geometry. The acceleration load can be applied on a part that is fixed on a side and moves rapidly along a direction. To use this tool, choose the **Acceleration** tool from the **Load Type** drop-down list of the **Loads and Conditions** group of the **Home** tab; the **Acceleration** dialog box will be displayed, as shown in Figure 9-36. To apply acceleration on the model, six options are available in the drop-down list of the **Type** rollout: **Whole Model-Components**, **Whole Model-Magnitude and Direction**, **Components**, **Magnitude and Direction**, **Normal** and **Node ID Table**. Applying acceleration using these options is discussed next.

Figure 9-36 *The **Acceleration** dialog box*

Whole Model-Components

Using this option, you can apply the acceleration on the entire model by specifying the components of acceleration along the individual coordinate system. When you choose this option, apart from the **Type** rollout, six other rollouts will be displayed in the dialog namely, **Name**, **Destination Folder**, **Model Object**, **Direction**, **Components**, and **Scaling**. As discussed earlier, you can specify name using the **Name** rollout. The **Destination Folder** rollout is used to define the load location in the **Simulation Navigator**. By using the options of the **Model Object** rollout, you can select the part of the model on which the force needs to be applied. Other rollouts of the dialog box are discussed next.

Direction: In this rollout, the **CSYS** drop-down list is available which can be used to select the coordinate system to apply the acceleration in the specified direction. You can select the **Global**, **Cartesian**, **Cylindrical**, or **Spherical** coordinate system from the drop-down list or you can also create the following coordinate systems: **Cartesian**, **Cylindrical**, and **Spherical**.

Components: Using the options of this rollout, you can specify the magnitude of acceleration along the individual axis of coordinate system. You can use the **Expression** or **Field** option to define acceleration.

Scaling: In this rollout, the **Direction** drop-down list is used to define the variation in the acceleration in the specified direction. Using the **Scale Factor** area, you can define a scale factor in terms of unitless field.

Whole Model-Magnitude and Direction

Using this option, you can apply the acceleration on the entire model by specifying the magnitude and direction of acceleration along individual coordinate system. When you choose this option, apart from the **Type** rollout, six other rollouts will be displayed in the dialog box, namely **Name**, **Destination Folder**, **Model Object**, **Magnitude**, **Direction**, and **Scaling**. Using the options of the **Acceleration** drop-down list in the **Magnitude** rollout, you can apply the magnitude for the selected part of the model. There are two options available in the **Acceleration** drop-down list, **Expression** and **Field**. The options of the **Direction** rollout are used to define the magnitude direction. Other rollout options of the dialog box have been discussed earlier.

Components

The **Components** option is similar to the **Whole Model-Components** option. In this option, you have to select individual geometry on which you want to define the acceleration load. When you choose this option, apart from the **Type** rollout, six rollouts will be displayed in the dialog box. These rollouts are **Name**, **Destination Folder**, **Model Objects**, **Direction**, **Components**, and **Distribution**. You have to select the part of the model using the options of the **Model Objects** rollout on which the acceleration needs to be applied. Other rollout options of the dialog box have been discussed earlier.

Magnitude and Direction

The **Magnitude and Direction** option is similar to the **Whole Model-Magnitude and Direction** option. In this option, you have to select individual geometry on which you want to define acceleration load. When you choose this option, six rollouts will be displayed in the dialog box apart from the **Type** rollout. These rollouts are **Name**, **Destination Folder**, **Model Objects**, **Magnitude**, **Direction**, and **Distribution**. You have to select the part of the model on which the acceleration needs to be applied using the options in the **Model Objects** rollout. Other rollout options of the dialog box work the same way as discussed earlier.

Normal

Using this option, you can apply the acceleration with magnitude along the normal direction of the selected face. You can specify negative or positive direction of the normal direction. When you choose this option, apart from the **Type** rollout, five rollouts will be displayed in the dialog box **Name**, **Destination Folder**, **Model Objects**, **Magnitude**, and **Distribution**. All these options work same as discussed earlier.

Node ID Table

Using this option, you can apply the acceleration value at each node. When you choose this option, the **Type**, **Name**, **Destination Folder**, **Nodes**, **Magnitude**, **Direction**, and **Scaling ID Table**. All these options are explained earlier.

After specifying all the values for the part of the model, the **OK** and **Apply** buttons in the dialog box will become active. Choose the **OK** button to specify the acceleration load and exit the dialog box. Figure 9-37 shows a model for applying the acceleration load on the top face of the model using the **Magnitude and Direction** option. Figure 9-38 shows the model after applying the acceleration load in downward direction.

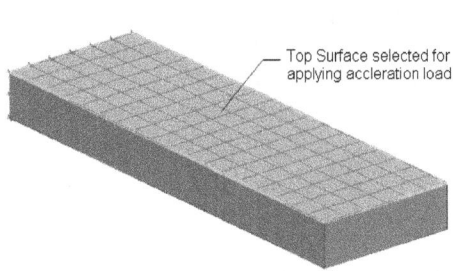

Figure 9-37 Face selected of the meshed model for applying acceleration load

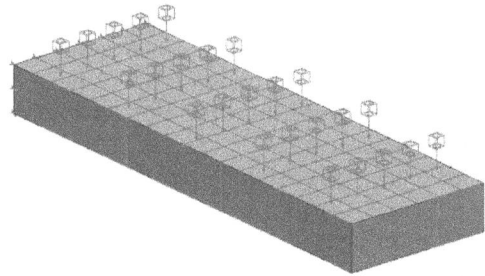

Figure 9-38 Model after applying acceleration load

Force

Ribbon: Home > Loads and Conditions > Load Type > Force

Using the **Force** tool, you can apply the magnitude and direction of force on the model. To apply the force on the model, choose the **Force** tool from the **Load Type** drop-down list of the **Loads and Conditions** group of the **Home** tab; the **Force** dialog box will be displayed, as shown in Figure 9-39. To apply the force on the model, five options, **Magnitude and Direction**, **Normal**, **Components**, **Node ID Table**, and **Edge-Face** are displayed in the **Type** rollout. These options are discussed next.

Figure 9-39 The **Force** dialog box

Magnitude and Direction

Using this option, you can apply the force with magnitude along a direction. When you choose this option, apart from the **Type** rollout, six rollouts will be displayed in the dialog box, namely, **Name**, **Destination Folder**, **Model Objects**, **Magnitude**, **Direction**, and **Distribution**. As discussed earlier, you can specify the name using the **Name** rollout. The option of the **Destination Folder** rollout is used to define the load location in the **Simulation Navigator**. You can select the part of the model using the option of the **Model Objects** rollout on which the force needs to be applied. Other options of this rollout are discussed next.

Note
*Working of the **Destination Folder** rollout option here is similar to constraint types options. The created node will be added in the **Load Container** in the **Simulation Navigator** rather than in the **Constraint Container**.*

Magnitude: You can apply the magnitude for the selected part of the model by using the two options available in the **Force** drop-down list, **Expression** and **Field**. On choosing the **Expression** option, an edit box will be displayed in which you can specify the constant value of force. On choosing the **Field** option, the **Specify Field** drop-down list will be displayed. You can use the options of this drop-down list for specifying the frequency, temperature or time dependent force magnitude. You will learn further in this chapter to use these field options.

Direction: In this rollout, three options are available in the **Method** drop-down list for defining the force direction, **Along Vector (FORCE)**, **Along 2 Nodes (FORCE1)**, and **Normal to 4 Nodes (FORCE2)**. Using the **Along Vector (FORCE)** option, you can specify the direction using triad or from the available vector defining options. Using the **Along 2 Nodes (FORCE1)** option, you can specify the direction using available nodes in the model. When you select this option from drop-down list, rollout options will be changed for nodes selection. Using the **Normal to 4 Nodes (FORCE2)** option, you can specify the direction using the four nodes. The derived direction vector would be parallel to crossed vector direction from Direction Node 1 to Direction Node 2, and from Direction Node 3 to Direction Node 4.

Distribution: In this rollout, three options are available in the **Method** drop-down list for defining the distribution of force load: **Total Per Object**, **Geometric distribution**, and **Spatial**. The **Total Per Object** option is used to apply total load on the selected geometry or face. The **Geometric distribution** option is used to apply the load on the selected faces in a distributed manner. For example, if you have selected two different surfaces then the load will be distributed in that area ratio. The **Spatial** option is used to apply the mapped load on the geometry.

Note
*You will learn to create mapped load using the **Spatial** option later in this chapter.*

Normal
Using this option, you can apply the force with magnitude along the normal direction of the selected face. You can specify negative or positive direction of the normal direction. When you choose this option, five rollouts will be displayed in the dialog box, besides the **Type** rollout, **Name**, **Destination Folder**, **Model Objects**, **Magnitude**, and **Distribution**. The functions of all these options are same as discussed earlier.

Components
Using this option, you can apply the force with magnitude along the different X, Y, and Z axes. When you choose this option, apart from the **Type** rollout, the **Name**, **Destination Folder**, **Model Objects**, **Direction**, **Components**, and **Distribution** rollouts will be displayed. You can define the criteria for component force using the **Direction** and **Components** rollouts. These rollouts are discussed next.

Direction: In this rollout, the **CSYS** drop-down list is available. You can select the type of coordinate system from this drop-down list to specify the magnitude of force.

Components: In this rollout, the **Magnitude** drop-down list is available using which you can specify force individually along each component of the coordinate system. Two options are available in the **Magnitude** drop-down list, **Expression** or **Field**. The coordinate component in this rollout will be changed according to the option selected from the **CSYS** drop-down list of the **Direction** rollout.

Node ID Table
Using this option, you can apply the force with magnitude along the X, Y, and Z axes. On choosing this option the **Type**, **Name**, **Destination Folder**, **Nodes**, **Magnitude**, **Direction**, and **Scaling ID Table** rollouts will be displayed. Usage of these options have been discussed earlier.

Edge-Face
This option is used to apply the force on an edge by using a face for orientation. The face defines the in/out direction of plane. You have to select an edge and adjacent face of the model along which you want to apply the force. When you choose this option the **Type**, **Name**, **Destination Folder**, **Model Objects**, **Associated Face**, **Components**, and **Distribution** rollouts will be displayed. Using the options of the **Model Objects** rollout, you can select the edges along which you want to apply force on the face. The option of the **Associated Face** rollout is used to select the face that will be affected due to force. Note that only the face that is attached to the selected edges can be selected. Working of the **Components** rollout is discussed next.

Components: In this rollout, the **Shear Force**, **In Plane Force**, and **Out of Plane Force** areas will be displayed. In each area, a drop-down list is available having the **Expression** or **Field** options. You can specify the force magnitude individually using these options. The **Shear Force** area is used to define the magnitude of force at a tangent to the selected edge, which will perform as a shear force. The **In Plane Force** area is used to define the force that will be normal to selected edge and is tangent to the selected face. The **Out of Plane Force** area is used to define the force that will be normal to the selected face.

After specifying all the values for the part of the model, the **OK** and **Apply** button in the dialog box will become active. Choose the **OK** button to specify the force load and exit the dialog box. Figure 9-40 shows a model for applying force load on the top face of the model using the **Magnitude and Direction** option. Figure 9-41 shows the model after applying force load in downward direction.

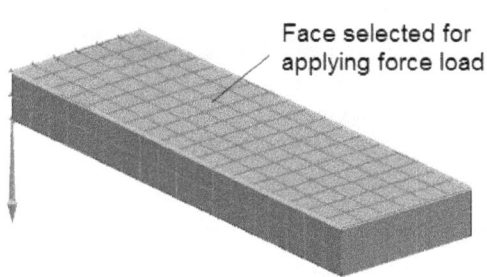

Figure 9-40 *Face selected of the meshed model for applying force load*

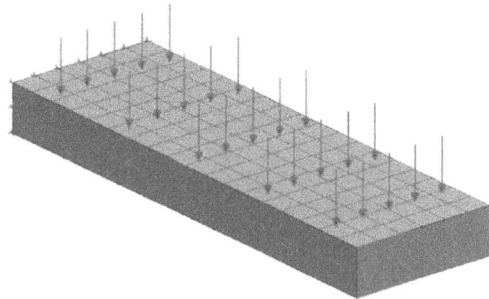

Figure 9-41 *Model after applying force load*

Moment

Ribbon: Home > Loads and Conditions > Load Type > Moment

Using the **Moment** tool, you can apply a magnitude of moment on the required part of the model. The moment load is applied as a rotational force on the selected part. To apply the moment on the model, choose the **Moment** tool from the **Load Type** drop-down list of the **Loads and Conditions** group of the **Home** tab; the **Moment** dialog box will be displayed, as shown in Figure 9-42. To apply moment on the model, there are five options available in the drop-down list of the **Type** rollout. They are **Magnitude and direction**, **Normal**, **Components**, **Node ID Table**, and **Edge-Face**. The method to apply the moment on a particular part of the body is the same as applying force.

Bearing

Ribbon: Home > Loads and Conditions > Load Type > Bearing

Using the **Bearing** tool, you can apply a magnitude of bearing load on cylindrical or circular part of the model. The bearing load is exerted where a cylindrical object is applied force on a cylindrical or circular part of the model. To apply the bearing load on the model, choose the **Bearing** tool from the **Load Type** drop-down list of the **Loads and Conditions** group of the **Home** tab; the **Bearing** dialog box will be displayed, as shown in Figure 9-43. In the dialog box, the **Name, Destination Folder, Cylindrical or Circular Object, Direction, Properties**, and **Distribution** rollouts are available. The functions of the **Name** and **Destination Folder** rollouts are same as discussed earlier. Working of other rollouts to apply bearing load is discussed next.

Cylindrical or Circular Object

Using the options of this rollout, you can select the cylindrical face or circular edge of the part on which the bearing load needs to be applied.

Direction

Using the options of this rollout, you can specify the direction for the bearing load. Note that the direction of the bearing load should be perpendicular to the selected cylindrical face or circular edge.

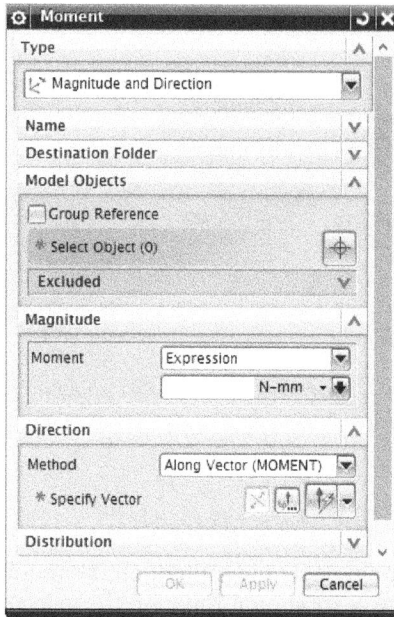

Figure 9-42 The **Moment** *dialog box*

Figure 9-43 The **Bearing** *dialog box*

Properties

Using the options of the **Properties** rollout, you can apply the force along a specific angle. In this rollout, two areas are available, **Force** and **Angle**. In the **Force** area, you can apply the magnitude of force using the **Expression** or **Field** options. The **Angle** area is used to specify the angular region of the bearing load. You can specify 0 degree to 360 degrees angle value. The intensity of the bearing load would be higher at the center of the angle and lower at the end of the angle range.

Distribution

You can distribute the applied bearing load on the selected geometry using two options available in the **Method** drop-down list of this rollout, **Parabolic** and **Sinusoidal**. These options become more specific in terms of mathematical equations.

After specifying all the values for the part of the model, the **OK** and **Apply** buttons in the dialog box will become active. Choose the **OK** button to specify the bearing load and exit the dialog box. Figure 9-44 shows a model for applying bearing load on the cylindrical face. Figure 9-45 shows model after applying the bearing load.

Note

If the specified direction is not perpendicular to the selected cylindrical face or circular edge then you will get an error message box, as shown in Figure 9-46.

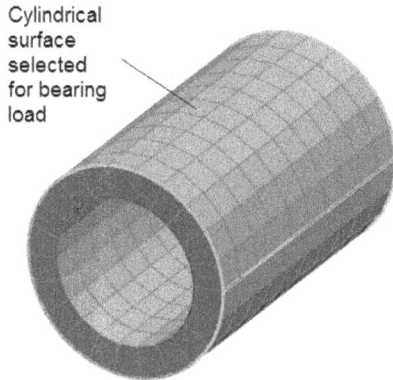

Figure 9-44 *Face selected of the meshed model for applying bearing load*

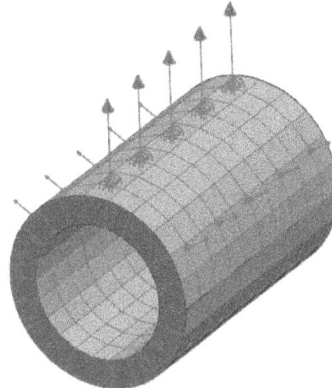

Figure 9-45 *Model after applying bearing load*

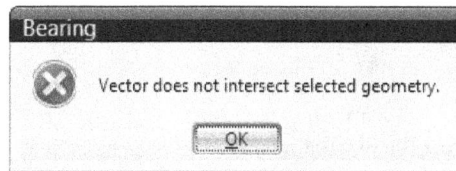

Figure 9-46 *The message box for vector direction error*

Torque

Ribbon: Home > Loads and Conditions > Load Type > Torque

Using the **Torque** tool, you can specify the magnitude of torque load for cylindrical or circular part of the model. Torque load follows the right hand rule for the selected object. To apply the torque load on the model, choose the **Torque** tool from the **Load Type** drop-down list of the **Loads and Conditions** group of the **Home** tab; the **Torque** dialog box will be displayed, as shown in Figure 9-47. You can apply torque on a particular part of the model using the following steps:

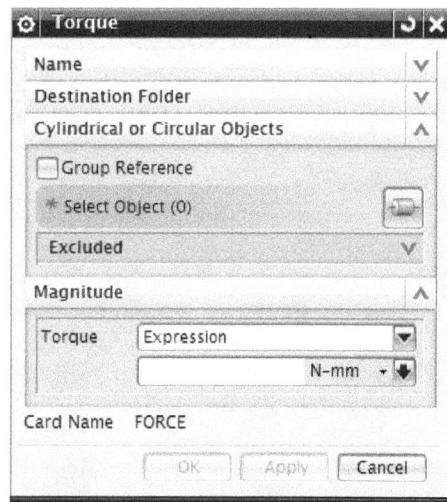

Figure 9-47 *The Torque dialog box*

1. Specify the name and destination of the torque load using the respective rollouts, as discussed earlier.
2. Select the cylindrical faces or circular edges of the model using the **Cylindrical or Circular Objects** rollout.
3. Specify the magnitude of torque in the **Magnitude** rollout. You can use negative or positive value for reversing the direction of load.
4. Choose the **Apply** button after specifying all the data.

Figure 9-48 shows a model for applying torque load on one side of the face. Figure 9-49 shows model after applying torque load.

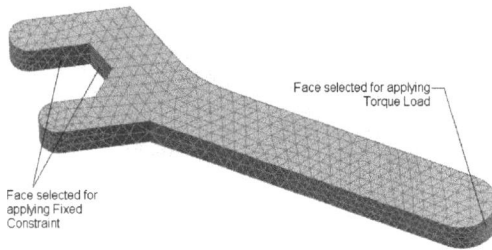

Figure 9-48   Face selected of the meshed model for applying torque load

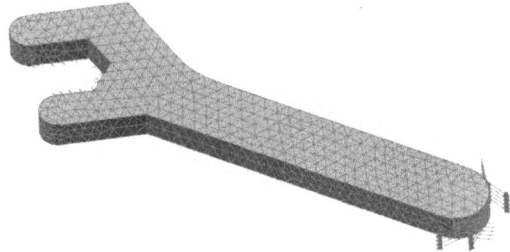

Figure 9-49   Model after applying torque load

Pressure

Ribbon:	Home > Loads and Conditions > Load Type > Pressure

Using the **Pressure** tool, you can apply pressure load on a flat or cylindrical surface of the model as force per unit area. To apply the pressure load on the model, choose the **Pressure** tool from the **Load Type** drop-down list of the **Loads and Conditions** group of the **Home** tab; the **Pressure** dialog box will be displayed, as shown in Figure 9-50. To apply pressure on the model, eight options are available in the drop-down list of the **Type** rollout, **Normal Pressure on 2D Elements or 3D Element Faces**, **Normal Pressure on 2D Elements**, **Components**, **On Beams**, **Normal Pressure on 2D Elements or 3D Element Faces – Spatial**, **Normal Pressure on 2D Elements – Spatial**, **Components – Spatial**, and **On Beams – Spatial**. These options are discussed next.

*Figure 9-50   The **Pressure** dialog box*

Normal pressure on 2D element or 3D element faces

Using this option, you can define pressure load normal to 2D or 3D element faces. Select the face using the options in the **Model Objects** rollout and specify the pressure value in the **Magnitude** rollout. Other rollouts work same as discussed earlier. Note that you can define the pressure direction using the positive or negative values for the magnitude.

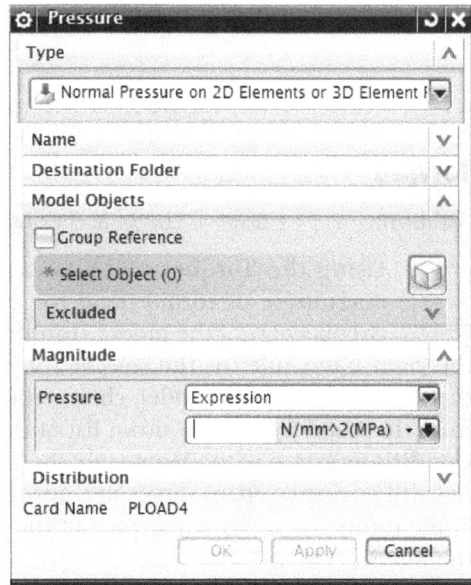

Normal pressure on 2D elements

Using this option, you can define pressure load normal to 2D element faces. Note that you can not apply pressure on a model having 3D elements using this option. Other rollouts work similar as discussed earlier.

Components

Using this option, you can define pressure load along a global or local coordinate system. After selecting the face for pressure load, select or create a coordinate system from the options in the **Direction** rollout. By default, the **Global** coordinate system is selected. After defining the coordinate system, specify magnitude using the **Components** rollout. You can apply pressure of different magnitudes along different axes of coordinate system. Other rollouts work same as discussed earlier.

On beams

Using this option, you can define pressure load on a beam element along the default **Global** coordinate system. When you choose this option, the **Properties** rollout will become available in the dialog box. The options in this rollout define the scale factor of the selected beam element. The value 0 **X1** box and the value 1 in the **X2** box show the nature of the fixed beam element. You cannot edit the values in this rollout. Using the options in the **Components** rollout, you can apply pressure of different magnitudes along different axes of coordinate system. The other rollouts work similar as discussed earlier.

Note

1. *The working of the **Normal Pressure on 2D Elements or 3D Element Faces – Spatial**, **Normal Pressure on 2D Elements – Spatial**, **Components – Spatial**, and **On Beams – Spatial** options is same as those of **Normal Pressure on 2D Elements or 3D Element Faces**, **Normal Pressure on 2D Elements**, **Components**, and **On Beams** options. You will use the unitless field data using the Spatial options to apply magnitude.*

2. *You will learn to use the **Spatial** option later in this chapter.*

Figure 9-51 shows a model for applying pressure on the top face. Figure 9-52 shows model after applying pressure.

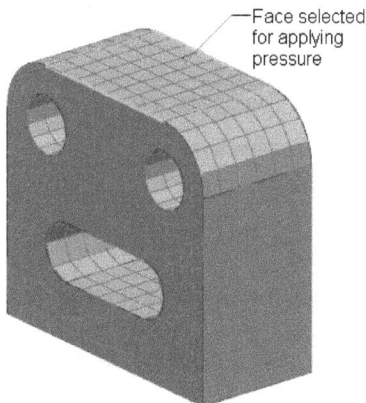

Figure 9-51 *Face selected of the meshed model for applying pressure*

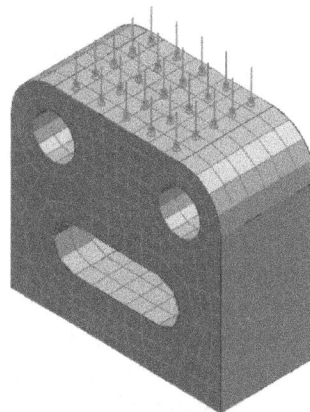

Figure 9-52 *Model after applying pressure*

Hydrostatic Pressure

Hydrostatic pressure is the pressure applied by a fluid at a certain depth within the fluid. It increases with the depth of the fluid. In NX Nastran, you can simulate this fluid property and pressure exerted at a certain depth. To apply hydrostatic pressure on a model surface, choose the **Hydrostatic Pressure** tool from the **Load Type** drop-down list of the **Loads and Conditions** group of the **Home** tab; the **Hydrostatic Pressure** dialog box will be displayed, as shown in Figure 9-53. Method to apply hydrostatic pressure on a surface of the model is discussed next.

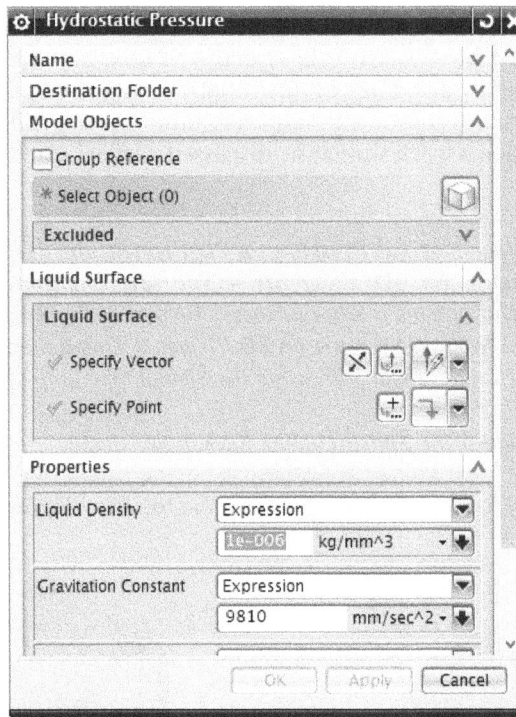

*Figure 9-53 The **Hydrostatic Pressure** dialog box*

1. Specify the name and destination of the load using the options of the **Name** and **Destination Folder** rollouts.
2. Select the face of the meshed and constraint model using the **Model Objects** rollout on which hydrostatic pressure need to the applied.
3. Specify the direction of the applied pressure using the options in the **Liquid Surface** rollout. By default -Z direction is applied.
4. You can specify the pressure and fluid property using the **Properties** rollout. The **Liquid Density** area is used to define the density of fluid. By default, the water density is specified. The **Gravitation Constant** area is used to define the gravity value. By default, the earth gravity is specified. The **Surface Pressure** area is used to define the surface pressure value. By default, atmospheric pressure on earth is specified.
5. After specifying all the options, choose the **Apply** button to apply the hydrostatic pressure.

Figure 9-54 shows a model for applying hydrostatic pressure on top of the face and in the vertically downward direction. Figure 9-55 shows model after applying hydrostatic pressure.

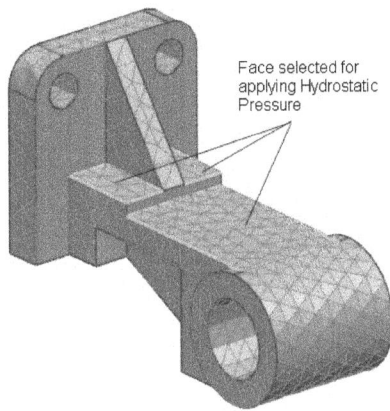

Figure 9-54 *Face to be selected for applying hydrostatic pressure*

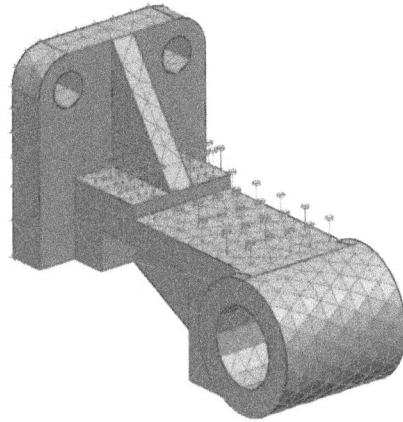

Figure 9-55 *Model after applying hydrostatic pressure*

Centrifugal Pressure

Ribbon: Home > Loads and Conditions > Load Type > Centrifugal Pressure

Centrifugal pressure is the radially applied pressure on the selected surface of the model. You can vary the pressure along the selected surface. To apply a centrifugal pressure on a model surface, choose the **Centrifugal Pressure** tool from the **Load Type** drop-down list of the **Loads and Conditions** group of the **Home** tab; the **Centrifugal Pressure** dialog box will be displayed, as shown in Figure 9-56. Method to apply the centrifugal pressure on the surface of the model is discussed next.

1. Specify the name and destination of the load using the options in the **Name** and **Destination Folder** rollouts.
2. Select the face of the meshed and constraint model using the options in the **Model Objects** rollout on which the centrifugal pressure needs to apply.
3. Specify the rotation axis for the pressure applied using the **Rotation Axis** rollout option. You can flip the rotation direction if needed. It defines the axis along which the centrifugal pressure will be applied.
4. Specify the pressure and fluid properties using the options in the **Centrifugal Loading Parameters** rollout. The **Liquid Density** area is used to define the density of the given fluid. The **Angular Velocity** area is used to define the angular velocity. The **Inlet Radius** area is used to define the radius value or the distance from the rotation axis that you define using the **Rotation Axis** rollout. The **Static Pressure** area is used to define the static pressure value.
5. After specifying all the options, choose the **Apply** button to apply the centrifugal pressure.

Figure 9-57 shows a model for applying the centrifugal pressure on top of the face and the rotation axis. Figure 9-58 shows model after applying centrifugal pressure.

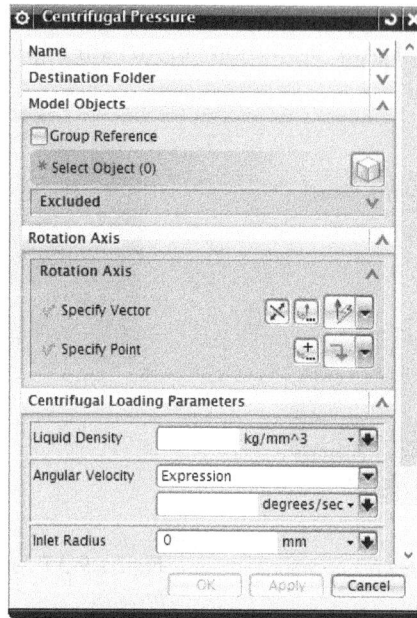

Figure 9-56 The **Centrifugal Pressure** *dialog box*

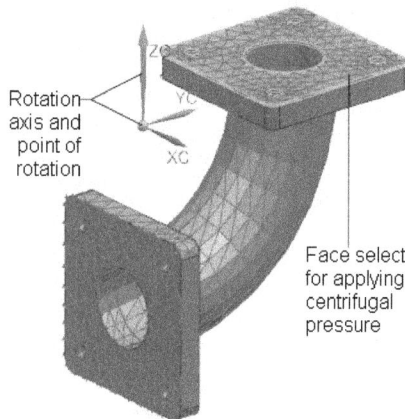

Figure 9-57 *Face of the meshed model selected for applying centrifugal pressure*

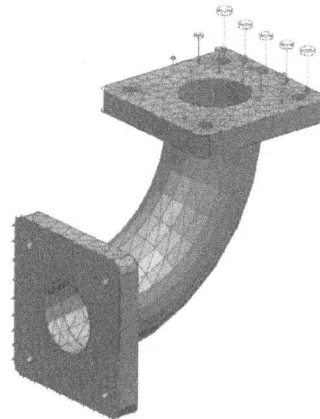

Figure 9-58 *Model after applying centrifugal pressure*

Gravity

Ribbon:	Home > Loads and Conditions > Load Type > Gravity

Gravity is termed as the force of attraction between two parts. For example, the Earth applies a gravitational force of unit 9.81m/s^2 on all bodies. In NX Nastran, you can apply the gravity on the whole model. To apply gravity on a model, choose the **Gravity** tool from the **Load Type** drop-down list of the **Loads and Conditions** group of the **Home** tab; the **Gravity** dialog box will be displayed, as shown in Figure 9-59. Two options are available in the drop-down list in the **Type** rollout, **Magnitude and Direction** and **Components**. Steps to apply the gravity load using these options are discussed next.

1. Specify the name and destination of the load using the options in the **Name** and **Destination Folder** rollout.
2. The gravity load is applied on the whole model so the option under the **Model Object** rollout is already selected.
3. If you select the **Magnitude and Direction** option from the drop-down list in the **Type** rollout; the **Magnitude** and **Direction** rollouts will be displayed in the dialog box. Specify the gravitational magnitude using the options in the **Magnitude** rollout and direction in which the gravitational force acts using the options of the **Direction** rollout.
4. If you select the **Components** option from the drop-down list of the **Type** rollout, the **Direction** and **Components** rollouts will be displayed in the dialog box. Using these options, you can define the gravity in different directions. Specify the coordinate type from the **Direction** rollout and gravitational magnitude along different axes using the options in the **Components** rollout.
5. After specifying all the options, choose the **Apply** button to apply gravity on the body.

Figure 9-60 shows the model after applying gravity load. As shown in figure, after applying gravity, a symbol is applied to show the direction of gravity.

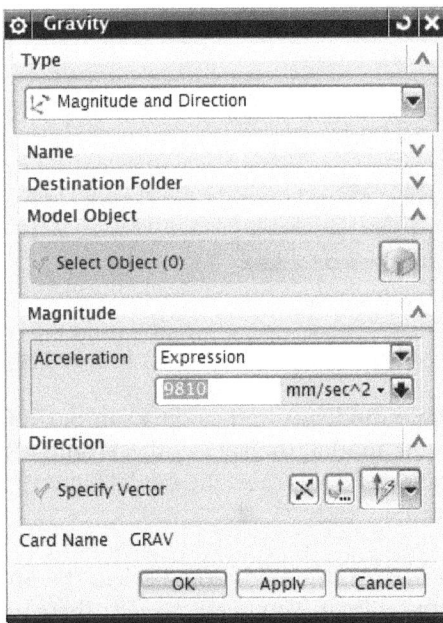

Figure 9-59 The **Gravity** dialog box

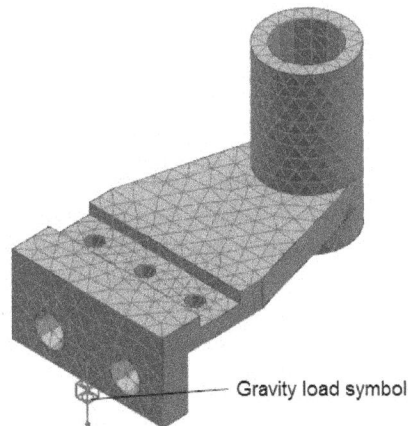

Figure 9-60 Model after applying gravity load

Rotation

Ribbon: Home > Loads and Conditions > Load Type > Rotation

The rotation load is applied on a model which is rotating along its center of axis. You can specify the angular velocity and acceleration using the **Rotation** tool. To apply rotation on a model, choose the **Rotation** tool from the **Load Type** drop-down list of the **Loads and Conditions** group of the **Home** tab; the **Rotation** dialog box will be displayed, as shown in Figure 9-61. Two options are available in the drop-down list in the **Type** rollout for rotation load, **Whole Model** and **Model Subset**. Using these options, you can apply the rotation load on the whole model or sub part of the model. Steps to apply the rotation load using these options are discussed next.

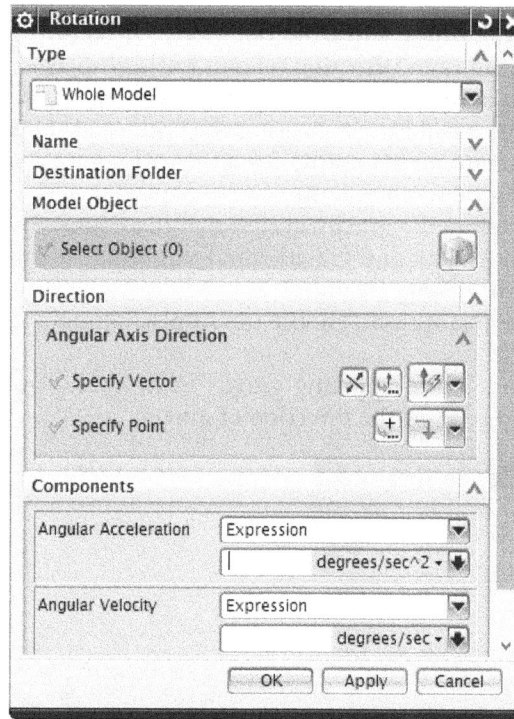

*Figure 9-61 The **Rotation** dialog box*

1. Specify the name and destination of the load using the options in the **Name** and **Destination Folder** rollout.
2. The rotation load can be applied on the whole model as well as on a sub part of the model. If you want to apply load on a complete model, then select the **Whole Model** option from the drop-down list in the **Type** rollout. In this case, the model object under the **Model Object** rollout will be already selected.

    If you want to apply rotation load on a sub part of the model then select the **Model Subset** option from the drop-down of the **Type** rollout. You need to select the edge or the face of the model that is to be affected by the rotation load.

3. Specify the direction vector and point of axis using the options available in the **Direction** rollout.
4. Specify the angular acceleration and angular velocity of the rotation load using the options available in the **Components** rollout.
5. After specifying all the options, choose the **Apply** button to apply rotation load on the body.

Bolt Pre-Load

In NX Nastran, you can specify the bolt pre-load of a modeled bolt in a model. In general, bolt pre-load is applied where bolt is extra screwed with nut. To apply this load, select the **Bolt Pre-Load** tool from the **Load Type** drop-down list in the **Loads and Conditions** group of the **Home** tab; the **Bolt Pre-Load** dialog box will be displayed, as shown in Figure 9-62. Two options are available in the drop-down list of the **Type** rollout for pre-bolt load, **Force on 1D Elements** and **Force on 3D Elements**. The **Force on 1D Elements** option is used to define the bolt pre-load on the bolts that are created using 1D elements like CBAR and CBEAM. While the **Force on 3D Elements** option is used to define load on bolts that are created using 3D elements like CTETRA or CPENTA elements. Steps to apply bolt pre-load using the options available in the dialog box are discussed next. Note that to apply bolt pre-load, you have to create a bolt connection in the model as discussed in the previous chapter.

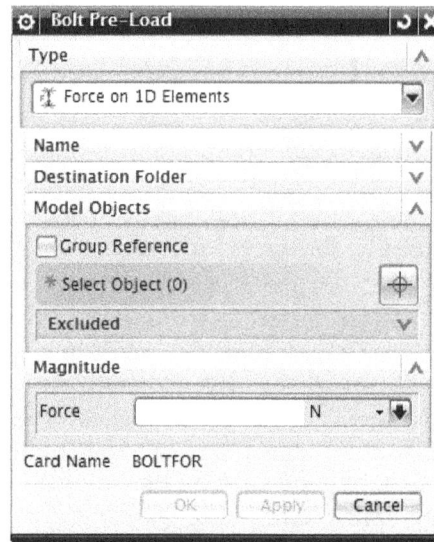

Figure 9-62 The Bolt Pre-Load dialog box

1. Select the required type from the **Type** rollout for bolt elements and specify the name and destination of the load using the options in the **Name** and **Destination Folder** rollout.
2. Select the created 1D or 3D bolt from the model using the **Model Objects** rollout.
3. Specify the bolt pre-load in the **Force** edit box available in the **Magnitude** rollout.
4. After specifying all the options, choose the **Apply** button to apply the bolt pre-load.

Figure 9-63 shows the model with 1D element bolt that is selected for pre-load. Figure 9-64 shows the bolt connection with applied bolt pre-load after hiding the meshed model.

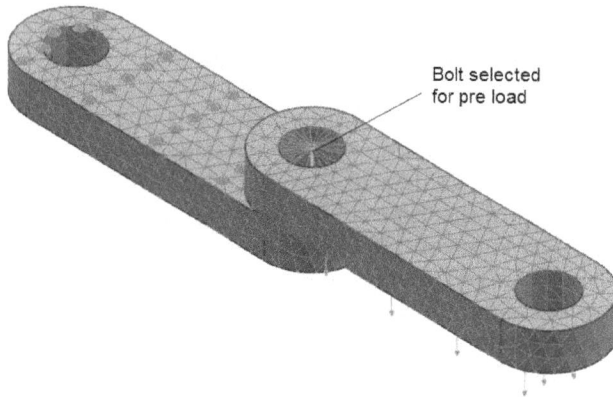

Figure 9-63   *Bolt connection selected for applying bolt pre-load*

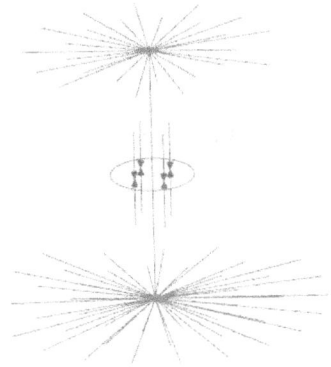

Figure 9-64   *Model after applying bolt pre-load*

FIELDS

In NX Nastran, you can use the fields options to specify a functional variable of a particular load, constraint, or their distribution on the selected geometry object. You can assign an equation variable or time or frequency variable value to the particular function. Next, you will learn to specify a functional value of pressure for the given object.

Applying Pressure using Spatial Field

Choose the **Pressure** tool from the **Load Type** drop-down list of the **Loads and Conditions** group of the **Home** tab; the **Pressure** dialog box will be displayed, refer to Figure 9-65. Select the **Normal Pressure on 2D Elements or 3D Element Faces – Spatial** option available in the drop-down list of the **Type** rollout; the dialog box will be modified, refer to Figure 9-65. You will apply the pressure using tangent expression on the cylindrical face of the model shown in Figure 9-66 using the following steps.

1. Specify the name and destination of the load using the options in the **Name** and **Destination Folder** rollout.
2. Choose the **Face to Create Normal Pressure On** button from the **Model Objects** rollout and select the top cylindrical face of the model, as shown in Figure 9-67.
3. Select the **Formula Constructor** option from the drop-down list available in the **Magnitude** rollout; the **Formula Field** dialog box will be displayed, as shown in Figure 9-68.
4. Select the **Angle** option from the **Independent** drop-down list available in the **Domain** rollout.
5. In the edit box available in the **Expressions** rollout, type **tan(** and then double click on **angle**, available below the **Filters** sub-rollout; the value in the edit box will be modified to **tan(ug_var("angle")**. Next, add the closing parenthesis to complete the expression, refer to Figure 9-69.
6. Now, choose the **Accept Edit** button available along the edit box to accept the expression; the expression will be assigned to pressure and the **OK** button will become active.
7. Choose the **OK** button from the **Formula Field** dialog box and the **Pressure** dialog box to apply the pressure on the selected face, as shown in Figure 9-70.

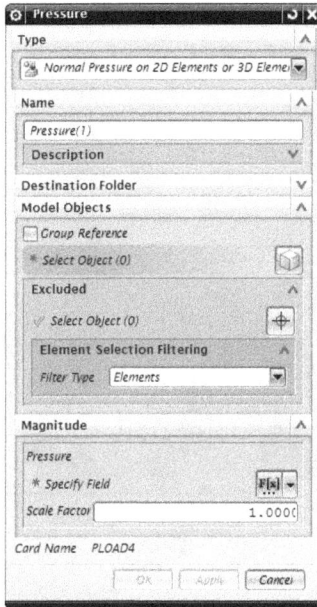

Figure 9-65 *The **Pressure** dialog box*

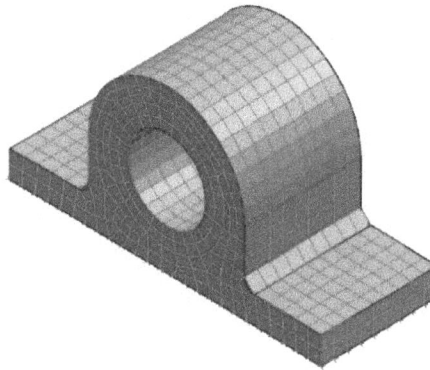

Figure 9-66 *Model for applying spatial pressure*

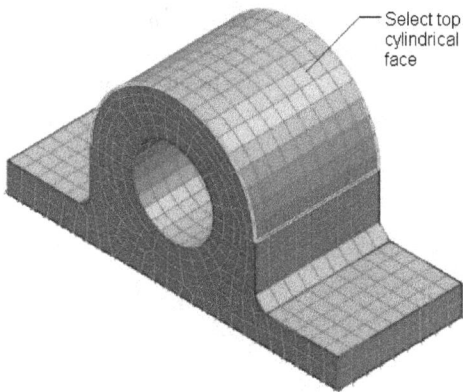

Figure 9-67 *Face of the meshed model selected for applying pressure using expression*

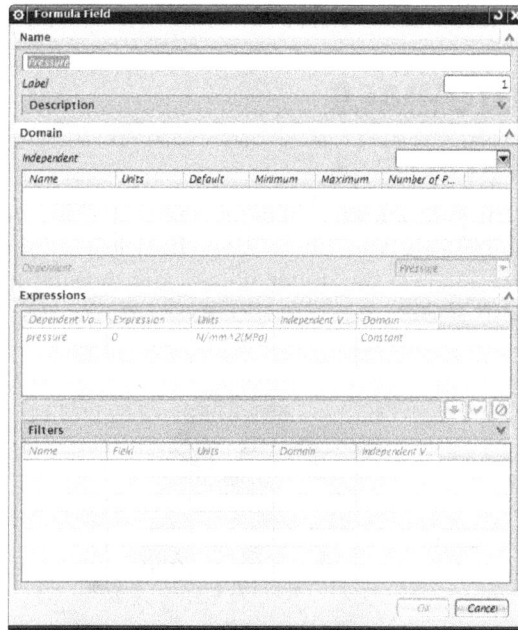

Figure 9-68 *The **Formula Field** dialog box*

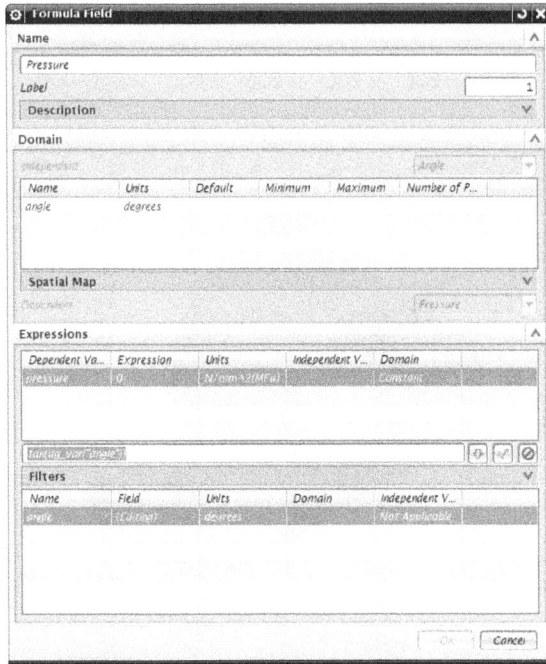

Figure 9-69 *The **Formula Field** dialog box with tangent expression*

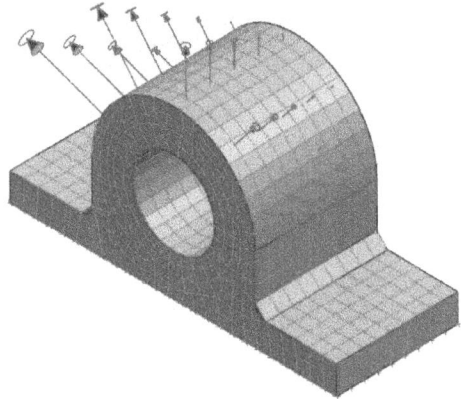

Figure 9-70 *Model after applying pressure using expression*

TUTORIALS

Tutorial 1

In this tutorial, you will define material using mesh collector in Exercise 1 of chapter 5, refer to Figure 9-71. Also, define edge to surface gluing contact, apply fixed contact on the hole, and force load on the upper hole of the meshed midsurface model. The final model is shown in Figure 9-72. **(Expected time: 45 min)**

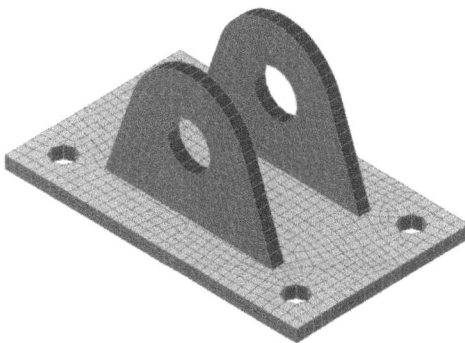

Figure 9-71 *Model for defining boundary conditions*

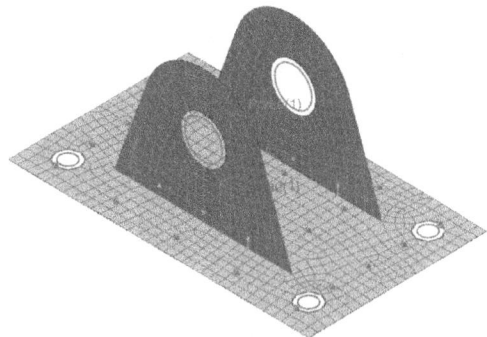

Figure 9-72 *Model after defining boundary conditions*

Note

While copying the meshed model, you also need to copy the part and idealized files from Exercise 1 of Chapter 4 in the same folder.

The following steps are required to complete this tutorial:

a. Open the FEM file into the NX and load the idealized file.
b. Hide the visibility of element thickness.
c. Apply the material using Mesh Collector.
d. Create a new Simulation file.
e. Apply edge to surface gluing contact.
f. Apply the fixed constraint on the base holes.
g. Apply the force load on the upper hole of the model.
h. Save and close the model.

Opening the Assembly File

1. Start NX 9.0 by double-clicking on the shortcut icon of NX 9.0 on the desktop of your computer.

2. Copy the **c05_exr01.fem** file from the location *C:\NX Nastran\c05\Exr01* and paste it to *C:\NX Nastran\c09\Tut01*. Also, you need to copy and paste the part and idealized files from *C:\NX Nastran\c04\Exr01*. You can also download these files from *www.cadcim.com*.

Note

You must create the Tut01 folder in c09 folder at the location C:\NX Nastran that was created in the previous chapter.

3. Choose **File > Open** from the **Ribbon**; the **Open** dialog box is displayed. Select the **c05_exr01.fem** file from the location and choose the **OK** button; the file is opened, as shown in Figure 9-73.

4. Choose the **Load** option from the shortcut menu displayed by right-clicking on the **c04_exr01_i.prt** node in the **Simulation Navigator**, as shown in Figure 9-74.

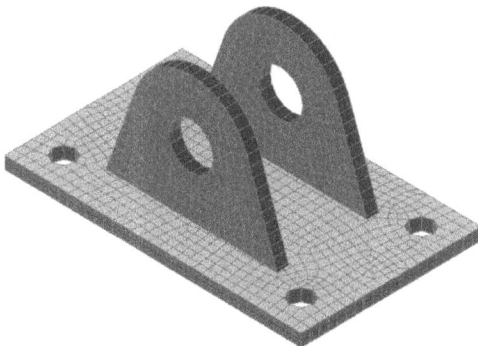

Figure 9-73 The model opened in FEM environment

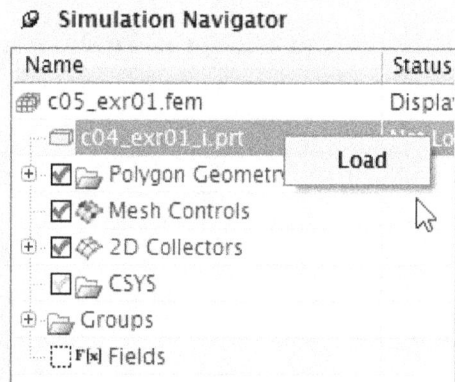

Figure 9-74 Options available on right-clicking the node

Hiding the Displayed Thickness of the Meshed Model

You need to hide the displayed thickness of the meshed model.

1. Choose **Menu > Preferences > Mesh Display** from the **Top Border Bar**; the **Mesh Display** dialog box will be displayed, as shown in Figure 9-75.

2. Clear the **Element Thickness and Offset** check box available in the **2D** tab of the **Display** rollout and choose the **OK** button from the dialog box; the meshed model without thickness is displayed, as shown in Figure 9-76.

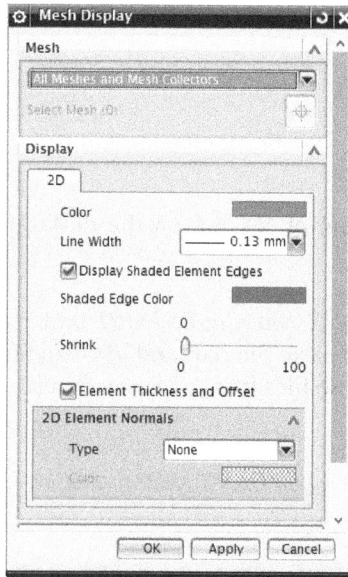

Figure 9-75 The **Mesh Display** *dialog box*

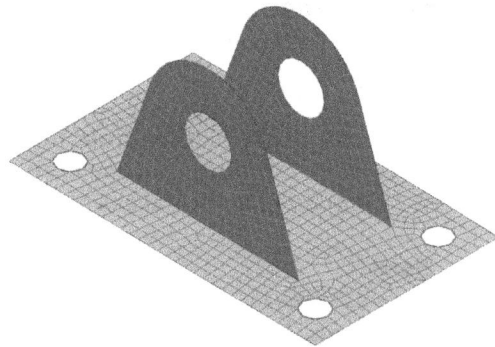

Figure 9-76 *The meshed midsurface model without thickness*

Note
You can also perform this tutorial with the displayed thickness but for better visibility of the model you need to hide the displayed thickness.

Applying the Material to the Model Using Mesh Collector

Next, you need to assign material to the model by using the Mesh Collector.

1. Expand the **2D Collectors** node from the **Simulation Navigator** and right-click on the **ThinShell(1)** node; a shortcut menu is displayed, refer to Figure 9-77.

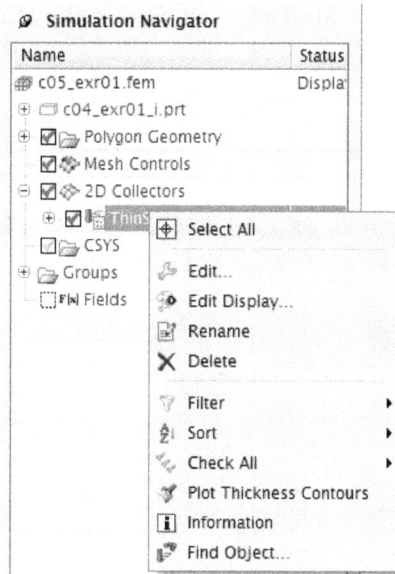

Figure 9-77 Shortcut menu

2. Choose the **Edit** option from the shortcut menu; the **Mesh Collector** dialog box is displayed, as shown in Figure 9-78.

3. Choose the **Edit** button available next to the **Shell Property** drop-down list of the **Physical Property** sub-rollout under the **Properties** rollout; the **PSHELL** dialog box is displayed, as shown in Figure 9-79.

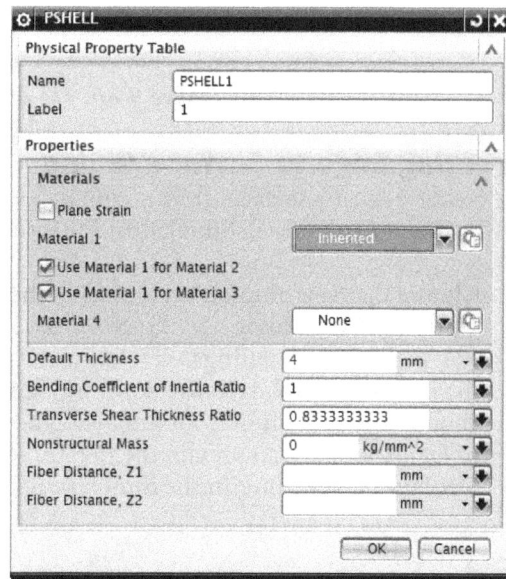

Figure 9-78 The **Mesh Collector** dialog box   *Figure 9-79* The **PSHELL** dialog box

4. Choose the **Choose Material** button available next to the **Material 1** drop-down list in the **Materials** sub-rollout of the **Properties** rollout; the **Material List** dialog box is displayed, refer to Figure 9-80.

5. Select the **AISI_Steel_4340** material from the **Material List** dialog box, refer to Figure 9-80, and choose the **OK** button from all the dialog boxes to accept the options and exit.

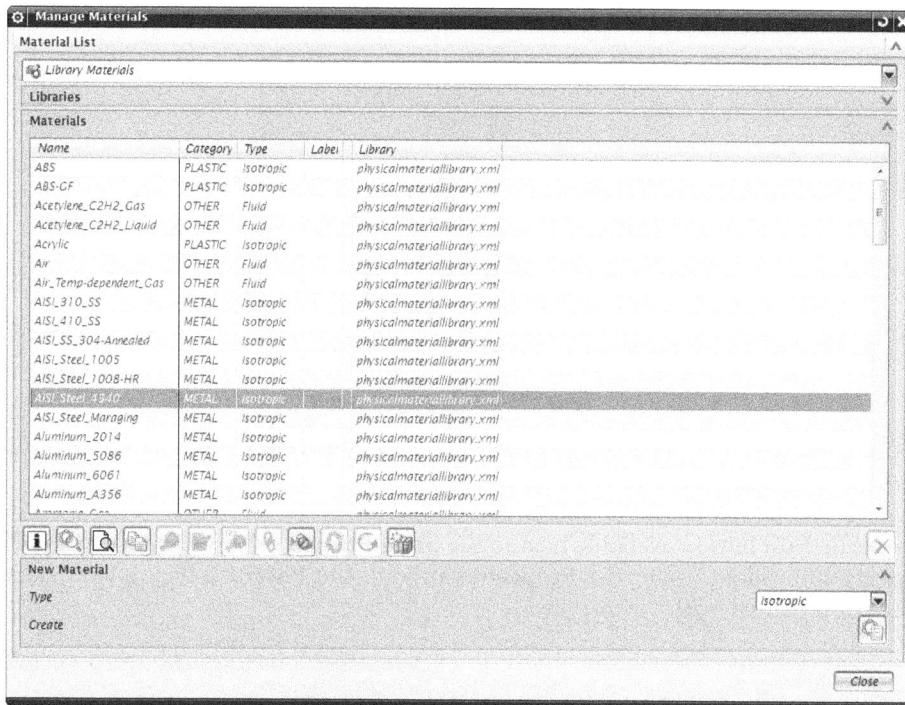

Figure 9-80 *The* **Material List** *dialog box*

Creating Edge to Surface Gluing

To create edge to surface gluing contact in the model, you need to be in the Simulation Environment. To enter in Simulation Environment, you need to create the Simulation file.

1. Choose the **New Simulation** tool from the **Change Window Drop-Down** list of the **Context** group from the **Home** tab; the **New Part File** window is displayed, as shown in Figure 9-81. Retain the default settings and choose the **OK** button from the window; the **New Simulation** dialog box is displayed, as shown in Figure 9-82. Do not change the default settings and choose the **OK** button from the dialog box; the **Solution** dialog box is displayed, as shown in Figure 9-83. Again, retain the default settings in the **Solution** dialog box and choose the **OK** button; you enter in the Simulation Environment.

Note
You will learn more about the Simulation Environment and its dialog boxes in the later chapters of this book.

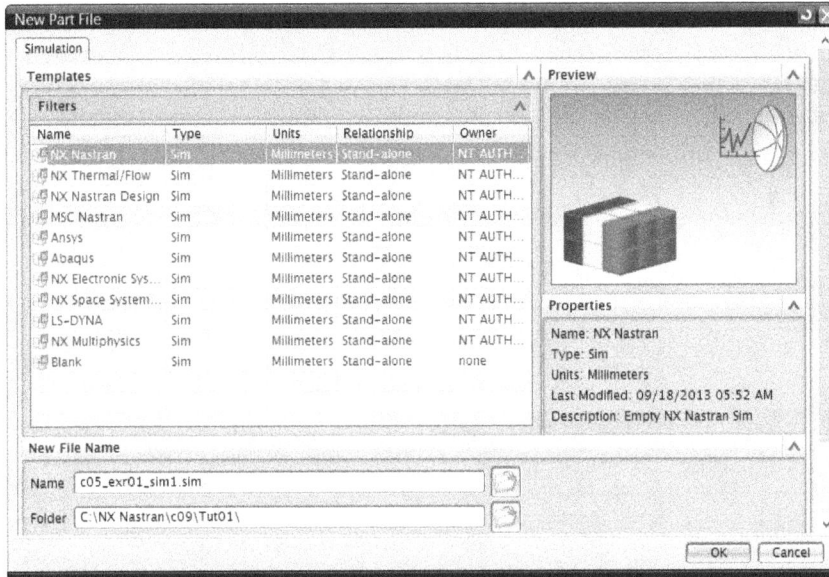

Figure 9-81 The **New Part File** *window*

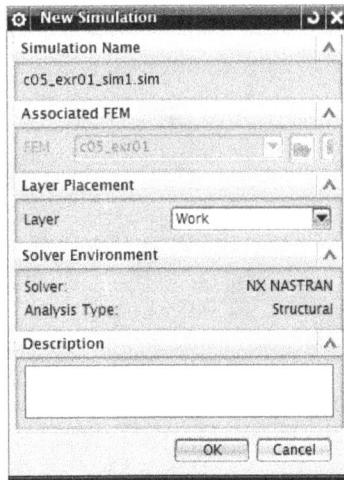

Figure 9-82 The **New Simulation** *dialog box*

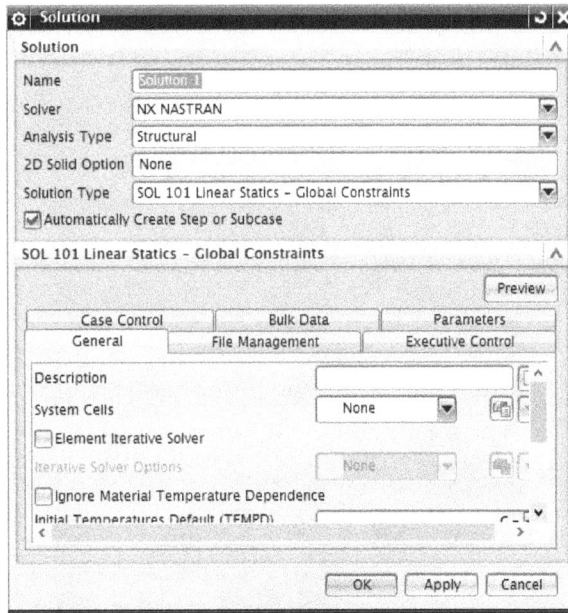

Figure 9-83 The **Solution** *dialog box*

Next, you will create edge to surface gluing in the model.

2. Choose the **Edge-to-Surface Gluing** tool from the **Simulation Object Type** drop-down list in the **Loads and Conditions** group of the **Home** tab; the **Edge-to-Surface Gluing** dialog box is displayed, as shown in Figure 9-84.

3. Choose the **Create Region** button available next to the **Edge Region** drop-down list in the **Source Region** rollout; the **Region** dialog box is displayed, as shown in Figure 9-85.

4. Select the two edges of the model, refer to Figure 9-86, and choose the **OK** button from the **Region** dialog box; the name of the created region is displayed in the **Edge Region** drop-down list in the **Source Region** rollout of the **Edge-to-Surface Gluing** dialog box.

5. Similarly, select the surface, as shown in Figure 9-87, for the **Target Region** rollout.

6. Retain the default settings in the dialog box and choose the **OK** button; the edge to surface gluing is created, as shown in Figure 9-88.

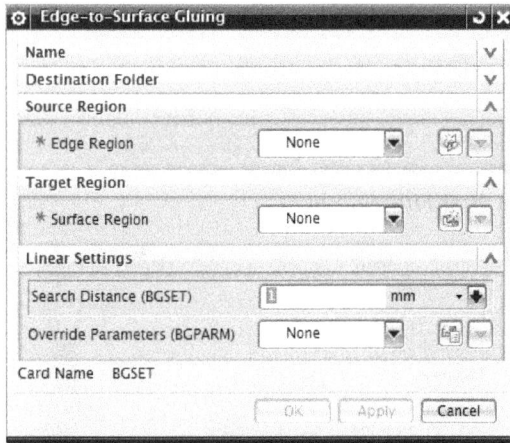

Figure 9-84 The **Edge-to-Surface Gluing** *dialog box*

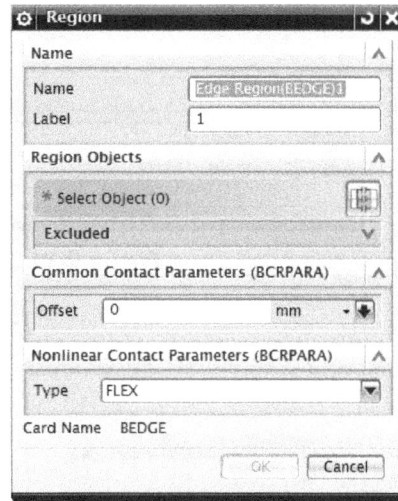

Figure 9-85 The **Region** *dialog box*

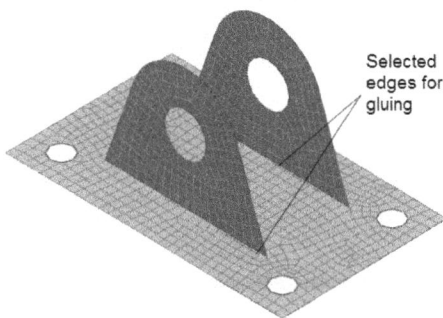

Figure 9-86 Edges selected for gluing

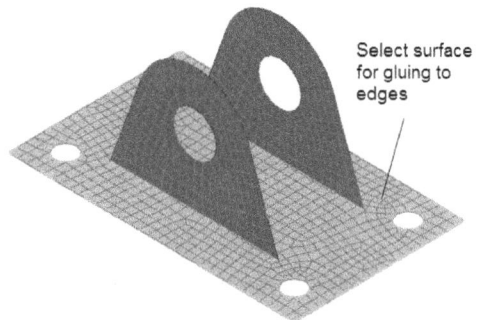

Figure 9-87 Surface selected for gluing

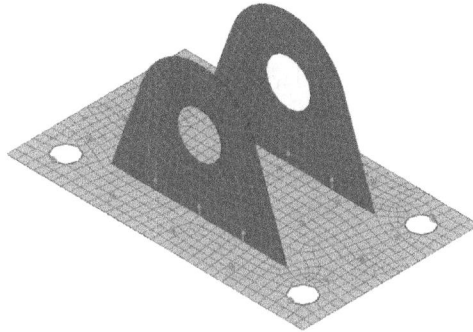

Figure 9-88 Model after edge to surface gluing

Applying Fixed Constraint to Model

Next, you need to apply the fixed constraint to the model.

1. Choose the **Fixed Constraint** tool from the **Constraint Type** drop-down list of the **Loads and Conditions** group of the **Home** tab; the **Fixed Constraint** dialog box is displayed, as shown in Figure 9-89.

2. Select all four edges of hole in the base surface, as shown in Figure 9-90.

3. Retain the default settings in the dialog box and choose the **OK** button; the fixed constraint using the holes edges is created.

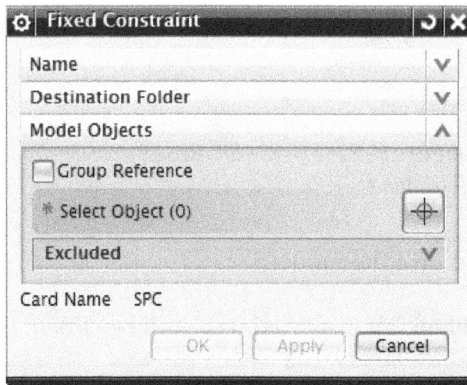

*Figure 9-89 The **Fixed Constraint** dialog box*   *Figure 9-90 Edges selected for fixed constraint*

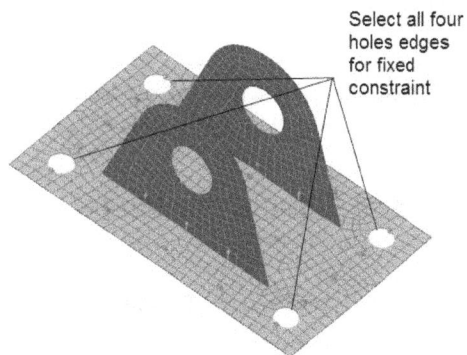

Applying Force Load on the Model

Next, you need to apply the force load on the model.

1. Choose the **Force** tool from the **Load Type** drop-down list of the **Loads and Conditions** group of the **Home** tab; the **Force** dialog box is displayed, as shown in Figure 9-91.

2. Make sure the **Magnitude and Direction** option is selected in the drop-down list in the **Type** rollout, refer to Figure 9-91.

3. Select the edges of holes in the vertical walls using the options in the **Direction** rollout, and specify downward direction as a direction of force, as shown in Figure 9-92.

4. Select the **Expression** option from the **Force** drop-down list in the **Magnitude** rollout and specify **20** N force load in the edit box.

5. Retain the default settings in the dialog box and choose the **OK** button; the force load is applied on the model.

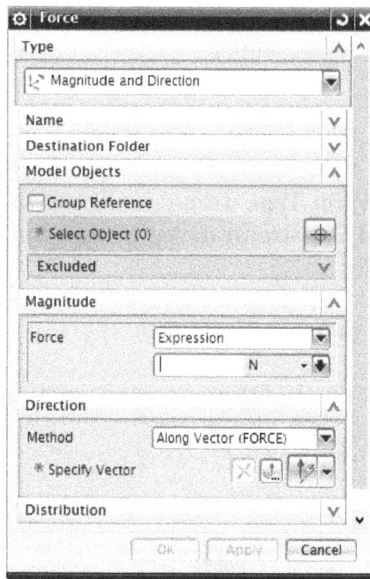

Figure 9-91 The *Force* dialog box

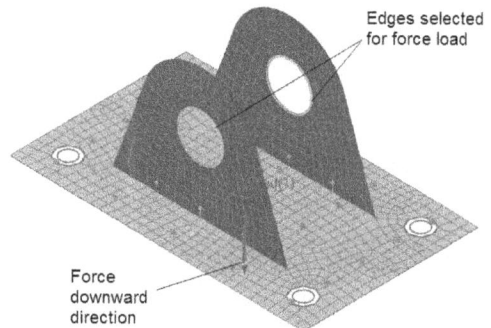

Figure 9-92 Edges and direction selected for force load

Saving and Closing the Model

1. Choose **File > Save> Save** from the **Ribbon** to save the model.

2. Choose **File > Close > All Parts** from the **Ribbon**; all displayed parts will be closed.

Tutorial 2

In this tutorial, you will define material in model of Tutorial 3 of chapter 5, refer to Figure 9-93. Also, apply fixed contact on the hole and pressure on the upper surface of cylinder. The final model is shown in Figure 9-94.   **(Expected time: 45 min)**

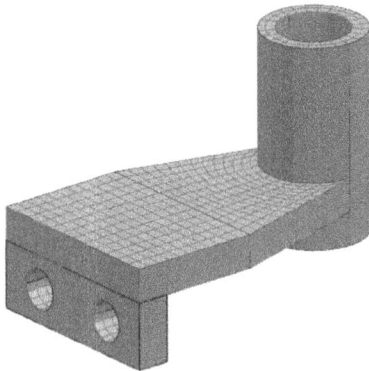

Figure 9-93 *Model for defining boundary condition*

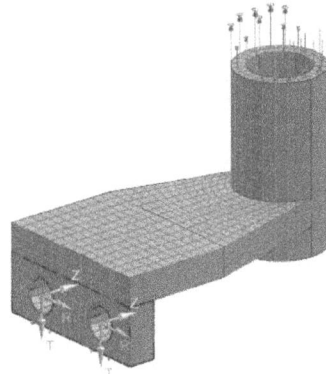

Figure 9-94 *Model after defining boundary condition*

The following steps are required to complete this tutorial:

a. Open the FEM file into NX.
b. Apply the material using the **Assign Materials** tool.
c. Create a new Simulation file.
d. Apply the pinned constraint on the side wall holes.
e. Apply the spatial pressure load on the upper face of the cylindrical body.
f. Save and close the model.

Opening the Assembly File

1. Start NX 9.0 by double-clicking on the shortcut icon of **NX 9.0** on the desktop of your computer.

2. Copy the **c05_tut03.fem** file from the location *C:\NX Nastran\c05\Tut03* and paste it to *C:\NX Nastran\c09\Tut02*. You can also download these files from *www.cadcim.com*.

Note
You must create the Tut02 folder in c09 folder at the location C:\NX Nastran that was created in the previous chapter.

3. To open the FEM file, choose **File > Open** from the **Ribbon**; the **Open** dialog box is displayed. Select the **c05_tut03.fem** file from the location and choose the **OK** button; the file is opened in the FEM environment.

Applying the Material to the Model

Next, you need to apply material to the model.

1. Choose **Menu > Tools > Materials > Assign Materials** from the **Top Border Bar**; the **Assign Materials** dialog box is displayed, as shown in Figure 9-95.

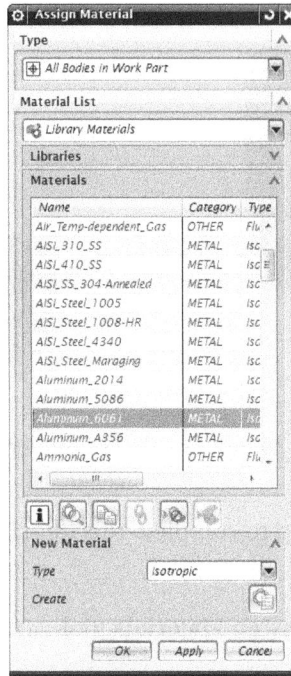

*Figure 9-95 The **Assign Material** dialog box*

2. Select the **All Bodies in Work Part** option from the drop-down list of the **Type** rollout, refer to Figure 9-95; the complete model is selected from the drawing area.

3. Select the **Aluminum_6061** material from the **Materials** sub-rollout, refer to Figure 9-95, and choose the **OK** button to apply the material on the body.

Applying Pinned Constraint to the Model

To apply pinned constraints to the model, you need to enter in the Simulation Environment. To do so, you need to create a new simulation file.

1. Choose the **New Simulation** tool from the **Change Window Drop-Down** list of the **Context** group under the **Home** tab; the **New Part File** window will be displayed. Retain the default settings and choose the **OK** button from the dialog box; the **New Simulation** dialog box is displayed. Do not change the default settings and choose the **OK** button from the dialog box; the **Solution** dialog box is displayed. Again, retain the default settings in the **Solution** dialog box and choose the **OK** button; you enter in the Simulation Environment.

Note
You will learn more about the Simulation Environment and its dialog boxes in the later chapters of this book.

Next, you need to apply pinned constraint to the model.

2. Choose the **Pinned Constraint** tool from the **Constraint Type** drop-down list of the **Loads and Conditions** group of the **Home** tab; the **Pinned Constraint** dialog box is displayed, as shown in Figure 9-96.

3. Select all the faces of both the holes, as shown in Figure 9-97.

4. Retain the default settings in the dialog box and choose the **OK** button; the pinned constraint is applied and all the displacement degrees of freedom of the model get fixed.

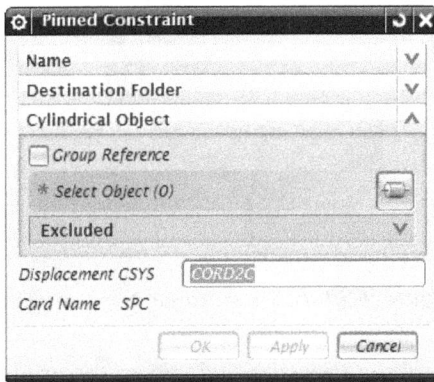

Figure 9-96 The Pinned Constraint dialog box

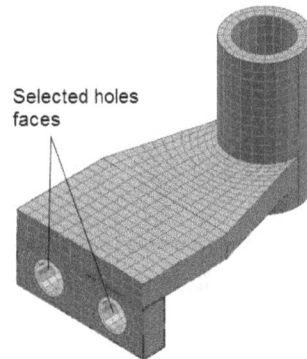

Figure 9-97 Holes faces selected for fixed constraint

Applying Spatial Pressure Load on the Model

Next, you need to apply the spatial pressure load on the model.

1. Choose the **Pressure** tool from the **Load Type** drop-down list of the **Loads and Conditions** group of the **Home** tab; the **Pressure** dialog box is displayed, as shown in Figure 9-98.

2. Select the **Normal Pressure on 2D Elements or 3D Element Faces – Spatial** option from drop-down list available in the **Type** rollout, refer to Figure 9-98.

3. Select the top face of the cylindrical geometry, as shown in Figure 9-99.

4. Select the **Formula Constructor** option from the drop-down list available in the **Magnitude** rollout; the **Formula Field** dialog box is displayed, as shown in Figure 9-100.

5. Select the **Length** option from the **Independent** drop-down list available in the **Domain** rollout.

6. In the edit box available in the **Expressions** rollout type **sin(** and then double click on **x**, available below the **Filters** sub-rollout, the value in the edit box is modified to **sin(ug_var("x")**. Next, add the closing parenthesis to complete the expression, refer to Figure 9-100.

7. Choose the **Accept Edit** button available next the edit box to accept the expression, the expression is assigned to pressure and the **OK** button becomes active.

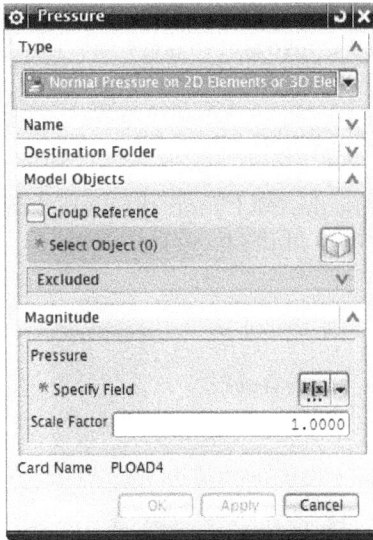

Figure 9-98 *The Pressure dialog box*

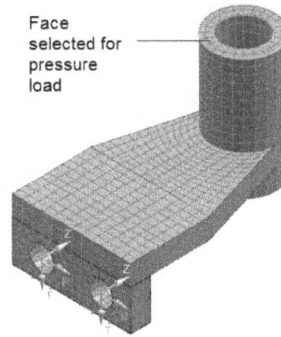

Figure 9-99 *Face selected for pressure load*

8. Choose the **OK** button from the **Formula Field** and the **Pressure** dialog boxes to apply the pressure on the selected face, as shown in Figure 9-101.

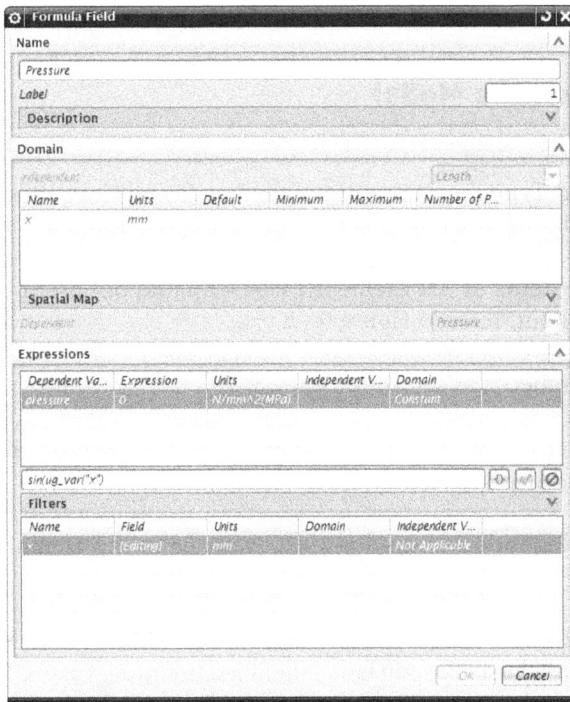

Figure 9-100 *The Formula Field dialog box*

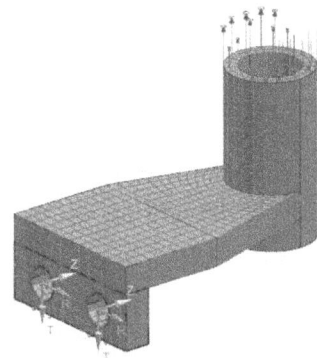

Figure 9-101 *Model after applying pressure load*

Saving and Closing the Model

1. Choose **File > Save > Save** from the **Ribbon** to save the model.

2. Choose **File > Close > All Parts** from the **Ribbon**; all the displayed parts will be closed.

Self-Evaluation Test

Answer the following questions and then compare them to those given at the end of this chapter:

1. The _____ button is used to view the material properties in the material dialog box.

2. The _____ tool is used to individually restrict the degrees of freedom of the selected part of the model.

3. The _____ tool is used to restrict the translation degrees of freedom of the selected part of the model.

4. The _____ tool is used to restrict the degrees of freedom for the cylindrical face of the model with respect to cylindrical coordinate system.

5. The _____ tool is used to apply gravity on the whole model.

6. The **Assign Materials** tool is used to define material to a body. (T/F)

7. You cannot import your user defined material library in NX Nastran. (T/F)

8. The **Mesh Collector** dialog box is also used to define the material. (T/F)

9. You can also create a user defined material in NX Nastran. (T/F)

10. You can not specify the functional variable of a particular load or constraint on the model in NX Nastran. (T/F)

Review Questions

Answer the following questions:

1. Which of the following tools is used to specify the angular velocity and acceleration on the model?

    (a) **Rotation**   (b) **User Defined Constraint**
    (c) **Force**   (d) None of these

2. Which of the following tools is used to radially applied pressure on the selected surface of the model?

    (a) **Centrifugal pressure**   (b) **Hydrostatic pressure**
    (c) **Pressure**   (d) None of these

3. Which of the following tools is used to define a displacement force on the selected geometry along the individual degree of freedom or in a specified direction?

    (a) **Enforced Displacement Constraint**   (b) **Manage Materials**
    (c) **User Defined Constraint**   (d) None of these

4. Which of the following tools is used to restrict all the degrees of freedom for the selected part of the model?

    (a) **Enforced Displacement Constraint**   (b) **Manage Materials**
    (c) **Fixed Constraint**   (d) None of these

5. Which of the following tools is used to restrict the rotational degree of freedom of the selected part of the model?

    (a) **Enforced Displacement Constraint**   (b) **Fixed Constraint**
    (c) **Fixed Rotation Constraint**   (d) None of these

6. The _____ tool is used to restrict the Z axis degree of freedom for the selected part of the model.

7. The _____ tool is used to restrict all the displacement DOFs and leave all rotational DOFs free for a selected cylindrical face of the model.

8. The _____ load can be applied on a part that is fixed on a side and moving rapidly along a direction.

9. The _____ tool is used to restrict all the DOFs except the displacement along X axis.

10. The _____ tool is used to apply symmetric loading conditions to a model.

EXERCISE

Exercise 1

In this exercise, you will define material in model of Exercise 1 of Chapter 8, refer to Figure 9-102. Also, apply fixed contact on the bottom face and acceleration. The final model is shown in Figure 9-103.   **(Expected time: 45 min)**

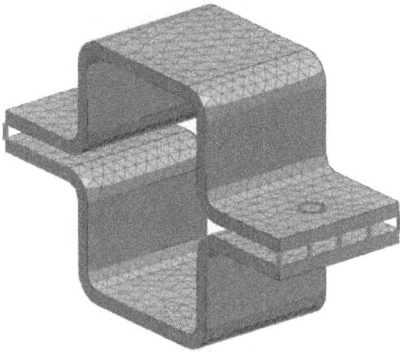

Figure 9-102 *Model for Exercise 1*

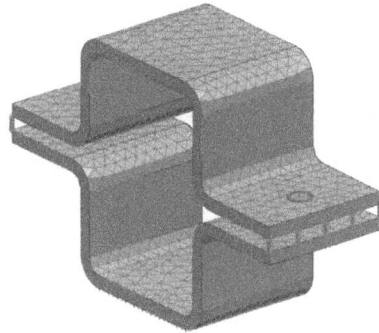

Figure 9-103 *Final model after applying boundary condition*

Hint

1. Open the simulation file; the model will be opened in the Simulation Environment.
2. Switch to FEM environment.
3. Expand the **1D Collectors** node from the **Simulation Navigator** and right-click on the **Spot Weld** node. Next, choose the **Edit** option from the shortcut menu displayed to open the **Mesh Collector** dialog box. Specify the material as **Aluminium_5086**.
4. Similarly, specify the **AISI_Steel_4340** material for **Bolt Connection** and **Aluminum_6061** sheets.
5. Switch to the Simulation environment.
6. Choose the **Fixed Constraint** tool and fix the bottom face of the model, refer to Figure 9-104.
7. Choose the **Acceleration** tool and select the **Whole Model-Components** option from the **Type** rollout.
8. Specify the acceleration value under the **Components** rollout using the **Expressions** option.
   Ax=20, Ay=5, Az=10.
9. Specify direction X from the **Direction** drop-down list and select the **Formula Constructor** option from drop-down list in the **Scale Factor** area drop-down list available in the **Scaling** rollout; the **Formula Field** dialog box will be displayed.
10. Specify expression **tan(ug_var("x"))** in the edit box available in the **Expressions** rollout, refer to Figure 9-105, and choose the **Accept Edit** button.
11. Choose **OK** from all the dialog boxes to apply the acceleration on the model.
12. Save and close the model.

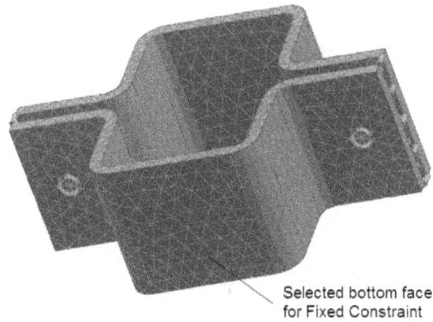

Figure 9-104 *Face selected for fixed constraint*

Figure 9-105 *Expression for acceleration scaling*

Answers to Self-Evaluation Test
1. Inspect Material, 2. User Defined Constraint, 3. Fixed Translation Constraint, 4. Cylindrical Constraint, 5. Gravity, 6. T, 7. F, 8. T, 9. T, 10. F

Chapter 10

Solving and Post-Processing

Learning Objectives

After completing this chapter, you will be able to:

- *Understand the Simulation Environment*
- *Understand Post-Processing Navigator*
- *Create graphs*
- *Create viewports*
- *Animate results*
- *Create reports*

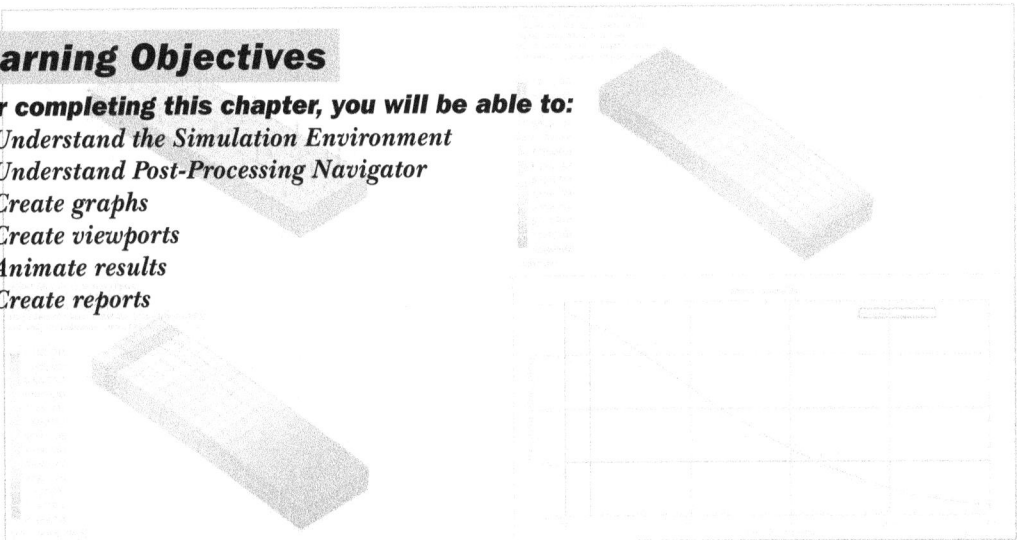

INTRODUCTION

In the previous chapters, you learned about preprocessing steps of the analysis. You also learned to apply boundary conditions in the Simulation Environment. In this chapter, you will learn about solving an analysis problem using post-processing methods and their related options.

SIMULATION ENVIRONMENT

As discussed in the earlier chapters, you can apply boundary condition and contacts in the Simulation Environment. You can further specify the analysis type from the dialog box of the Simulation Environment.

After specifying meshing, material, and connection, you need to switch to the Simulation Environment. To enter in the Simulation Environment, right-click on the simulation file in the **Simulation File View** area; a shortcut menu will be displayed, as shown in Figure 10-1. Choose the **Make Work Part** option from the shortcut menu; simulation file will be activated. You can also create a new simulation file. To do so, choose the **New Simulation** tool from the **Change Window Drop-down** list of the **Context** group in the **Home**

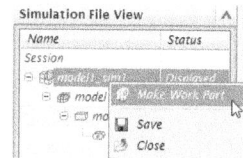

Figure 10-1 *Shortcut menu displayed on right-clicking*

tab; the **New Part File** window will be displayed, as shown in Figure 10-2. In this window, different templates for different solvers and analysis types are available in the **Templates** rollout. In this book, only structural analysis using NX Nastran is covered so you need to select NX Nastran to proceed, refer to Figure 10-2. Using the options in the **New File Name** rollout, you can enter name and folder location in their respective text boxes.

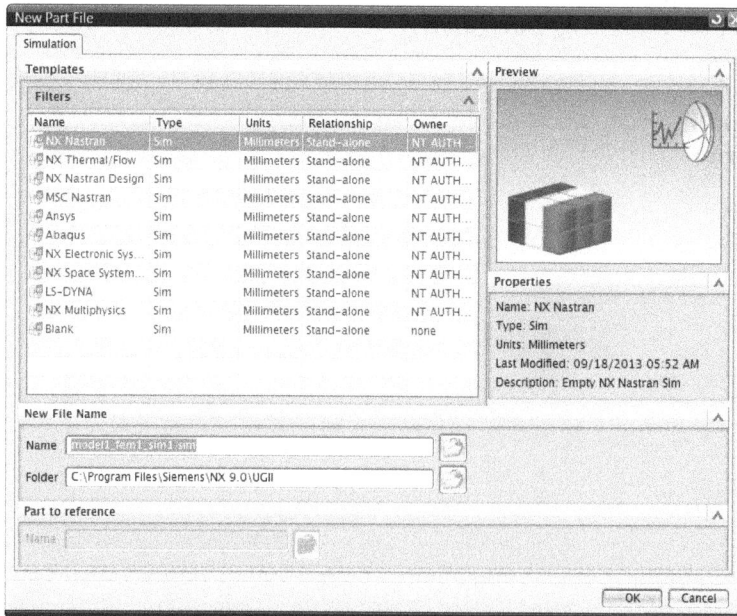

Figure 10-2 *The **New Part File** window*

After specifying the name and location, choose the **OK** button from the window; the **New Simulation** dialog box will be displayed, as shown in Figure 10-3. In this dialog box, simulation name, FEM file name, and solver environment are displayed in their respective rollouts. You can also change the type of layer from the **Layer Placement** rollout and provide additional description under the **Description** rollout. Choose the **OK** button from the dialog box; the **Solution** dialog box will be displayed, as shown in Figure 10-4.

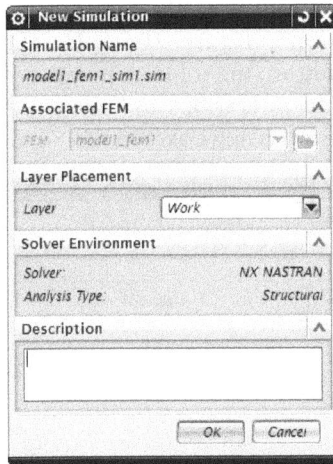

Figure 10-3 *The **New Simulation** dialog box*

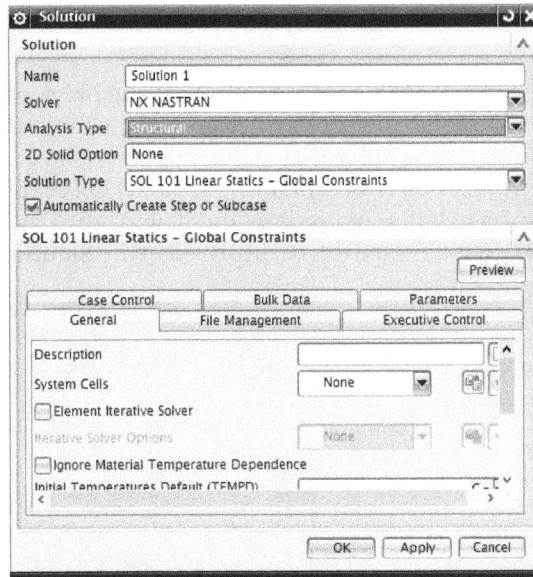

Figure 10-4 *The **Solution** dialog box*

The **Solution** dialog box is used to specify other analysis criteria. Using the **Solution** rollout, you can select the name, solver type and analysis type from their respective drop-down lists. You can choose a more specific solution type from the options available in the **Solution Type** drop-down list, refer to Figure 10-5. Options below the **Solution** rollout change according to the option selected from the **Solution Type** drop-down list. These options are used to make more custom solutions for a solution type.

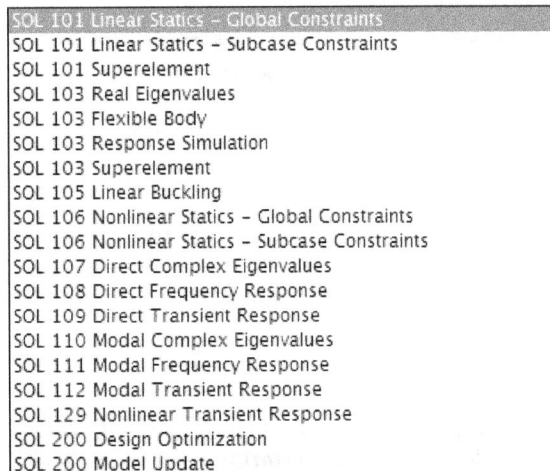

Figure 10-5 *Preview of solution type in NX Nastran*

After specifying all the settings, choose the **OK** button; you will enter in the Simulation Environment and the created solution node will be displayed in the **Simulation Navigator**.

Tip
For the same model, you can create another solution to compare between two or more solutions. You will learn about this comparison method in the later chapter of this book.

After specifying all the steps of preprocessing like specifying boundary conditions and materials, you need to solve the problem. To do this, choose the **Solve** tool from the **Solution** group of the **Home** tab; the **Solve** dialog box will be displayed, as shown in Figure 10-6. Select the **Solve** option from the **Submit** drop-down list and choose the **OK** button; the solution process will start and display in different windows and dialog boxes. After all the operations and calculations, the **END OF JOB** message will be displayed in the **Solution Monitor** dialog box, refer to Figure 10-7.

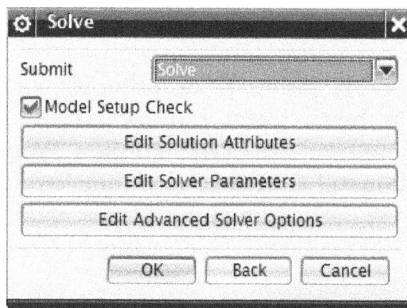

*Figure 10-6 The **Solve** dialog box*

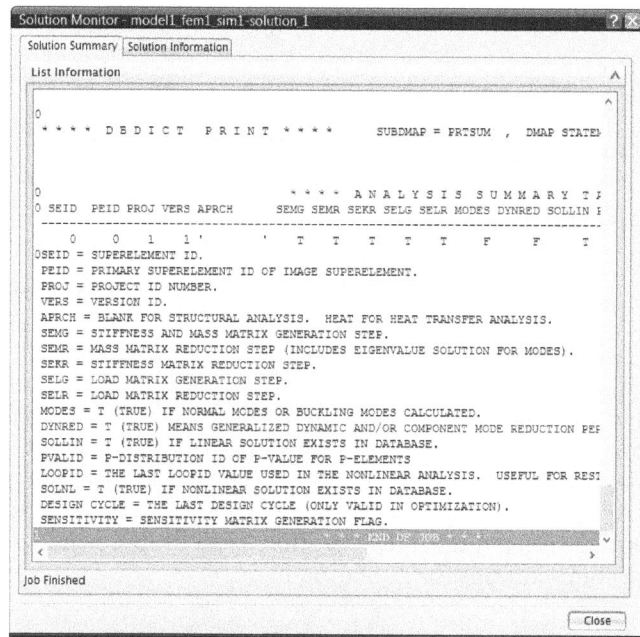

*Figure 10-7 The **Solution Monitor** dialog box*

Close all the open windows and dialog boxes; the result will be generated and saved in the file saving location. To open the result, right-click on the **Results** node under the **Solution** node and choose the **Open** option from shortcut menu, refer to Figure 10-8; the result of the analysis will be displayed in the **Post Processing Navigator**, refer to Figure 10-9.

With the help of these results, you can analyze the model for different parameters which is known as the post-processing and is discussed next.

Note
*If there is any step missing in the preprocessing like meshing, material, load, constraint, then a warning message displayed in the **Information** window while solving. You are suggested to complete that step and restart the solving process.*

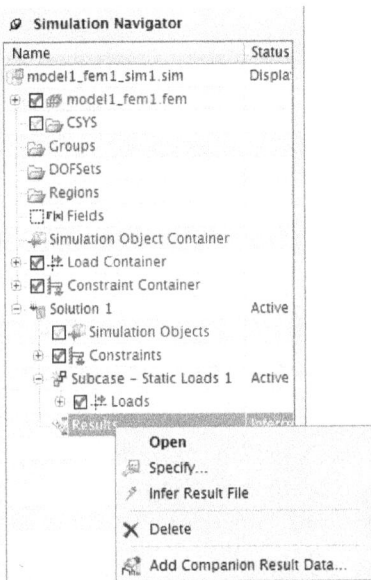

Figure 10-8 *The* **Results** *node in the* **Simulation Navigator**

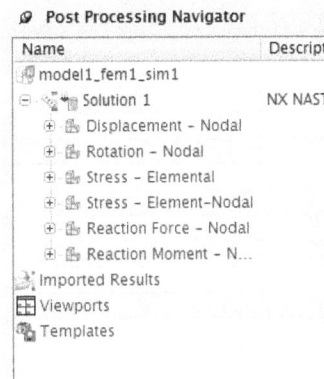

Figure 10-9 *Displayed result in the* **Post Processing Navigator**

POST-PROCESSING

Post-processing includes the complete study of the model with the defined boundary conditions. After solving the model, the results will be loaded in the **Post Processing Navigator**. The main steps of post-processing are as follows:

- Checking and comparing result parameters like displacement, rotation, or stress.
- Creating report of the analysis.
- Comparing analysis in viewport.
- Creating graphics and animating results.
- Creating graph of the variation of results along with different nodes.

Next, you will learn different tools to perform post-processing operations.

Post Processing Navigator

After the results are loaded, different parameters of the solution are loaded in the **Post Processing Navigator**, refer to Figure 10-9. You can activate the required solution by double-clicking on the respective node of the solution. You can also activate the result by right-clicking and choosing the **Plot** option from the shortcut menu. The active results are displayed with varied model in the graphics window. Also, the result are loaded in the **Post View** node under the **Fringe Plots** nodes in the **Post Processing Navigator**. Figure 10-10 shows the solved model with the **Displacement-Nodal** option active. You can check the load variation through colors as well as by the data displayed at the left side of the model.

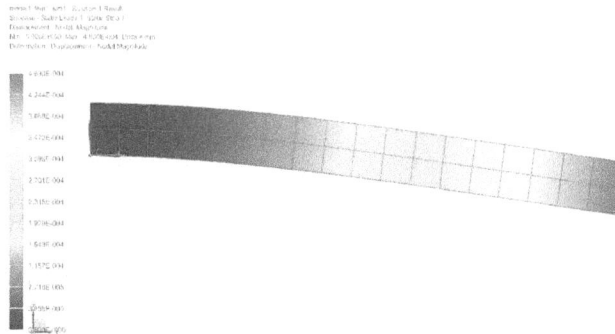

*Figure 10-10 Solved model with the **Displacement-Nodal** option active*

Note
*You can also plot the result using the **Set Result** tool. To do this, choose the **Set Result** tool from the **Post Processing** group of the **Results** tab; the **Smooth Plot** dialog box will be displayed, as shown in Figure 10-11. You can select a result from the drop-down list available in the dialog box.*

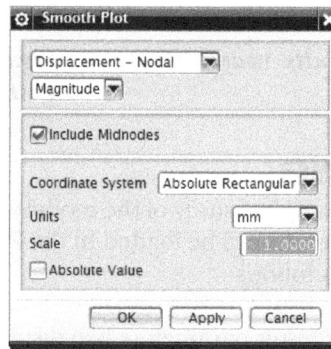

*Figure 10-11 The **Smooth Plot** dialog box*

Creating the Graph

Ribbon: Results > Post Processing > Create Graph
Menu: Tools > Results > Graph

You can create a graph showing the variation of results along different selected nodes. To create a graph, choose **Menu > Tools > Results > Graph** from the **Top Border Bar**; the **Graph** dialog box will be displayed, refer to Figure 10-12. The method to create graph using this dialog box is discussed next.

1. Invoke the dialog box, a selector will replace the mouse cursor in the environment. Also, you will be prompted to select the nodes along which you want to create the graph.
2. Select the nodes of the solved model to create graph; the selected nodes of the model will be highlighted, refer to Figure 10-13. The nodes will be added under the **Select Entities** sub-rollout in the **Y Axis** rollout.
3. You can individually change the selection method of X and Y using the **X Axis** and **Y Axis** rollouts respectively.
4. You can also change the graph name from the **Graph Title** rollout.

5. After specifying all the options, choose the **OK** button; the **Viewport** dialog box will be displayed, refer to Figure 10-14 and you are prompted to select the viewport to place the graph. Select the desired viewport.

6. You can also choose the **Create New Window** button from the **Viewport** dialog box to create the graph in a new window.

Figure 10-15 shows the plotted graph of the selected nodes of the model.

Figure 10-13 Nodes selected for plotting the graph

Figure 10-12 The **Graph** dialog box

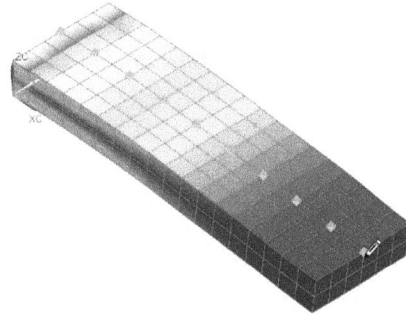

Figure 10-14 The **Viewport** dialog box

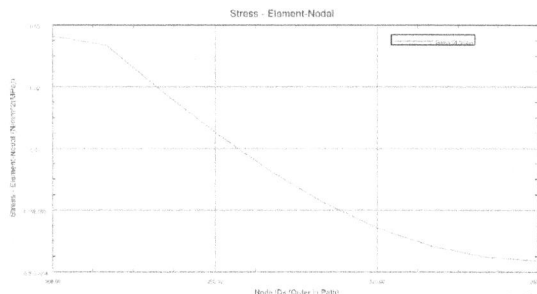

Figure 10-15 The graph plotted from selected nodes

Creating the Viewport

Viewport is the division of the screen in which you can individually show different results at the same time. You can create multiple viewports in the Simulation Environment for result comparison. Different viewport settings are available in the **Layout** group of the **Results** tab in the **Ribbon**. Next you will create four viewports to display model with boundary condition, displacement result, Von-Mises Stress and the graph of the Von-Mises Stress in individual viewports. The method to create four viewports is discussed next.

1. Select the **Four Views** button from the **Layout** group of the **Results** tab in the **Ribbon**; the screen will be divided into four parts with random part preview in each viewport.
2. Right-click on the **Displacement-Nodal** node from the **Post Processing Navigator** and choose the **Plot** option from the shortcut menu; the **Viewport** dialog box will be displayed and you will be prompted to select the viewport to plot the result.
3. Select the upper right viewport; the result will be plotted in the viewport.
4. To plot the Von-Mises Stress, expand the **Stress-Elemental** node from the **Post Processing Navigator**. Plot the Von-Mises Stress using the **Von-Mises** node in the lower left viewport.
5. To create graph of the Von-Mises Stress result, choose **Menu > Tools > Results > Graph** from the **Top Border Bar**; the **Graph** dialog box will be displayed and the cursor will change into a selector. Note that the selector will only appear in the lower left viewport as this is the active viewport. You can activate the viewport by selecting that result node from the **Post Processing Navigator**.
6. Select the elements of the model to create the graph for them and choose the **OK** button from the dialog box; the **Viewport** dialog box will be displayed. Select the bottom right viewport to create the graph.

Figure 10-16 shows the different viewports with plotted results and graph.

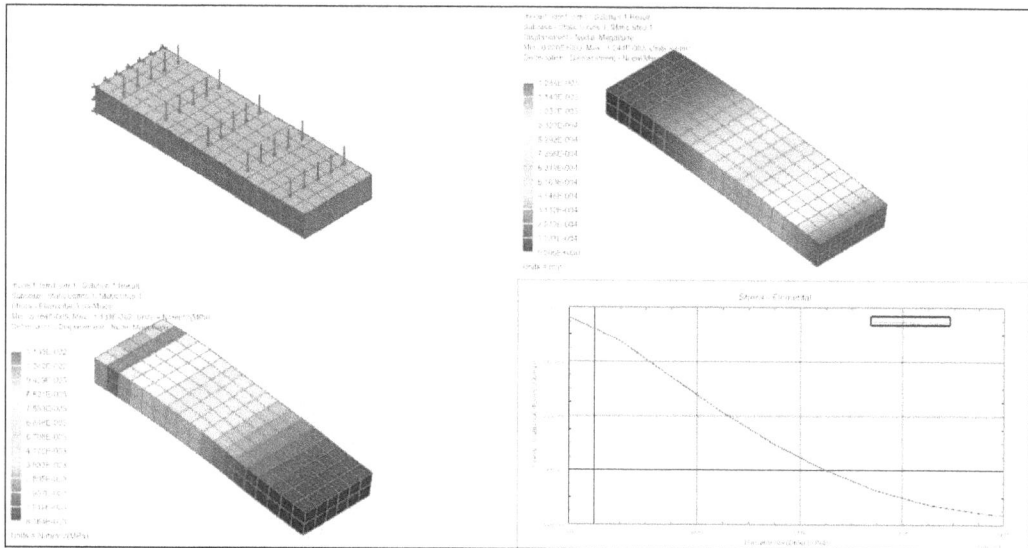

Figure 10-16 Four viewports with plotted results and graph

Tip
*You can use the **View Synchronize** tool available in the **Layout** group of the **Results** tab to synchronize the operations such as zooming, panning, or rotating in the viewports, to synchronize the viewport, choose the **View Synchronize** tool; the **Viewport Settings** dialog box will be displayed and you will be prompted to select the viewport to synchronize. Select the viewports and choose the **OK** button from the **Viewport Settings** dialog box. The selected viewport will synchronize and act simultaneously.*

Animating the Result

Ribbon: Results > Animation > Animate
Menu: Tools > Results > Animation

You can animate the result to visualize how the model will perform under the specified boundary condition. To animate the result, choose **Menu > Tools > Results > Animation** from the **Top Border Bar**; the **Animation** dialog box will be displayed, as shown in Figure 10-17.

Figure 10-17 *The Animation dialog box*

In the **Animation** dialog box, you can choose the animation style from the **Style** drop-down list. The **Linear** style option is used to animate the result from zero to maximum while the **Modal** style option is used to animate the result from negative maximum to positive maximum. The **Number of Frames** edit box is used to specify frame count. More the frame count, smoother the result will be or vice versa.

By using the **Full-cycle** check box, you can animate the result in complete range of forward and backward motion. If the check box is clear, the part will move upto the maximum forward range and then again start from the beginning. The **Synchronize frame change for all post views** option is used to specify frame delay. You can specify a value in the edit box available below this option. The lower is the value, faster the animation will be or vice versa.

In this dialog box, there are animation control buttons such as **Play**, **Pause**, **Stop**, **Step Backward** and **Step**. You can also create a GIF image using the **Capture animated GIF** button.

Creating the Report

Ribbon: Home > Solution > Create Report
Menu: Tools > Create Report

You can create a report of the analysis that may include the material, boundary condition values, or developed stress. To create a report, choose **Menu > Tools > Create Report** from the **Top Border Bar**; the **Report** node will be added under the **Simulation Navigator**. Expand the **Report** node; different parameters related to analysis will be shown, refer to Figure 10-18. You can edit the parameters name or description using the node. You can also add the images of the solution result.

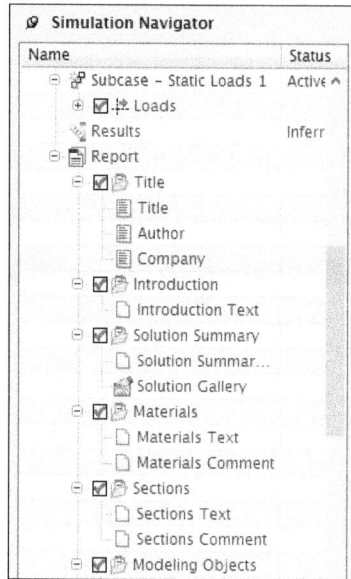

Figure 10-18 *Expanded **Report** node in the **Simulation Navigator***

TUTORIALS

Tutorial 1

In this tutorial, you will solve the model of Tutorial 1 of Chapter 9, refer to Figure 10-19, and then create a report. **(Expected time: 45 min)**

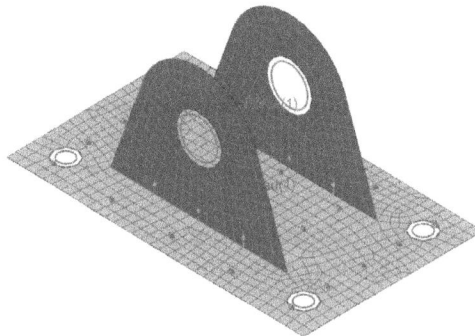

Figure 10-19 *Model to be solved*

The following steps are required to complete this tutorial:

a. Open the Simulation file in NX Nastran.
b. Solve the model and load results in **Post Processing Navigator**.
c. Create report and take snapshot of results.

d.   Export and save the report.
e.   Save and close the model.

Opening the Simulation File

1.   Start NX 9.0 by double-clicking on its shortcut on the desktop of your computer.

2.   Copy the *Tut01* folder from the location *C:\NX Nastran\c09* and paste it at *C:\NX Nastran\c10*. You can also download this folder from *www.cadcim.com*.

Note
You must create the c10 folder at the location C:\NX Nastran\ as created in the previous chapter.

3.   Choose **File > Open** from the **Ribbon**; the **Open** dialog box is displayed. Select the **c05_exr01_sim1.sim** file from the location where it is saved and choose the **OK** button; the file is opened in the Simulation Environment, refer to Figure 10-19. Also, the FEM file is loaded.

Solving and Loading the Result

Next, you will solve the model and load the result in the **Post Processing Navigator**.

1.   Choose the **Solve** tool from the **Solution** group of the **Home** tab; the **Solve** dialog box is displayed, as shown in Figure 10-20.

2.   Make sure the **Solve** option is chosen in the **Submit** drop-down list and then choose the **OK** button; the solution process gets started. Once the solution process is completed, a **END OF JOB** message is displayed in the **Solution Monitor** dialog box, refer to Figure 10-21.

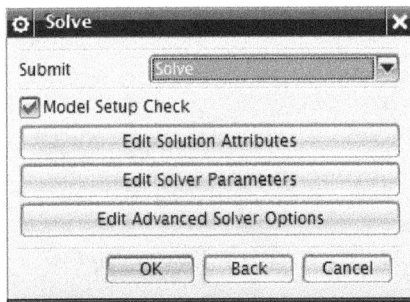

*Figure 10-20 The **Solve** dialog box*

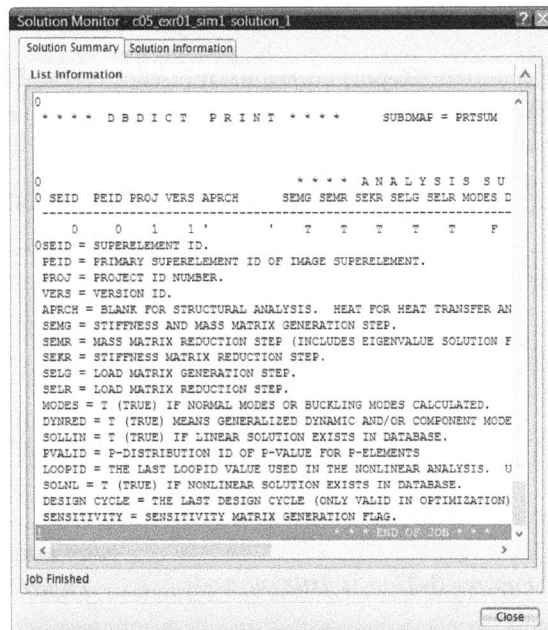

*Figure 10-21 The **Solution Monitor** dialog box*

3.　Close all the opened dialog boxes and window.

4.　Right-click on the **Results** node in the **Simulation Navigator**; a shortcut menu is displayed, refer to Figure 10-22. Choose the **Open** option from the shortcut menu; the result is loaded in the **Post Processing Navigator**.

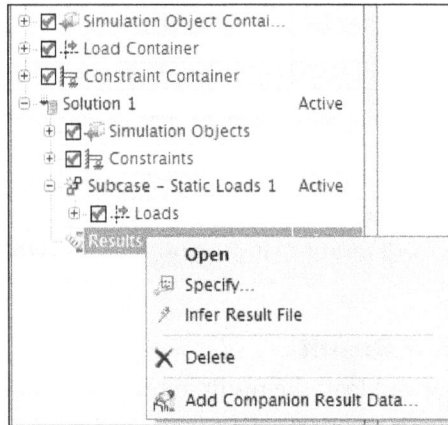

*Figure 10-22 Shortcut menu displayed on right-clicking on the **Results** node*

Creating the Report

Next, you will create a report of the analysis and add images of the results.

1.　Choose the **Create Report** tool from the **Solution** group of the **Home** tab; the **Report** node is added to the **Simulation Navigator**.

2.　Expand the **Report** node and right-click on the **Title** sub-node; a shortcut menu is displayed, refer to Figure 10-23. Choose the **Edit** option from the shortcut menu; the **Report Editor** dialog box is displayed, as shown in Figure 10-24.

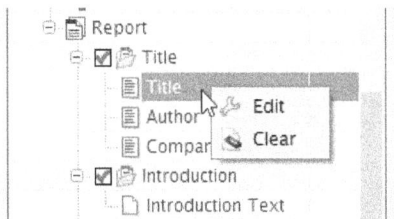

*Figure 10-23 The **Title** sub-node*

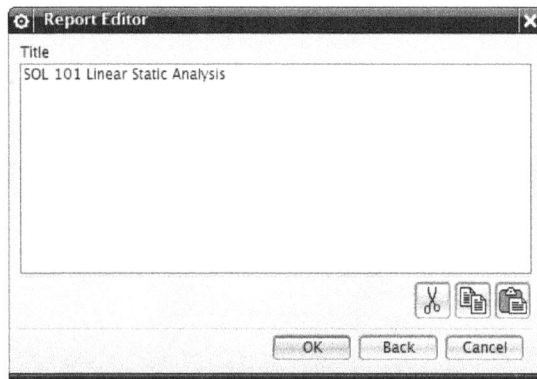

*Figure 10-24 The **Report Editor** dialog box*

3. Specify the title **SOL 101 Linear Static Analysis** in the **Report Editor** dialog box, refer to Figure 10-24, and choose the **OK** button; model with the Boundary condition is shown in the environment.

Note that you can also change the Author and Company name using the same method.

4. Right-click on the **Images** sub-node, refer to Figure 10-25, under the **Report** node and choose **Snapshot** from the shortcut menu; the **Image1** node is added under the **Images** node.

5. Right-click on the **Image1** node and choose **Edit Description** from the shortcut menu; a text box is displayed in the **Simulation Navigator**. Specify the description as **Model with Boundary Condition** in the text box, refer to Figure 10-26, and press ENTER.

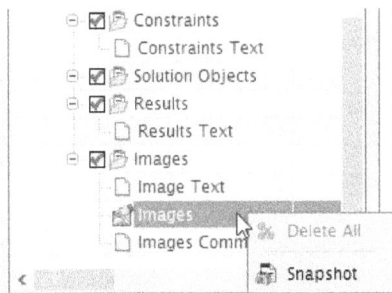

Figure 10-25 *The shortcut menu for snapshot*

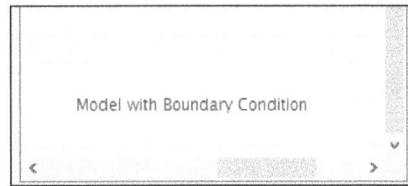

Figure 10-26 *The description text box*

Now, you need to plot the **Displacement - Nodal** result using the **Post Processing Navigator**.

6. Right-click on it and select the **Plot** option from the shortcut menu; the result is displayed on the model in the environment.

7. In the **Simulation Navigator**, right-click on the **Images** node and choose **Snapshot** from the shortcut menu; the **Image2** node is added under the **Images** node.

8. Right-click on the **Image2** node and choose **Edit Description** from the shortcut menu; a text box is displayed in the **Simulation Navigator**. Specify the description as **Displacement-Nodal Result** in the text box and press ENTER.

9. Similarly, add the Von-Mises and Stress-Elemental results as well as add their description.

10. Next, right-click on the **Report** node in the **Simulation Navigator** and choose the **Export** option from the shortcut menu, refer to Figure 10-27; the report gets opened in your default browser.

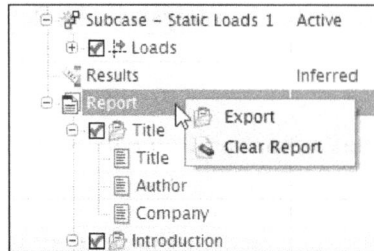

Figure 10-27 *The shortcut menu displayed on right-clicking the **Report** node*

Saving and Closing the Model and Report

1.   Choose **File > Save > Save** from the **Ribbon** to save the model.

2.   Choose **File > Close > All Parts** from the **Ribbon**; all displayed parts are closed.

3.   Also save and close the report opened in the browser.

Tutorial 2

In this tutorial, you will solve the model of Tutorial 2 of Chapter 9, refer to Figure 10-28 and then create a report. You will also create the viewport for comparing results. (**Expected time: 45 min**)

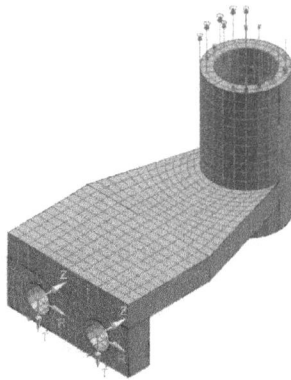

Figure 10-28 *Model to be solved*

The following steps are required to complete this tutorial:

a.   Open the Simulation file into NX Nastran.
b.   Solve the model and load the result in **Post Processing Navigator**.
c.   Create report and take snapshot of different results.
d.   Create viewport to compare the results.
e.   Export and save the report.
f.   Save and close the model.

Opening the Simulation File

1. Start NX 9.0 by double-clicking on its shortcut icon on the desktop of your computer, if it is not already started.

2. Copy the *Tut02* folder from the location *C:\NX Nastran\c09* and paste it to *C:\NX Nastran\c10*. You can also download these files from *www.cadcim.com*.

Note
You must create the c10 folder at the location C:\NX Nastran\, as created in the previous chapter.

3. Choose **File > Open** from the **Ribbon**; the **Open** dialog box is displayed. Select the **c05_tut03_sim1.sim** file from the location where it is saved and choose the **OK** button; the file is opened in the Simulation Environment, refer to Figure 10-28. Also, the FEM file is loaded.

Solving and Loading the Result

Next, you will solve the model and load the result in the **Post Processing Navigator**.

1. Choose the **Solve** tool from the **Solution** group of the **Home** tab; the **Solve** dialog box is displayed.

2. Make sure the **Solve** option is selected in the **Submit** drop-down list and choose the **OK** button; the solution process starts. After completion of the solution process, the **END OF JOB** message is displayed in the **Solution Monitor** dialog box.

3. Close all the opened dialog boxes and window.

Now you need to load the result in the **Post Processing Navigator**.

4. Right-click on the **Results** node from the **Simulation Navigator**; a shortcut menu is displayed, refer to Figure 10-29. Choose the **Open** option from the shortcut menu; the result is loaded in the **Post Processing Navigator**.

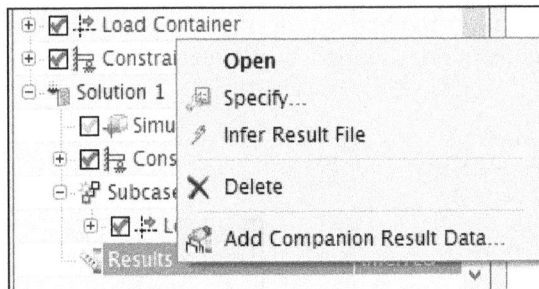

Figure 10-29 Shortcut menu displayed on right-clicking

Creating the Report

Next, you will create the report for the analysis and add the images of the results.

1. Choose the **Create Report** tool from the **Solution** group of the **Home** tab; the **Report** node is added to the **Simulation Navigator**.

2. Expand the **Report** node and right-click on the **Title** sub-node; a shortcut menu is displayed, refer to Figure 10-30. Choose the **Edit** option from the shortcut menu; the **Report Editor** dialog box is displayed, as shown in Figure 10-31.

3. Specify **Linear Static Analysis with Spatial Pressure Load** as the title in the **Title Bar** in the **Report Editor** dialog box, refer to Figure 10-31, and choose the **OK** button to change the title.

Note that you can also change the Author and Company name in the same way.

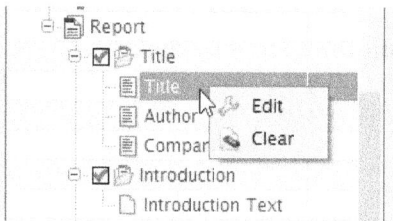

*Figure 10-30 The **Title** sub-node*   *Figure 10-31 The **Report Editor** dialog box*

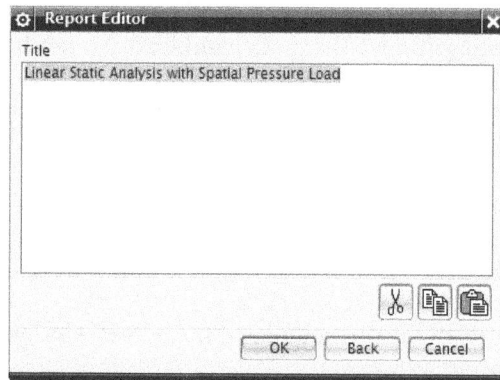

4. Right-click on the **Images** sub-node under the **Report** node and choose the **Snapshot** from the shortcut menu, refer to Figure 10-32; the **Image1** node is added under the **Images** node.

5. Right-click on the **Image1** node and choose the **Edit Description** from the shortcut menu; a text box in the **Simulation Navigator** is displayed. Specify the description as **Model with Boundary Condition** as the description in the text box, refer to Figure 10-33, and press ENTER.

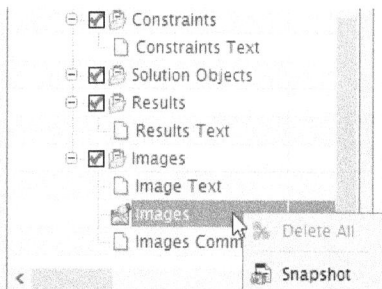

Figure 10-32 The shortcut menu for snapshot   *Figure 10-33 The description text box*

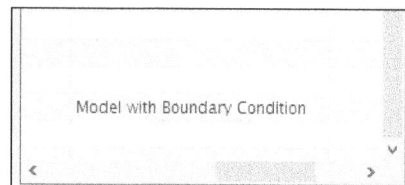

6. Using the **Post Processing Navigator**, plot the **Displacement - Nodal** result by right-clicking on the respective node and choosing the **Plot** option from the shortcut menu displayed. The results will affect the model in the environment.

7. In the **Simulation Navigator**, right-click on the **Images** node and choose **Snapshot** from the shortcut menu displayed; the **Image2** node is added under the **Images** node.

8. Right-click on the **Image2** node and choose the **Edit Description** from the shortcut menu; the text box in the **Simulation Navigator** is displayed. Specify **Displacement-Nodal Result** as the description in the text box and press ENTER.

9. Similarly, add the von-Mises and Stress-Elemental results as well as their description.

Creating the Viewports

Next, you will create side by side viewport for comparing the model deformation due to spatial pressure load. You will also synchronize both the views and add image to the report. First, choose the **Return to Home** button from the **Context** group of the **Home** tab; the model with boundary condition becomes available in the environment.

1. Select the **Side by Side** button from the **Layout** group of the **Results** tab in the **Ribbon**; the screen is divided into two parts with random part preview in each viewport.

2. Right-click on the **Displacement-Nodal** node from the **Post Processing Navigator** and choose the **Plot** option from the shortcut menu; the **Viewport** dialog box is displayed and you are prompted to select the viewport to plot the result.

3. Select the right viewport; the result is plotted in the right viewport.

4. Choose the **View Synchronize** tool from the **Layout** group of the **Results** tab; the **Viewport Settings** dialog box is displayed and you are prompted to select the two viewports to be synchronized. Select the viewports and choose the **OK** button from the **Viewport Settings** dialog box; the selected viewports get synchronized and start acting simultaneously.

5. Choose the **Trimetric** button to orient both the views in trimetric position, refer to Figure 10-34.

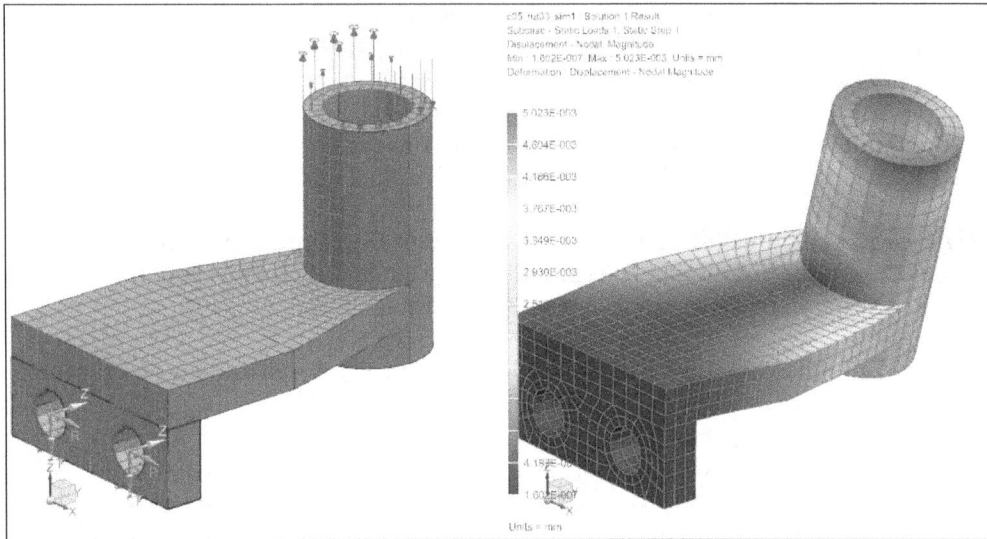

Figure 10-34 *Displacement comparison in the viewport*

6. In the **Simulation Navigator**, right-click on the **Images** node and choose **Snapshot** from the shortcut menu displayed; the **Image4** node is added under the **Images** node.

7. Right-click on the **Image4** node and choose the **Edit Description** from the shortcut menu; the text box in the **Simulation Navigator** is displayed. Specify **Displacement comparison** as the description in the text box and press ENTER.

8. Next, right-click on the **Report** node and choose the **Export** option from the shortcut menu displayed, refer to Figure 10-35; the report is opened in your default browser.

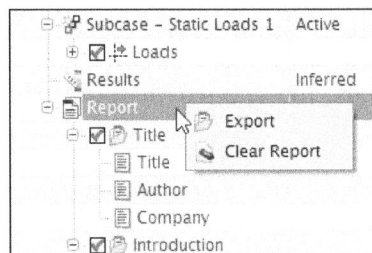

Figure 10-35 **Export** *option in the shortcut menu*

Saving and Closing the Model and Report

1. Choose **File > Save > Save** from the **Ribbon** for saving the model.

2. Choose **File > Close > All Parts** from the **Ribbon**; all the displayed parts are closed.

3. Also save and close the report that opened in the browser.

Self-Evaluation Test

Answer the following questions and then compare them to those given at the end of this chapter:

1. Which of the following environments is used to assign boundary condition to the model?

    (a) **Simulation Environment**   (b) **Modeling Environment**
    (c) **FEM Environment**   (d) None of these

2. Which of the following tools is used to create a graph of the result in the environment?

    (a) **Create Graph**   (b) **Set Result**
    (c) **Viewport**   (d) None of these

3. Which of the following tools is used to create a report of the analysis?

    (a) **Graph**   (b) **Viewport**
    (c) **Create Report**   (d) None of these

4. The **Solve** tool is used to solve the parameters defined in the model. (T/F)

5. You can create multiple viewports in the Simulation Environment for result comparison. (T/F)

Review Questions

Answer the following questions:

1. You can choose specific solution type from the _____ drop-down list available in the **Solution** dialog box.

2. The result of the analysis will be displayed and loaded in the _____ .

3. You can plot the result using the _____ tool.

4. You can animate the result to visualize the performance of the model under some specified boundary condition. (T/F)

5. You can apply boundary condition and contacts in the FEM Environment. (T/F)

EXERCISE

Exercise 1

In this exercise, you will solve and create the report of the model of Exercise 1 of Chapter 9, refer to Figure 10-36. You will add the Displacement-Nodal, von Mises, and Stress Elemental results to the report.   **(Expected time: 45 min)**

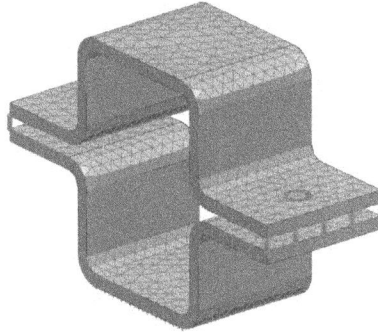

Figure 10-36 *Model to be solved*

Answers to Self-Evaluation Test
1. a, **2.** a, **3.** c, **4.** T, **5.** T

Chapter 11

Projects

Learning Objectives

After completing this chapter, you will be able to:

- *Perform structural analysis*
- *Perform buckling analysis*
- *Perform responsive analysis*

STRUCTURAL ANALYSIS

The Static Structural analysis is one of the important analyses in NX Nastran. This system analyses the structural components for displacements (deformation), stresses, strains, and forces under different loading conditions. In this analysis, the loads are assumed to have no damping characteristics (time dependent) and only steady loading and damping conditions are assumed.

Project 1

In this project, you will do structural analysis of a support sheet. You can download this file from *www.cadcim.com*. For analyzing and comparing the support sheet, you will create two groups, one with center rib and the other with two side ribs having same boundary conditions. Figure 11-1 shows the model having all the ribs and boundary conditions applied. Figure 11-2 shows the model with one center rib and Figure 11-3 shows the model with two side ribs.

(Expected time: 3 hr)

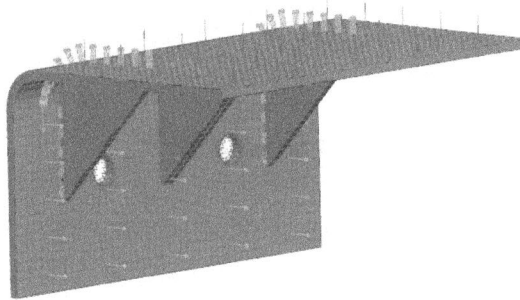

Figure 11-1 *Rotated view of model with all ribs and boundary conditions*

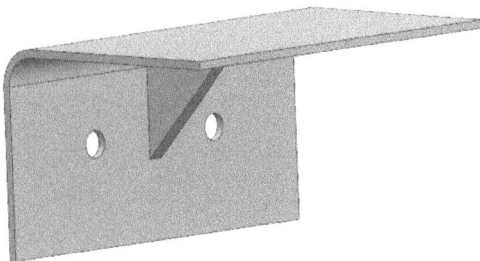

Figure 11-2 *Model with center rib*

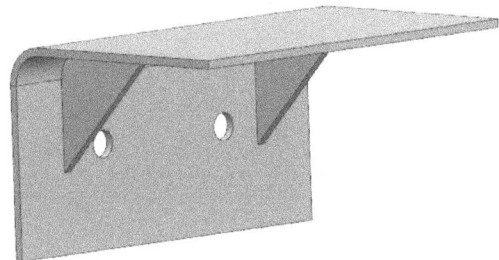

Figure 11-3 *Model with two side ribs*

The following steps are required to complete this tutorial:

a. Open the model in NX and create FEM and idealized file.
b. Split the model for 3D swept meshing.
c. Mesh the model.
d. Apply material.

e. Enter the Simulation Environment.
f. Create surface to surface gluing
g. Apply fixed constraint on the model.
h. Apply force on the model.
i. Solve the analysis.
j. Create group of one rib and two rib model.
k. Create Solution Card for one rib model.
l. Apply boundary condition and glue for one rib model.
m. Solve the one rib model analysis.
n. Create Solution Card for two rib model.
o. Apply boundary condition and glue for two rib model.
p. Solve the Two Rib Model analysis.
q. Create Solution Card for two rib model.
r. Post process the results.
s. Save and close the model.

Opening the Model in NX and Creating the FEM and Idealized Files

1. Start NX 9.0 by double-clicking on the shortcut icon of NX 9.0 on the desktop of your computer.

2. Download the *c11_P1* zipped folder file from *www.cadcim.com*. Extract the folder to the location *C:\NX Nastran\c11\P1*.

Note
You must create the c11 folder at C:\NX Nastran as created in the previous chapter.

3. Choose **File > Open** from the **Ribbon**; the **Open** dialog box is displayed. Select the **Sheet_support.prt** file from the location it is saved at and then choose the **OK** button; the file is opened in the Modeling Environment.

4. Now you need to create the FEM and Idealized file in the Advanced Simulation environment. Choose **Advanced Simulation** from the **Applications** area of the **File** tab in the **Ribbon**; the Advanced Simulation environment is invoked.

5. Choose the **New FEM** tool from the **Change Window Drop-down** list in the **Context** group under the **Home** tab; the **New Part File** window is displayed. Select **NX Nastran** from the **Name** column for the **FEM** type from the window and choose the **OK** button; the **New FEM** dialog box is displayed with the preview of the imported part in the **CAD Part** window. Select the **Create Idealized Part** check box if it is not selected by default and choose the **OK** button from the dialog box; the part is opened in the FEM environment.

Splitting the Model for 3D Swept Meshing

For uniform meshing on the sheet, you need to split the sheet around the rib in Ideal Environment using the created idealized part.

1. In the **Simulation File View**, right-click on the **Sheet_support_fem1_i** file name; a shortcut menu is displayed. Choose the **Make Displayed Part** option from the shortcut menu;

the **Idealized Part Warning** message box is displayed. Choose the **OK** button from the message box; the part is opened in the Ideal Environment.

2. Choose **Menu > Insert > Associative Copy > Promote** from the **Top Border Bar**; the **Promote Body** dialog box is displayed. Select the part from the environment and choose the **OK** button; an associative copy of the part selected is created.

3. Choose **Menu > Insert > Model Preparation > Trim > Split Body** from the **Top Border Bar**; the **Split Body** dialog box is displayed, refer to Figure 11-4 and you are prompted to select the target bodies to be split. Select the model to be split in the window.

4. Select the **Extrude** option from the **Tool Option** drop-down list in the **Tool** rollout; the **Select Curve (0)** area is highlighted in the **Section** sub-rollout and you are prompted to select the curve for extrusion. Select the curved edge of the model, refer to Figure 11-5 and select the **-YC-axis** from the **Direction** sub-rollout.

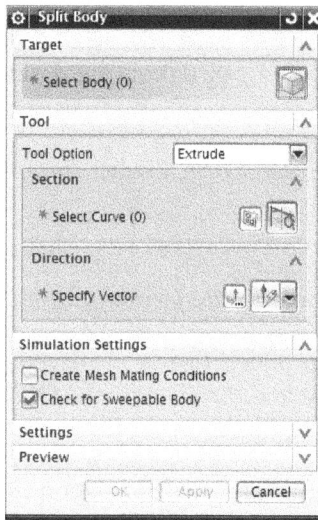

*Figure 11-4 The **Split Body** dialog box*   *Figure 11-5 Curve selected to split the model*

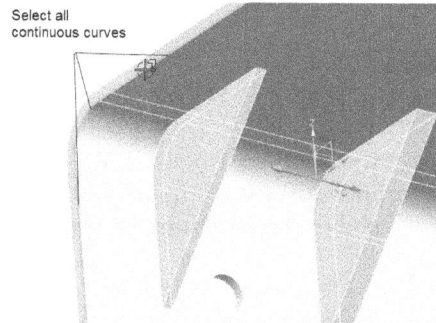

5. Select the **Check for Sweepable Body** check box from the **Simulation Settings** rollout. Also, make sure that the **Create Mesh Mating Conditions** check box is clear. Next, choose the **Apply** button from the dialog box; the model gets split between rib and support.

6. Split the support sheet by selecting the edge, as shown in Figure 11-6 and select **-XC-axis** from the **Direction** sub-rollout.

7. Choose the **Apply** button from the dialog box; the model splits and turns green. Close the dialog box.

Meshing the Model

Next, you will mesh the model using the **3D Swept Mesh** tool. To create mesh, you need to switch into the FEM Environment for meshing.

1. Right-click on **Sheet_support_fem1** from the **Simulation File View** and choose the **Make Displayed Part** option from the shortcut menu; the FEM Environment is invoked and also the **Information** window is displayed. Close this window.

2. Choose **Menu > Insert > Mesh > 3D Swept Mesh** from the **Top Border Bar**; the **3D Swept Mesh** dialog box is displayed, refer to Figure 11-7.

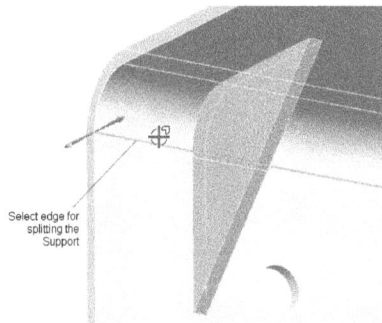

Figure 11-6   *Curve selected for splitting the model*

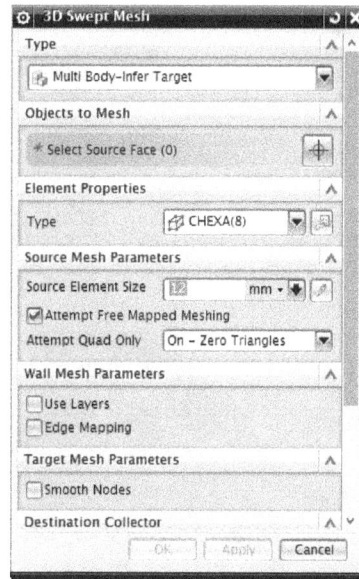

Figure 11-7   *The **3D Swept Mesh** dialog box*

3. Select the side face of the upper split model, as shown in Figure 11-8.

4. Select the CHEXA(8) element type from the **Type** drop-down list in the **Element Properties** rollout. Specify **12** as the element size in the **Source Element Size** edit box and choose the **On - Zero Triangles** option from the **Attempt Quad Only** drop-down list available in the **Source Mesh Parameters** rollout. Also, select the **Attempt Free Mapped Meshing** check box.

5. Retain the rest of the values and choose the **Apply** button; the swept mesh is created on the model.

6. Select the front face of the model, refer to Figure 11-9. Retain the settings of the previous step and choose the **Apply** button; the swept mesh is created on the support.

7. Select the side face of the rib, as shown in Figure 11-10. Also, select the **Off - Allow Triangles** option from the **Attempt Quad Only** drop-down list in the **Source Mesh Parameters** rollout. Retain the other values and choose the **Apply** button; the swept mesh is created on the first rib.

8. Similarly, create mesh on the other ribs. The model after creating the mesh is shown in Figure 11-11.

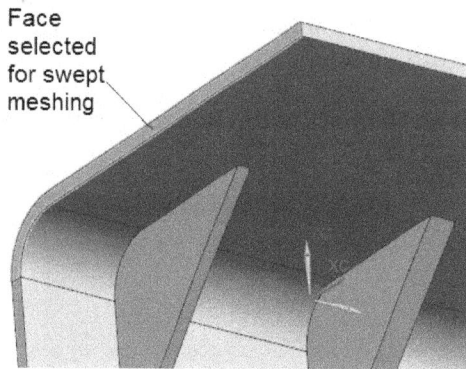

Figure 11-8 Selected face to create swept mesh

Figure 11-9 Front face selected to create swept mesh

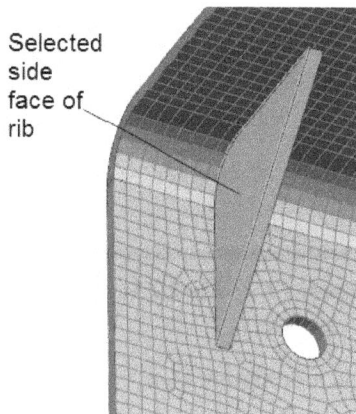

Figure 11-10 Side face of rib selected to create swept mesh

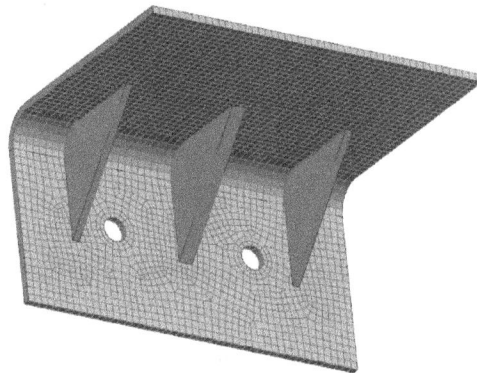

Figure 11-11 Model after creating swept mesh

Applying Material

Next, you will apply the **Aluminum_A356** material on the model.

1. Choose **Menu > Tools > Materials > Assign Materials** from the **Top Border Bar**; the **Assign Materials** dialog box is displayed, refer to Figure 11-12.

2. Select the **All Bodies in Work Part** option from the **Type** rollout and choose **Aluminum_A356** from the **Materials** sub-rollout in the **Material List** rollout. Choose the **OK** button to apply the material on the model.

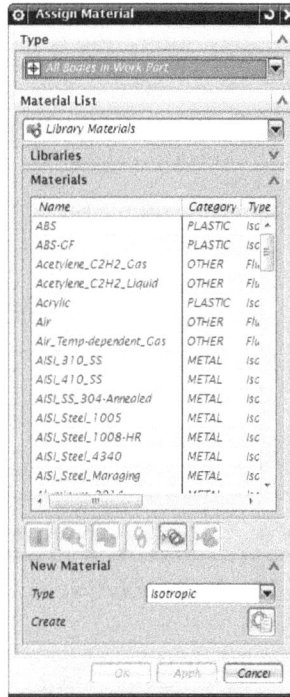

Figure 11-12 *The **Assign Material** dialog box*

Entering the Simulation Environment

To perform the analysis, you need to enter Simulation Environment and create a solution with Linear Buckling analysis solution type.

1. Choose the **New Simulation** tool from the **Change Window Drop-down** list in the **Context** group under the **Home** tab; the **New Part File** window is displayed. Select **NX Nastran** from the **Name** column for the **Sim** type and choose the **OK** button from the dialog box; the **New Simulation** dialog box is displayed. Retain the default settings and choose the **OK** button from the dialog box; the **Solution** dialog box is displayed.

2. Specify **Complete Model** in the **Name** edit box and select the **SOL 101 Linear Statics - Global Constraints** option from the **Solution Type** drop-down list, refer to Figure 11-13.

3. Retain the rest of settings in the dialog box and choose the **OK** button; you enter into Simulation Environment.

> **Note**
> *You will create three different Solution Cards for comparing the model: when the model has one rib, when the model has two ribs, and when the model has all the ribs. However, in this solution the card name is **Complete Model** and you will analyze the model with all three ribs.*

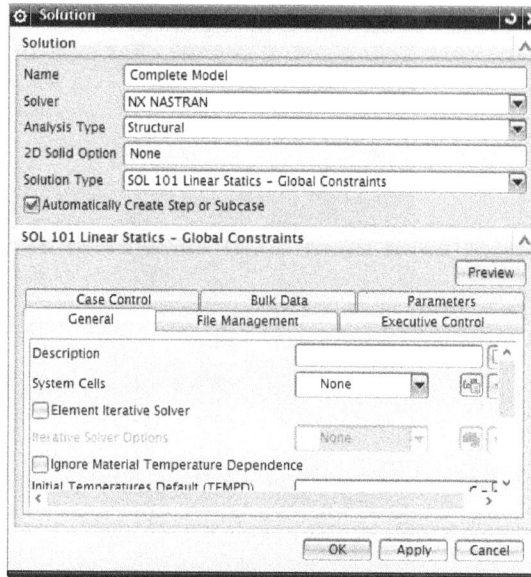

Figure 11-13 The *Solution* dialog box

Creating Surface to Surface Gluing

Now, you will create surface to surface gluing in the model to make a contact between rib and support.

1. Choose the **Surface-to-Surface Gluing** tool from the **Simulation Object Type** drop-down list of the **Loads and Conditions** group of the **Home** tab; the **Surface-to-Surface Gluing** dialog box is displayed, as shown in Figure 11-14.

2. Choose the **Create Face Pairs** button available in the **Automatic Face Pair Creation** rollout; the **Create Automatic Face Pairs** dialog box is displayed, as shown in Figure 11-15.

3. Specify **0.2** mm as the value in the **Distance Tolerance** edit box and select all the surfaces of the model. Choose the **OK** button to select all the surfaces that lie within the specified range.

4. Retain the rest of the settings in the **Surface-to-Surface Gluing** dialog box and choose the **OK** button; the surface to surface gluing is created, as shown in Figure 11-16.

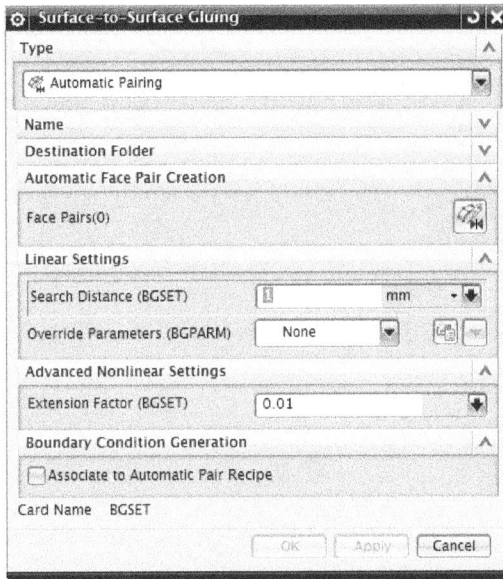

Figure 11-14 The *Surface-to-Surface Gluing* dialog box

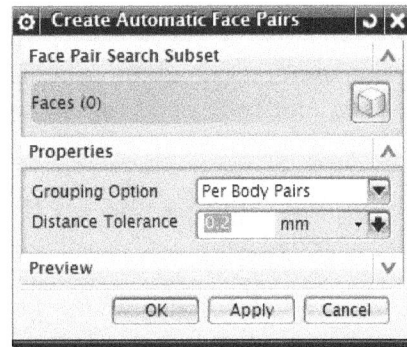

Figure 11-15 The *Create Automatic Face Pairs* dialog box

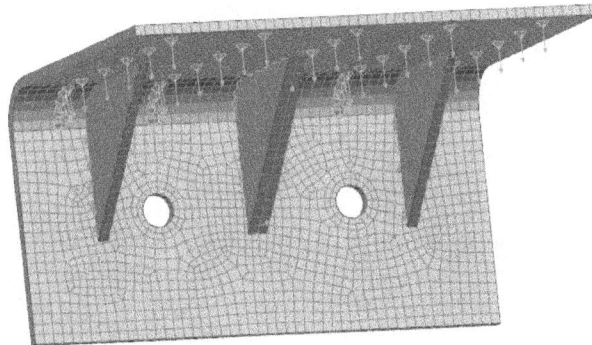

Figure 11-16 *Model after creating surface to surface gluing*

Applying Fixed Constraint to the Model

Next, you need to apply fixed constraint on the holes of the support.

1. Choose the **Fixed Constraint** tool from the **Constraint Type** drop-down list in the **Loads and Conditions** group of the **Home** tab; the **Fixed Constraint** dialog box is displayed, as shown in Figure 11-17.

2. Select the faces of the holes in the model, as shown in Figure 11-18.

3. Retain the default settings in the dialog box and choose the **OK** button; the fixed constraint is applied on the model.

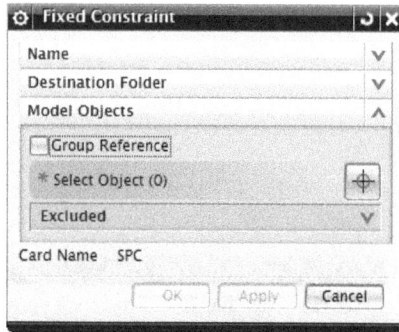

Figure 11-17 The *Fixed Constraint* dialog box

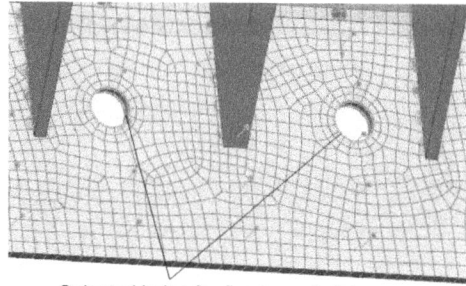

Selected holes for fixed constraint

Figure 11-18 *Surface selected for fixed constraint*

Applying the Force Load on the Model

Next, you need to apply load on the support.

1. Choose the **Force** tool from the **Load Type** drop-down list in the **Loads and Conditions** group of the **Home** tab; the **Force** dialog box is displayed.

2. Select the **Magnitude and Direction** option from the drop-down list in the **Type** rollout, refer to Figure 11-19.

3. Select the top face of the support, as shown in Figure 11-20.

Figure 11-19 The **Force** dialog box

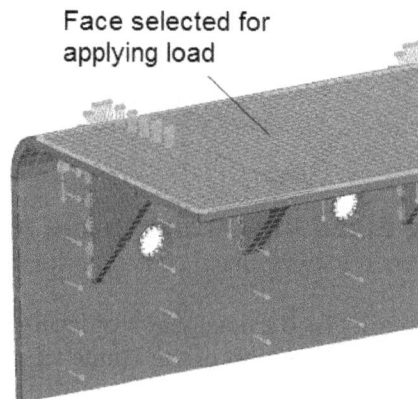

Face selected for applying load

Figure 11-20 *Face selected for load applying*

4. Select the **Expression** option from the **Force** drop-down list in the **Magnitude** rollout if it is not selected by default and specify **800** N as the value in the edit box.

5. Specify the direction of load as **-ZC-axis** by using the options available in the **Direction** rollout, refer to Figure 11-19.

6. Retain the default settings in the dialog box and choose the **OK** button; the force load is applied on the model.

Solving the Analysis

Next, you need to solve the model.

1. Choose the **Solve** tool from the **Solution** group in the **Home** tab; the **Solve** dialog box is displayed.

2. Make sure the **Solve** option is selected in the **Submit** drop-down list and choose the **OK** button; the solution process is started. Once the solution process is complete, the **END OF JOB** message is displayed in the **Solution Monitor** dialog box.

3. Close all the opened dialog boxes and the window.

After the solution process is complete, a node with the name **Complete Model** is added under the **Post Processing Navigator**, refer to Figure 11-21. This contains the result of the set boundary condition. You will use this result later.

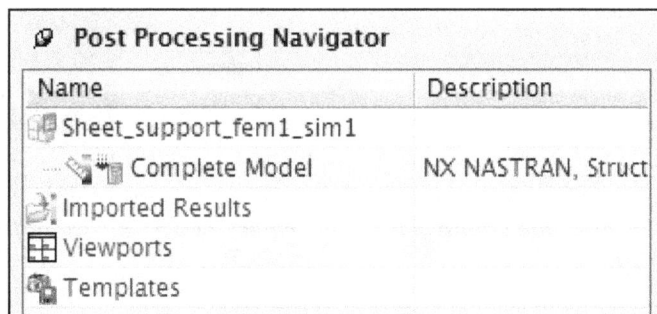

*Figure 11-21 Result node added to the **Post Processing Navigator***

Creating a Group of One Rib and Two Rib Models

Next, you will create groups of model with one and two ribs.

1. Right-click on the **Groups** node in the **Simulation Navigator**, refer to Figure 11-22 and choose the **New Group** option from the shortcut menu; the **New Group** dialog box is displayed, refer to Figure 11-23.

2. Specify **One Rib Model** in the edit box available in the **Name** rollout and select the **Related Elements** option from the **Method** drop-down list available in the **Selection Group** of the **Top Border Bar**, refer to Figure 11-24.

3. Select both the side faces of the center rib, refer to Figure 11-25 and similarly select all faces of support, refer to Figure 11-26.

4. Choose the **OK** button from the dialog box; all the elements of center rib and support are added to the **One Rib Model** group and also get displayed under the **Group** node.

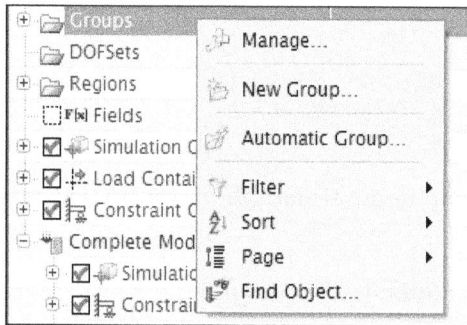

Figure 11-22 Shortcut menu for creating group

Figure 11-23 The New Group dialog box

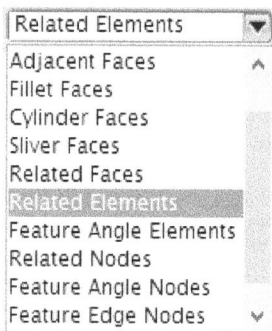

Figure 11-24 Selection Group in the Top Border Bar

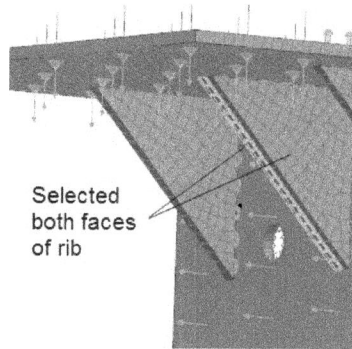

Figure 11-25 Selected surface of rib

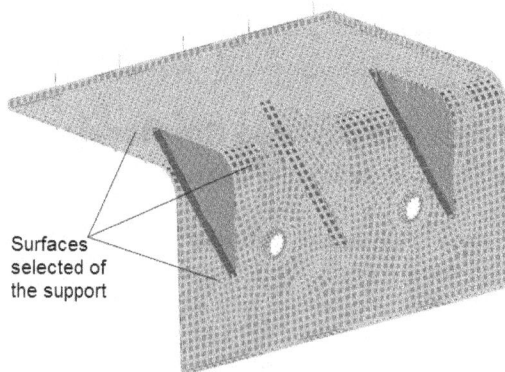

Figure 11-26 Surface selected of support

5. Similarly, create the **Two Rib Model** group and select the support and the rib faces on both sides, refer to Figure 11-27.

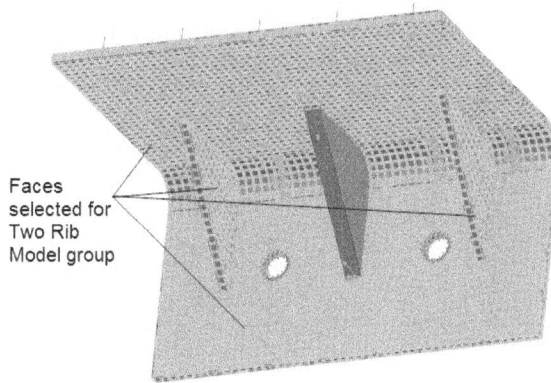

Figure 11-27 *Faces selected for creating the **Two Rib Model** group node*

> **Tip**
> *You can verify your selection by double-clicking on the respective node created under the **Group** node in the **Simulation Navigator**.*

Creating Solution Card for One Rib Model

Next, you will create a Solution Card for One Rib Model and solve the card after applying boundary condition and gluing.

1. Choose the **Solution** tool from the **Environment Drop-down** list in the **Solution** group of the **Home** tab; the **Solution** dialog box is displayed.

2. Specify the name of the Solution Card as **One Rib Model** in the **Name** edit box of the **Solution** rollout.

3. Retain the default settings in the dialog box and choose the **OK** button; a Solution Card named **One Rib Model** is added to the **Simulation Navigator**.

Note that this Solution Card has no boundary condition or gluing applied. So you will now apply boundary condition and gluing before solving the model.

Applying Boundary Conditions and Gluing

You do not need to create separate boundary conditions and gluing for this contact but can drag and drop the node object from the previous Solution Card to this card.

1. Expand the **Constraints** node in the **Complete Model** card, select the **Fixed(1)** node and drag it to the **Constraints** node in the **One Rib Model** card, refer to Figure 11-28. If after dragging a node, the corresponding node has not become active then you need to select the check box manually.

2. Similarly, apply the force load on the model.

For the Surface to Surface gluing, you only need to drag **Face Gluing (1)**, **Face Gluing (4)**, and **Face Gluing (5)** node from the **Simulation Object Container** node, refer to Figure 11-29, since only these contacts are between center rib and support.

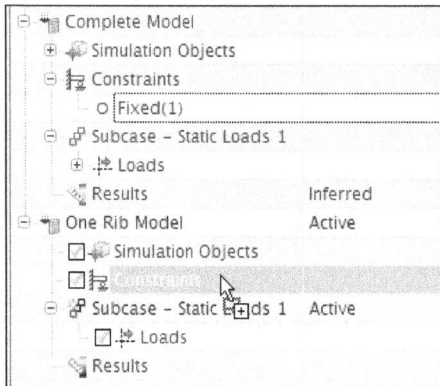

Figure 11-28 Applying constraint using drag and drop

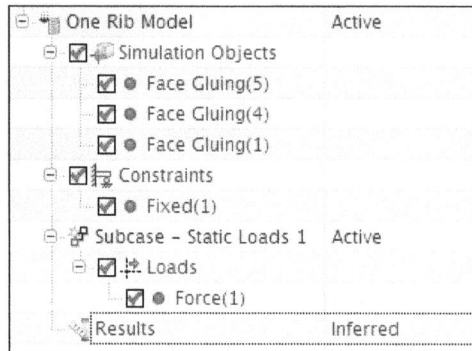

*Figure 11-29 Gluing node in the **Simulation Objects***

Note
*Gluing nodes in the **Simulation Object Container** node may differ in your case. So, you need to check individual nodes in **Simulation Navigator** for their gluing contacts. In case of One Rib Model, you need to drag the gluing contacts that are applied between the surface of centre rib and support surfaces. If you choose a wrong node, then there would be an error message displayed while solving.*

Solving the One Rib Model Analysis

Next, you need to solve the card with **One Rib Model** group that was created before.

1. Choose the **Solve** tool from the **Solution** group of the **Home** tab; the **Solve** dialog box is displayed, refer to Figure 11-30.

2. Clear the **Model Setup Check** check box and choose the **Edit Advanced Solver Options** button; the **Advanced Solver Options** dialog box is displayed, refer to Figure 11-31.

3. Choose the **Export Options** tab and expand the **Subset Export** rollout. Next, choose the **Selected groups** radio button; other options in the rollout become active.

4. Select **One Rib Model** from the **Available groups** area and choose the **Add** button to add it to the **Selected groups** area, refer to Figure 11-31.

5. Retain all the default settings and choose the **OK** button; the **Solve** dialog box is displayed again.

6. Make sure the **Solve** option is chosen in the **Submit** drop-down list and choose the **OK** button; the solving process is started. When the process is complete, the **'END OF JOB'** message is displayed in the **Solution Monitor** dialog box.

7. Close all the opened dialog boxes and window.

After the solution process is complete, another node name **One Rib Model** is added in the **Post Processing Navigator**. This contains the result of the set boundary condition. You will use this result later on.

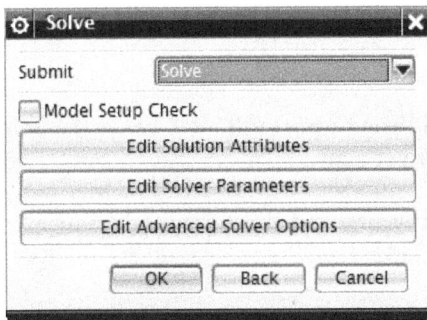

Figure 11-30 *The* **Solve** *dialog box*

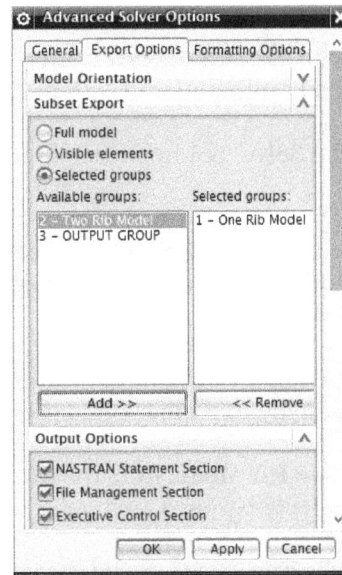

Figure 11-31 *The* **Advanced Solver Options** *dialog box*

Creating Solution Card for the Two Rib Model

Next, you will create a Solution Card for the Two Rib model and solve it after applying the boundary condition and gluing.

1. To create Solution Card, choose the **Solution** tool from the **Environment Drop-down** list in the **Solution** group of the **Home** tab; the **Solution** dialog box is displayed.

2. Specify the name of the Solution Card as **Two Rib Model** in the **Name** edit box in the **Solution** rollout.

3. Retain the default settings in the dialog box and choose the **OK** button; another Solution Card named **Two Rib Model** is added to the **Simulation Navigator**.

Applying Boundary Conditions and Gluing

Apply force and fixed constraint to the Solution Card created using the drag and drop method.

1. In this card, for the Surface to Surface gluing, you only need to drag the **Face Gluing (1)**, **Face Gluing (2)**, **Face Gluing (3)**, **Face Gluing (6)** and **Face Gluing (7)** nodes from the **Simulation Object Container** node, refer to Figure 11-32, since only these contacts are between side ribs and support.

Note
Gluing nodes in the **Simulation Object Container** *node may differ in your case. So, you need to check individual nodes in* **Simulation Navigator** *for their gluing contacts. In case of Two*

Rib Model, you need to drag the gluing contacts that are applied between the surfaces of side ribs and support surfaces. If you choose a wrong node, then there would be an error message displayed while solving.

Solving the Two Rib Model

Next, you need to solve the card with the **Two Rib Model** group that is created.

1. Choose the **Solve** tool from the **Solution** group of the **Home** tab; the **Solve** dialog box is displayed.

2. Make sure that the **Model Setup Check** check box is cleared and choose the **Edit Advanced Solver Options** button; the **Advanced Solver Options** dialog box is displayed, refer to Figure 11-33.

3. Choose the **Export Options** tab and expand the **Subset Export** rollout. Choose the **Selected groups** radio button; other options in the rollout get activated.

4. Select **2- Two Rib Model** from the **Available groups** area and choose the **Add** button to add it to the **Selected groups** area, refer to Figure 11-33.

5. Retain all the default settings and choose the **OK** button; the **Solve** dialog box is displayed again.

6. Make sure the **Solve** option is chosen in the **Submit** drop-down list and then choose the **OK** button; the solving process starts. After the solution process, the 'END OF JOB' message is displayed in the **Solution Monitor** dialog box.

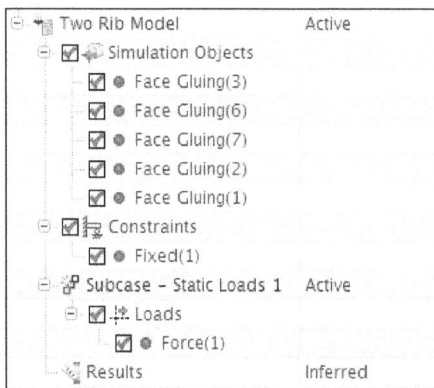

*Figure 11-32 Gluing node in **Simulation Objects***

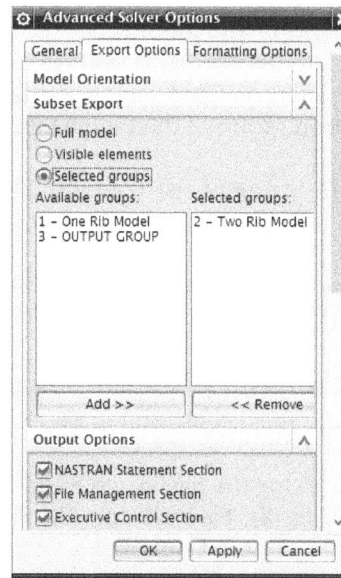

*Figure 11-33 The **Advanced Solver Options** dialog box*

7. Close all the opened dialog boxes and windows.

After the solution process is completed, another node name **Two Rib Model** is added under the **Post Processing Navigator**. This contains the result of the set boundary condition.

Post Processing the Results

Next, you will load the result of every Solution Card in the **Post Processing Navigator**. You will also create viewports and synchronize them to compare the results.

1. Right-click on the **Complete Model** node in the **Post Processing Navigator** and choose the **Load** option from the shortcut menu; all the result gets loaded in the **Post Processing Navigator**.

2. Similarly, load the result of **One Rib Model** and **Two Rib Model** Solution Cards.

3. Select the **Four Views** button from the **Layout** group of the **Results** tab in the **Ribbon**; the screen gets divided into four parts, with each part having model with boundary conditions applied in each viewport.

4. In the **Post Processing Navigator**, expand the **Complete Model** node, refer to Figure 11-34.

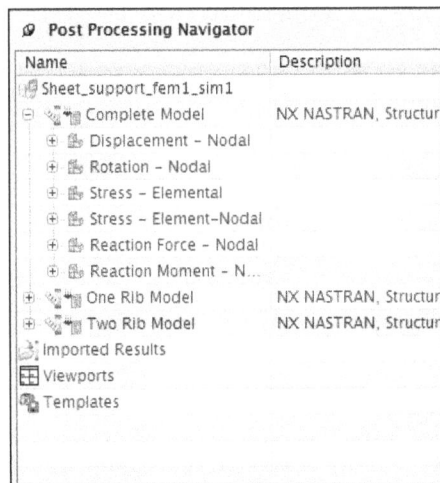

Figure 11-34 The **Complete Model** *node expanded*

5. Right-click on the **Displacement-Nodal** node and choose the **Plot** option from the shortcut menu; you are prompted to select a viewport for plotting the result. Select the second viewport to plot the result.

6. Similarly, plot the **Displacement-Nodal** result using the **One Rib Model** and **Two Rib Model** results.

Next, you will synchronize the orientation of the views.

7. Choose the **View Synchronize** tool from the **Layout** group to synchronize the views; the **Viewport Settings** dialog box is displayed and you are prompted to select all viewports to

synchronize. After selecting the viewports, choose the **OK** button from the **Viewport Settings** dialog box; selected viewports get synchronized.

Figure 11-35 shows the comparison of results in the viewports after the views are synchronized.

Saving and Closing the Model and Report
1. Choose **File > Save > Save** from the **Ribbon** to save the model.

2. Choose **File > Close > All Parts** from the **Ribbon**; all the displayed parts get closed.

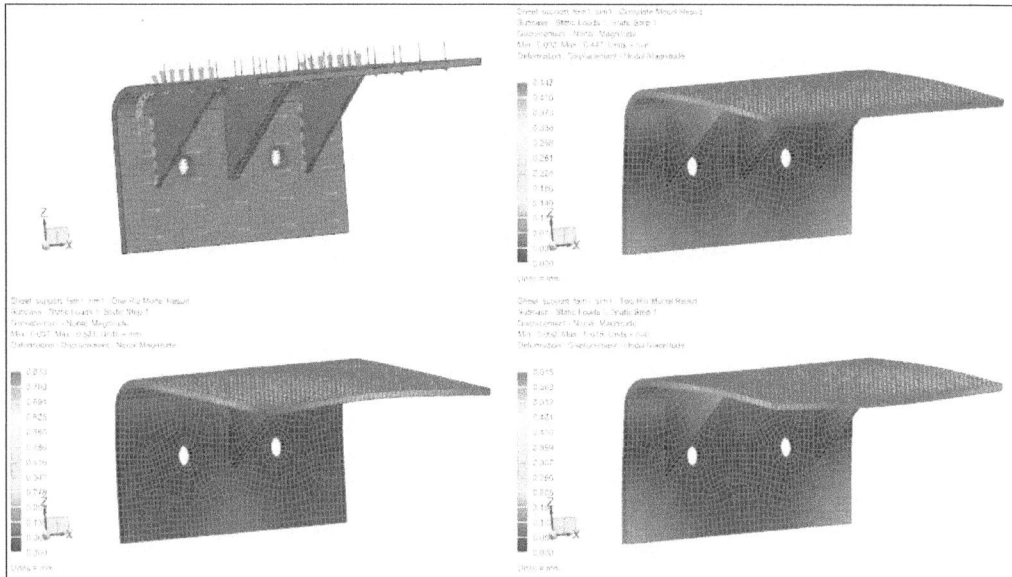

Figure 11-35 *Viewports showing results*

BUCKLING ANALYSIS
The load that creates axial tension or axial compression on the model is called buckling load and the method to analyze the tension or compression is called buckling analysis. Figure 11-36 shows the column deformation under the concentric axial load.

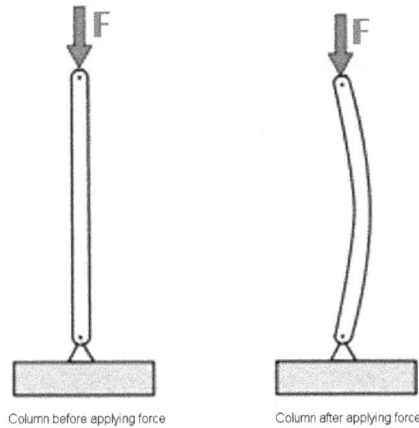

Figure 11-36 Column deformation due to buckling load

Project 2

In this tutorial, you will perform buckling analysis on the pillar shown in Figure 11-37. You can download this file from *www.cadcim.com*. You will idealize the small radii and lastly, you will analyze the result for different modes. Figure 11-38 shows the model with applied mesh and boundary conditions.   **(Expected time: 3 hr)**

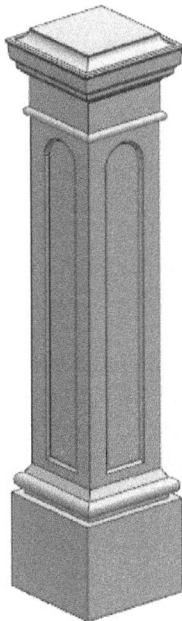

Figure 11-37 Pillar model for buckling load

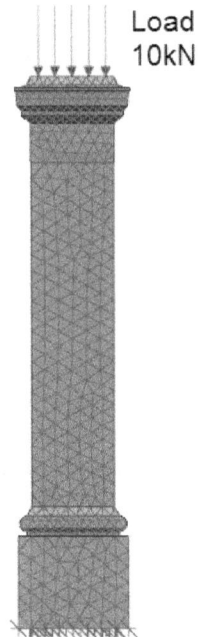

Figure 11-38 Meshed model with Boundary condition

The following steps are required to complete this tutorial:

a. Open the model in NX and create FEM and idealized file.
b. Idealize geometry and remove radii.
c. Defeature geometry and remove round faces.
d. Stitch free edges.
e. Create 3D tetrahedral mesh.
f. Apply material on the model.
g. Enter in Simulation Environment.
h. Apply constraint on the model.
i. Apply force on the model.
j. Solve and load the result.
k. Post process the results.
l. Save and close the model.

Opening the Model in NX and Creating the FEM and Idealized Files

1. Start NX 9.0 by double-clicking on the shortcut icon of NX 9.0 on the desktop of your computer.

2. Download the *c11_P2* zipped folder file from *www.cadcim.com*. Extract the folder to the location *C:\NX Nastran\c11\P2*.

> **Note**
> *You must create the c11 folder at C:\NX Nastran\ as created in the previous chapter.*

3. To open the file, choose **File > Open** from the **Ribbon**; the **Open** dialog box is displayed. Select the **pillar.prt** file from the location where the file is saved and choose the **OK** button; the file is opened in the Modeling environment.

Next, you need to invoke Advanced Simulation environments to create the FEM and Idealized files.

4. Choose **Advanced Simulation** from the **Applications** area of the **File** tab; you are entered in Advanced Simulation environment.

5. Choose the **New FEM** tool from the **Change Window Drop-down** list of the **Context** group under the **Home** tab; the **New Part File** window is displayed. Select the **NX Nastran** option from the **Name** column for the **FEM** type from the window and choose the **OK** button; the **New FEM** dialog box is displayed with the preview of the imported part in the **CAD Part** window. Make sure that the **Create Idealized Part** check box is selected and choose the **OK** button from the dialog box; the part is opened in FEM Environment.

Idealizing the Geometry

Next, you need to idealize the geometry. For Geometry Idealization, you need to enter in the Ideal Environment using the created idealized part.

1. Right-click on the **pillar_fem1_i** file in the **Simulation File View**; a shortcut menu is displayed. Choose the **Make Displayed Part** option from the shortcut menu; the **Idealized**

Part Warning message box is displayed. Choose the **OK** button from the message box; the part is opened in the Ideal Environment.

2. Choose **Menu > Insert > Associative Copy > Promote** from the **Top Border Bar**; the **Promote Body** dialog box is displayed. Select the part from environment and choose the **OK** button; an associative copy of the part selected is created.

Next, you need to remove the small radii.

3. Choose **Menu > Insert > Model Preparation > Idealize** from the **Top Border Bar**; the **Idealize Geometry** dialog box is displayed, refer to Figure 11-39. Select the model from the window to idealize it. Other options of the dialog box get activated.

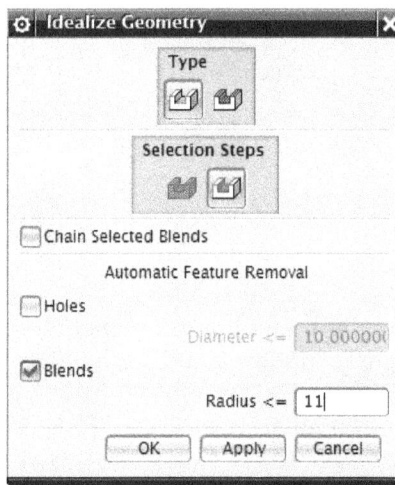

*Figure 11-39 The **Idealize Geometry** dialog box*

4. Select the **Blends** check box; the **Radius <=** edit box becomes available. Specify **11** in the edit box; the parts of the model that have radius value less than this are highlighted.

5. Choose the **OK** button from the dialog box; the **Failing Wounds Exist** window is displayed, refer to Figure 11-40. This warning is displayed as some of the parts having less than specified radius failed to idealize, refer to Figure 11-41. Choose the **Yes** button from the window to leave these radii and proceed further for idealization. Model after idealization is shown in Figure 11-42.

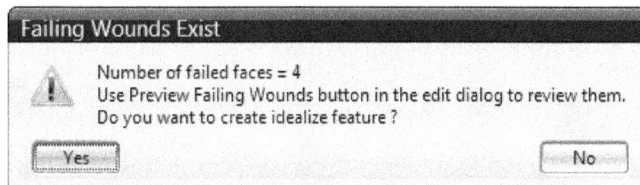

*Figure 11-40 The **Failing Wounds Exist** window*

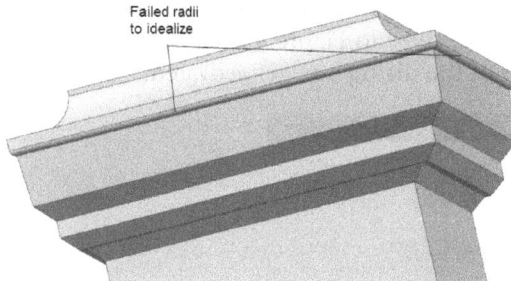

Figure 11-41 Failed radii from idealization

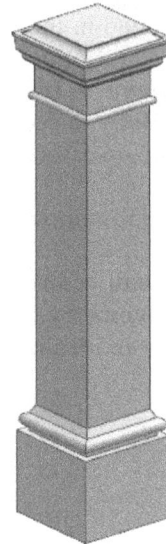

Figure 11-42 The model after idealization

Defeaturing Geometry

Next, you need to remove the upper round portion of the pillar as it will not be considered in the analysis and also take more time in solving and meshing.

1. Choose the **Defeature** tool from **Menu > Insert > Model Preparation** in the **Top Border Bar**; the **Defeature Geometry** dialog box is displayed, as shown in Figure 11-43.

2. Select all four round faces of the model, refer to Figure 11-44 and choose the **OK** button; all the round faces get deleted. Exit the dialog box.

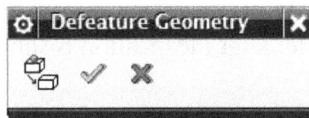

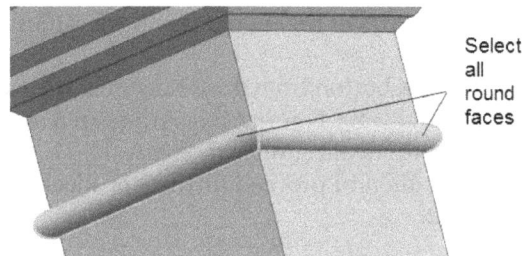

Figure 11-43 The **Defeature Geometry** dialog box

Figure 11-44 Selected round faces

Abstracting Geometry

Next, you will abstract the model by removing radial faces that were left in the model during idealization process. To abstract the geometry, you need to switch to FEM Environment. First you need to delete the round faces and then by using the **Face from Boundary** tool, you will create the face there.

1. Choose the **Delete Face** tool from the **Polygon Geometry** group of the **Home** tab; the **Delete Polygon Face** dialog box is displayed, refer to Figure 11-45.

2. Select all the four faces of the model, refer to Figure 11-46 and choose the **OK** button; all faces get deleted.

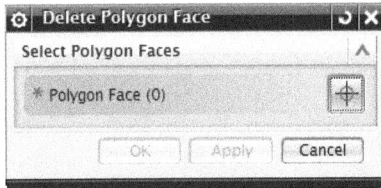

Figure 11-45 The **Delete Polygon Face** *dialog box*

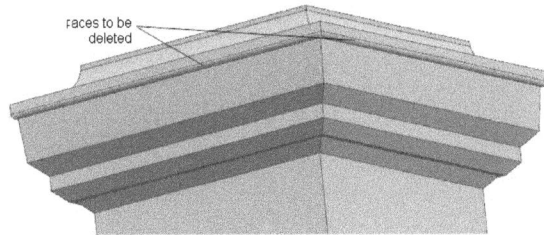

Figure 11-46 *Radii faces selected from the model*

3. Choose the **Face from Boundary** tool from **Menu > Insert > Model Cleanup** in the **Top Border Bar**; the **Face From Boundary Edges** dialog box is displayed. The **Select Edges(0)** area is highlighted in the **Select Boundary** rollout and you are prompted to select edges.

4. Select the two edges from the model, refer to Figure 11-47.

5. Choose the **Inferred Point** button available in the **Add Boundary** rollout and choose the corner point of the selected edges in a sequence to form a close loop, refer to Figure 11-47.

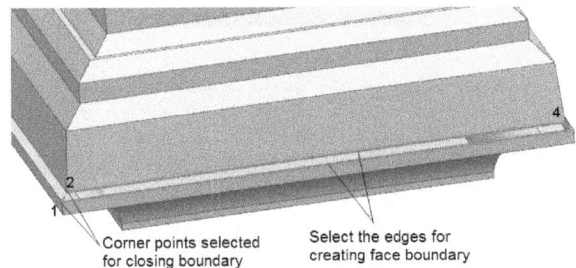

Figure 11-47 *Selected edges and point from the model*

6. Choose the **OK** button to create the face from the loop. Similarly, create faces from other loops.

Stitching Free Edges

Next, you need to stitch the free edges created by using the **Face from Boundary** tool. To stitch the free edges, first you need to display them in the model.

1. Choose **Menu > Preferences > Model Display** from the **Top Border Bar**; the **Model Display** dialog box is displayed, refer to Figure 11-48.

2. Choose the **Polygon Geometry** tab and select the **Display Free Edges** check box from

the **Free Edges** rollout and choose the **OK** button, refer to Figure 11-48; free edges get highlighted in the model.

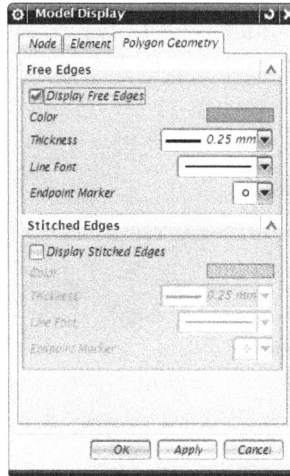

*Figure 11-48 The **Model Display** dialog box*

3. Next, choose **Menu > Insert > Model Cleanup > Stitch Edge** from the **Top Border Bar**; the **Stitch Edge** dialog box is displayed, refer to Figure 11-49.

4. Choose the **Manual** option from the **Method** drop-down list and **Edge to Face** from the **Geometry to Stitch** drop-down list, refer to Figure 11-49.

*Figure 11-49 The **Stitch Edge** dialog box*

5. Next, select the free edges from the model, refer to Figure 11-50 and click in the **Select Face(0)** area of the **Target Geometry** rollout; you are prompted to select the target face.

6. Select the adjacent face as the target face, refer to Figure 11-50 and choose the **Apply** button; the selected face and edge get stitched.

7. Similarly, stitch all four edges and exit the dialog box.

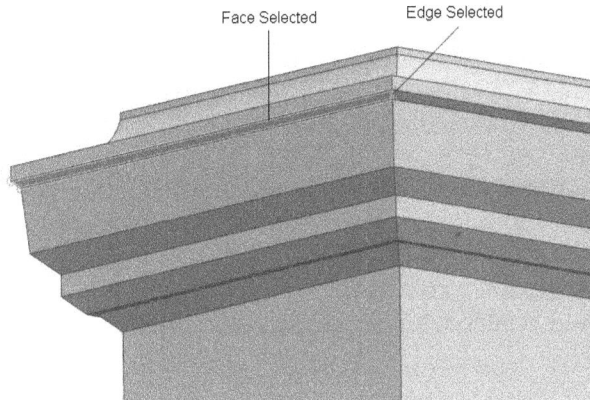

Figure 11-50 *Selected edge and face from the model to stitch*

Creating a 3D Tetrahedral Mesh

Next, you need to create mesh on the pillar. Since the model is a 3D solid, so you will apply 3D mesh using the **3D Tetrahedral Mesh** tool.

1. Choose **Menu > Insert > Mesh > 3D Tetrahedral Mesh** from the **Top Border Bar**; the **3D Tetrahedral Mesh** dialog box is displayed, refer to Figure 11-51.

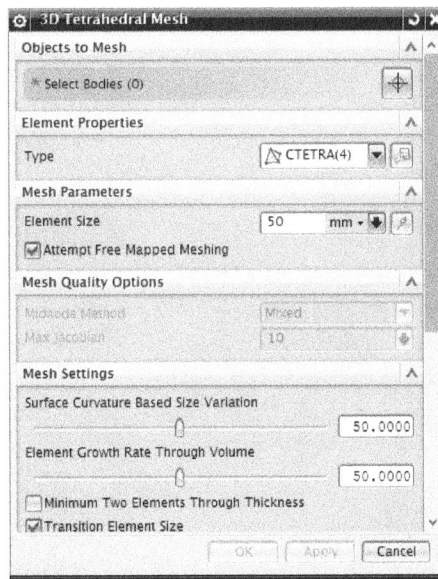

Figure 11-51 *The **3D Tetrahedral Mesh** dialog box*

2. Specify the **CTETRA(4)** element type using the **Type** drop-down list of the **Element Properties** rollout and **50** mm in the **Element Size** edit box of the **Mesh Parameters** rollout, refer to Figure 11-51.

3. Select the model. Retain the default settings and choose the **OK** button to create mesh on the model. The model after creating mesh is shown in Figure 11-52.

Applying Material

Next, you need to apply the material on the model. You will apply the **Iron_Cast_G60** material.

1. Choose **Menu > Tools > Materials > Assign Materials** from the **Top Border Bar**; the **Assign Materials** dialog box is displayed, refer to Figure 11-53.

2. Select the body and **Iron_Cast_G60** from the **Materials** sub-rollout of the **Material List** rollout, refer to Figure 11-53. Choose the **OK** button to apply material on the model.

Entering in Simulation Environment

For buckling analysis, you need to enter in Simulation Environment and create Linear Buckling analysis solution type.

1. Choose the **New Simulation** tool from the **Change Window Drop-down** list of the **Context** group of the **Home** tab; the **New Part File** window is displayed. Select **NX Nastran** from the **Name** column from the **Sim** type and choose the **OK** button from the dialog box; the **New Simulation** dialog box is displayed. Leave the default settings same and choose the **OK** button from the dialog box; the **Solution** dialog box will be displayed.

2. Specify the name **Linear Buckling** in the **Name** edit box and select the **SOL 105 Linear Buckling** option from the **Solution Type** drop-down list, refer to Figure 11-54.

Figure 11-52 The model after meshing

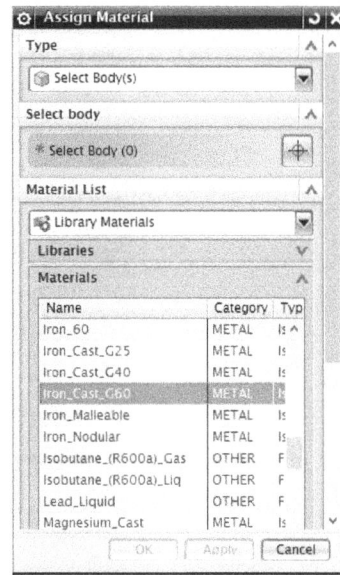

*Figure 11-53 The **Assign Material** dialog box*

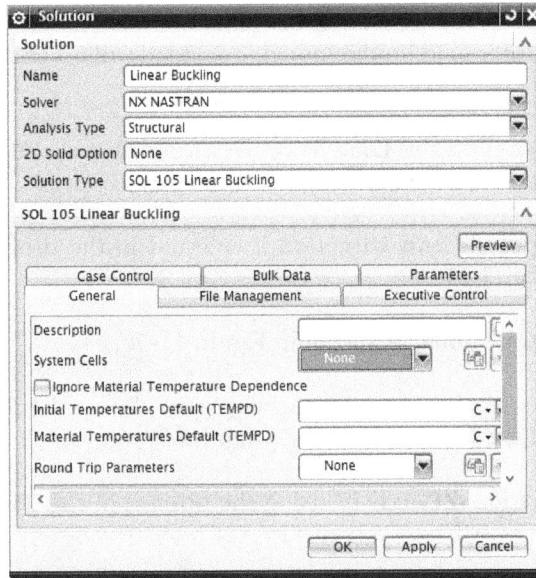

Figure 11-54 The Solution dialog box

3. Retain the rest of the settings in the dialog box and choose the **OK** button; you enter in Simulation Environment.

Applying Fixed Constraint to the Model

Next, you need to apply fixed constraint to the base of the pillar to fix it.

1. Choose the **Fixed Constraint** tool from the **Constraint Type** drop-down list of the **Loads and Conditions** group in the **Home** tab; the **Fixed Constraint** dialog box is displayed, as shown in Figure 11-55.

2. Select the bottom surface of the pillar, as shown in Figure 11-56.

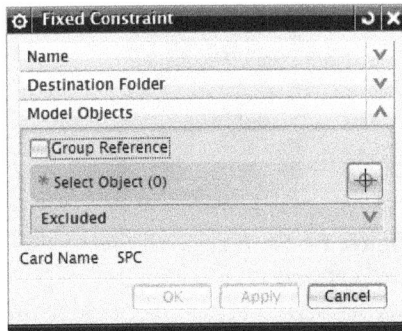

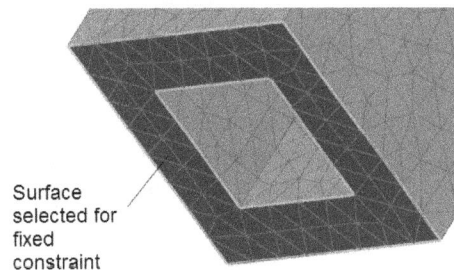

Surface selected for fixed constraint

Figure 11-55   The Fixed Constraint dialog box

Figure 11-56   Surface selected for fixed constraint

3. Retain the rest of the settings in the dialog box and choose the **OK** button; the fixed constraint is applied on the base of the pillar.

Applying Force Load On the Model

Next, you need to apply force load on the model. This load will act as the buckling load on the model.

1. Choose the **Force** tool from the **Load Type** drop-down list of the **Loads and Conditions** group in the **Home** tab; the **Force** dialog box is displayed.

2. Make sure that **Magnitude and Direction** is selected in the drop-down list of the **Type** rollout, refer to Figure 11-57.

3. Select the top face of the pillar, as shown in Figure 11-58.

4. Specify **10000** N force load in the edit box when you choose the **Expression** option from the drop-down list of the **Force** area of the **Magnitude** rollout.

5. Specify the load direction **-ZC-axis** by using the options of the **Direction** rollout, refer to Figure 11-57.

6. Retain the rest of the settings in the dialog box and choose the **OK** button; the force load is applied on the model.

Figure 11-57 *The **Force** dialog box*   *Figure 11-58* *Face selected for load applying*

Solving the Analysis

Next, you will solve the model.

1. Choose the **Solve** tool from the **Solution** group of the **Home** tab; the **Solve** dialog box is displayed.

2. Make sure the **Solve** option is chosen in the **Submit** drop-down list and choose the **OK** button; the solving process is started. After completion of the solution process, the **'END OF JOB'** message is displayed in the **Solution Monitor** dialog box.

3. Close all the opened dialog boxes and windows.

Post Processing the Result

Next, you need to load the result in the **Post Processing Navigator** and create the report.

1. Right-click on the **Result** node from the **Simulation Navigator**, refer to Figure 11-59 and choose the **Open** option from the shortcut menu; all result are loaded in the **Post Processing Navigator**, refer to Figure 11-60.

Note
The result contains the 10 modes along which the model is deformed by the buckling load. You can individually check the particular modes. Next, you will create report of Von-Mises Stress from first three modes of the analysis using Viewport.

2. Choose the **Create Report** tool from the **Solution** group of the **Home** tab; the **Report** node is added in the **Simulation Navigator**.

3. Expand the **Report** node and right-click on the **Title** sub-node, refer to Figure 11-61 and choose the **Edit** option; the **Report Editor** dialog box is displayed.

4. Specify the title **Linear Buckling Analysis** in the **Report Editor** dialog box, refer to Figure 11-62 to change the title. Choose the **OK** button from the dialog box.

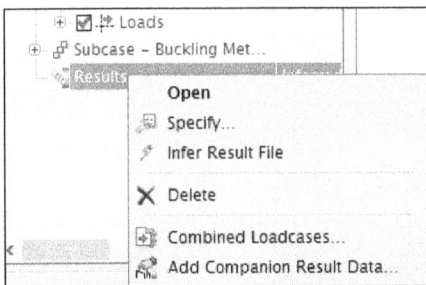

Figure 11-59 Shortcut menu for loading results

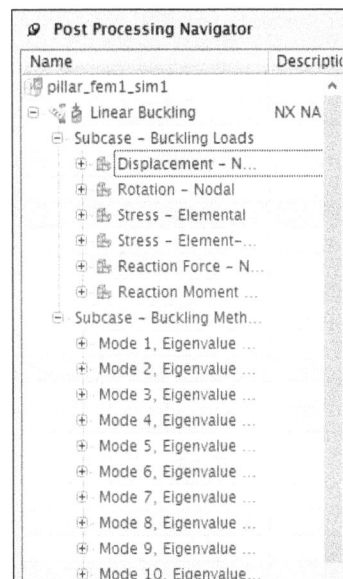

Figure 11-60 Expanded nodes of the uploaded results

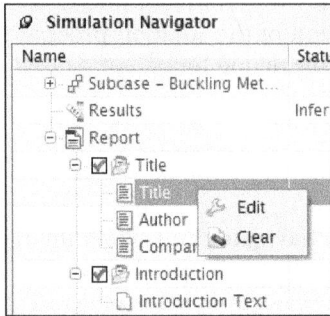

Figure 11-61 *The **Title** sub-node*

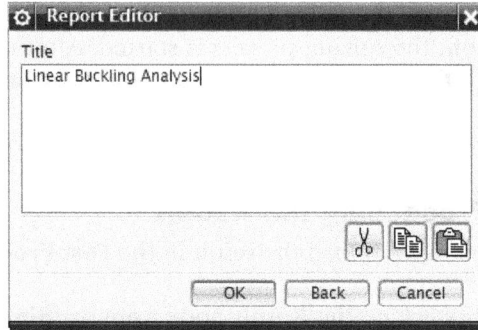

Figure 11-62 *The **Report Editor** dialog box*

Next, you will create four viewports.

5. Select the **Four Views** button from the **Layout** group of the **Results** tab in the **Ribbon**; the screen is divided into four parts with random part preview in each viewport.

6. In the **Post Processing Navigator**, expand the **Stress-Elemental** node under the **Mode1, Eigen Value=6.405e+003** node in the **Subcase-Buckling Method** node, refer to Figure 11-63.

7. Right-click on the **Von-Mises** and choose the **Plot** option from the shortcut menu; you will be prompted to select viewport for result plot. Select the first viewport to plot the result.

8. Similarly, plot the Von-Mises of Stress Elemental for other three modes.

Next, you will synchronize these views to orient correspondingly.

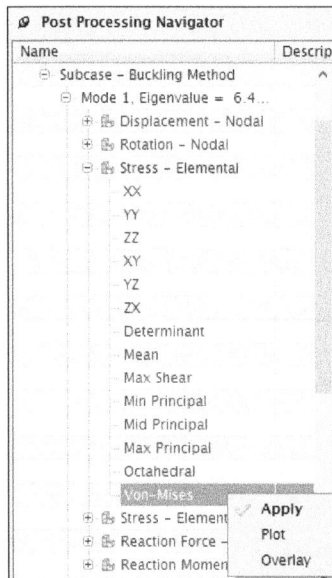

Figure 11-63 *The plotting of results*

9. Choose the **View Synchronize** tool from the **Layout** group of the **Result** tab to synchronize the view; the **Viewport Settings** dialog box is displayed and you are prompted to select all viewport to synchronize. After selecting the viewports, choose the **OK** button from the **Viewport Settings** dialog box; selected viewport are synchronized and act simultaneously.

10. Choose the **Front** button to orient the all the views in the front view position, refer to Figure 11-64.

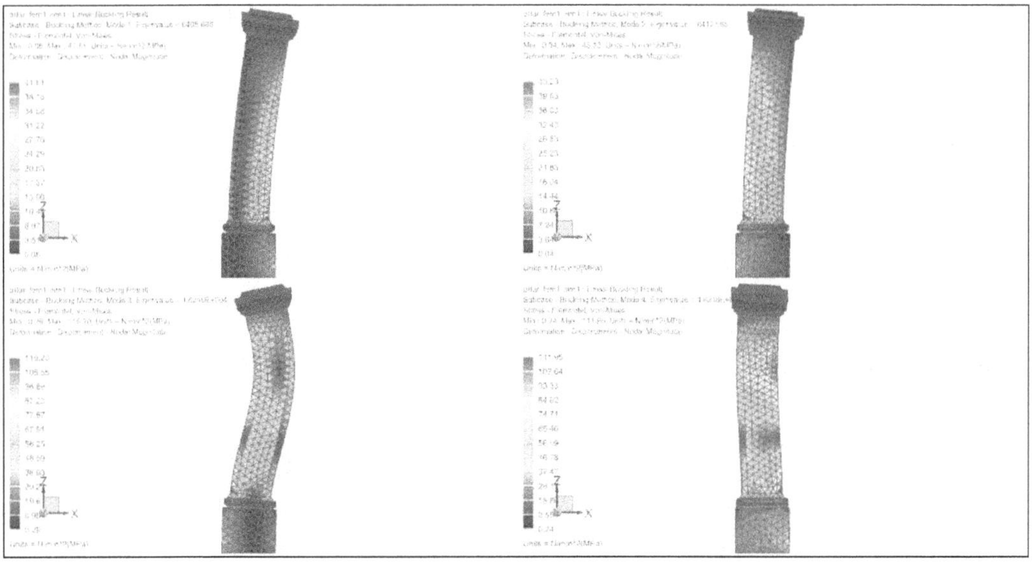

Figure 11-64 Results plotted in viewport

11. Right-click on the **Images** sub-node under the **Images** node of the **Report** node from the **Simulation Navigator**, and choose **Snapshot** from the shortcut menu; the **Image1** node is added under the **Images** node.

12. Right-click on the **Image1** node and choose **Edit Description** from the shortcut menu; the edit box in the **Simulation Navigator** is displayed. Specify description **Von-Mises Stress comparison** in the edit box.

13. Next, to export the report, right click on the **Report** node and choose the **Export** option from the shortcut menu; the report is opened in your default browser.

Saving and Closing the Model and Report

1. Choose **File > Save > Save** from the **Ribbon** to save the model.

2. Choose **File > Close > All Parts** from the **Ribbon**; all displayed parts are closed.

3. Also, save and close the report that is opened in the browser.

RESPONSIVE ANALYSIS

The responsive analysis is used on the models which are affected by the dynamic forced response.

Project 3

In this tutorial, you will do Response Simulation on the hook shown in Figure 11-65. You can also download this file from *www.cadcim.com*. Figure 11-66 shows the model with applied mesh and boundary condition.   **(Expected time: 3 hr)**

The following steps are required to complete this tutorial:

a. Open the model in NX and create the FEM and idealized files.
b. Split the model for meshing.
c. Create mesh on the hook.
d. Apply material on the model.
e. Enter in Simulation Environment.
f. Apply constraint on the model.
g. Apply force on the model.
h. Solve and load the result.

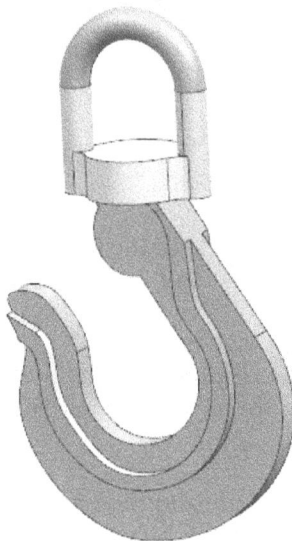

Figure 11-65 *Hook model for Response analysis*

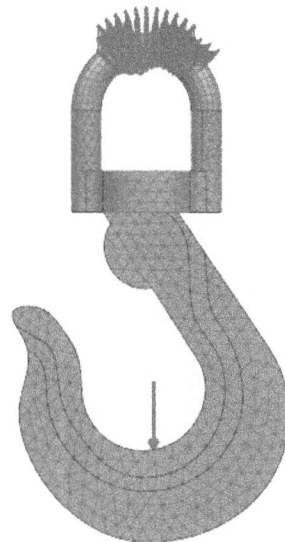

Figure 11-66 *Hook with boundary condition*

Opening the Model in NX and Creating the FEM and Idealized Files

1. Start NX 9.0 by double-clicking on the shortcut icon of NX 9.0 on the desktop of your computer.

2. Download the *c11_P3* zipped file from *www.cadcim.com*. Extract the folder to the location *C:\NX Nastran\c11\P3*.

> **Note**
> *You must create the c11 folder at the location C:\NX Nastran\ created in previous chapter.*

3. To open the file, choose **File > Open** from the **Ribbon**; the **Open** dialog box is displayed. Select the **Hook.prt** file from the location and choose the **OK** button; the file is opened in the Gateway Environment.

Next, you need to invoke Advanced Simulation environments to create the FEM and the Idealized files.

4. Choose the **Advanced Simulation** from the **Applications** area of the **File** tab; you enter in Advanced Simulation environments.

5. Choose the **New FEM** tool from the **Change Window Drop-down** list of the **Context** group in the **Home** tab; the **New Part File** window is displayed. Select **NX Nastran** from the **Name** column for the **FEM** type from the window and choose the **OK** button; the **New FEM** dialog box is displayed with the preview of the imported part in the **CAD Part** window. Select the **Create Idealized Part** check box, if it is not selected by default, and choose the **OK** button from the dialog box; the part is opened in FEM Environment.

Splitting the Model For Meshing

Before meshing, you need to split the hook for better mesh quality. To split the part, you must be in Ideal Environment and promote the part as discussed earlier.

1. Choose **Split Body** tool from **Menu > Insert > Model Preparation > Trim** from the **Top Border Bar**; the **Split Body** dialog box is displayed, refer to Figure 11-67.

2. Select the body using the option of the **Target** rollout.

3. Select the **New Plane** option from the **Tool Option** drop-down list and the **Point and Direction** method from the **Inferred** drop-down list, refer to Figure 11-67.

4. Select the round edge and Y-axis direction for creating the plane, refer to Figure 11-68.

5. Make sure the **Create Mesh Mating Conditions** and **Check for Sweepable Body** check boxes are selected.

6. Choose the **OK** button to split the upper part of the hook and exit the dialog box. Also, switch to FEM Environment using the **Simulation File View**.

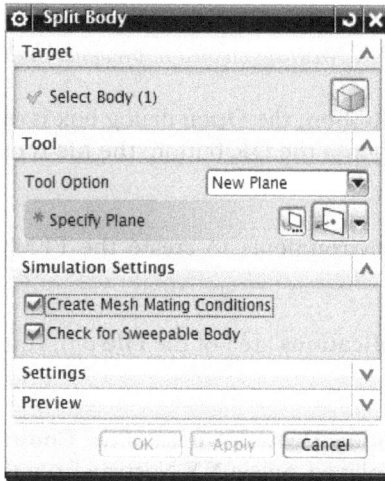

Figure 11-67 The **Split Body** dialog box

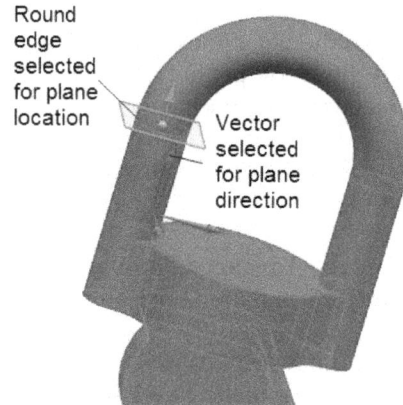

Figure 11-68 *Edge and vector selected for plane creation*

Creating Mesh On the Hook

Next, you need to create mesh on the hook. The upper part will have the 3D swept mesh while on the remaining part of hook, you will create a 3D tetrahedral mesh.

1. Choose **Menu > Insert > Mesh > 3D Swept Mesh** from the **Top Border Bar**; the **3D Swept Mesh** dialog box is displayed, as shown in Figure 11-69.

2. Select the inner face of the upper part of the hook. You can select this surface using the **QuickPick** dialog box, refer to Figure 11-70, or by hiding the bottom part. After selecting the face, a message is prompted to create the Hex meshing on the highlighted part. Choose the **No** button.

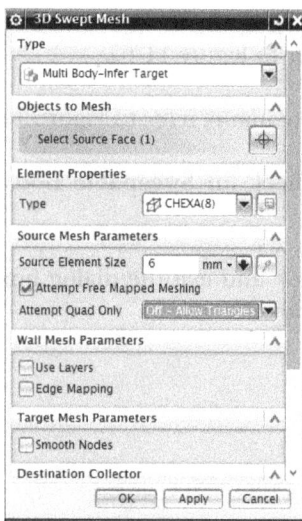

Figure 11-69 The **3D Swept Mesh** dialog box

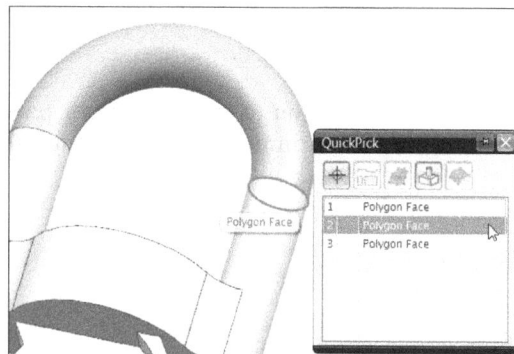

Figure 11-70 *Selecting surface using the **QuickPick** dialog box*

3. Select the **CHEXA(8)** element type and **6** mm element size in their respective rollouts. Also, select the **Attempt Free Mapped Meshing** check box and the **Off - Allow Triangles** option from the **Attempt Quad Only** drop-down list in the **Source Mesh Parameters** rollout.

4. Choose the **Apply** button to create swept mesh on the split part of the hook and exit the dialog box.

Next, create the 3D tetrahedral mesh on the remaining part of the hook.

5. Choose **Menu > Insert > Mesh > 3D Tetrahedral Mesh** from the **Top Border Bar**; the **3D Tetrahedral Mesh** dialog box is displayed, refer to Figure 11-71.

6. Select the **CTETRA(10)** element type from the **Type** drop-down list of the **Element Properties** rollout and specify **15** mm in the **Element Size** edit box of the **Mesh Parameters** rollout. Also, select the **Attempt Free Mapped Meshing** check box, refer to Figure 11-71.

7. Select the remaining part of the model from the environment. Retain the rest of the settings and choose the **OK** button to create the mesh on the model. The model after creating the mesh is shown in Figure 11-72.

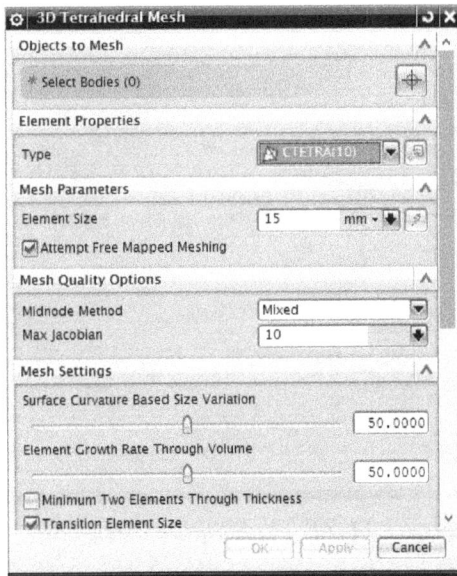

Figure 11-71 The **3D Tetrahedral Mesh** dialog box

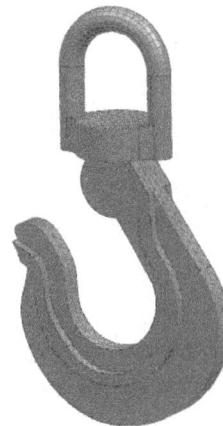

Figure 11-72 Hook after meshing

Applying Material

Next, you need to apply the material on the model. You will apply the **AISI_Steel_1005** material.

1. Choose **Menu > Tools > Materials > Assign Materials** from the **Top Border Bar**; the **Assign Materials** dialog box is displayed, refer to Figure 11-73.

2. Select the complete model and **AISI_Steel_1005** from the **Materials** sub-rollout in the **Material List** rollout, refer to Figure 11-73. Choose the **OK** button to apply the material on the model.

Entering in Simulation Environment

For response analysis, you have to enter in Simulation Environment and create Response analysis solution type.

1. Choose the **New Simulation** tool from the **Change Window Drop-down** list of the **Context** group under the **Home** tab; the **New Part File** window is displayed. Select **NX Nastran** from the **Name** column for the **Sim** type and choose the **OK** button from the dialog box; the **New Simulation** dialog box is displayed. Retain the rest of the settings and choose the **OK** button from the dialog box; the **Solution** dialog box is displayed.

2. Specify **Response Simulation** in the **Name** edit box and select the **SOL 103 Response Simulation** option from the **Solution Type** drop-down list, refer to Figure 11-74.

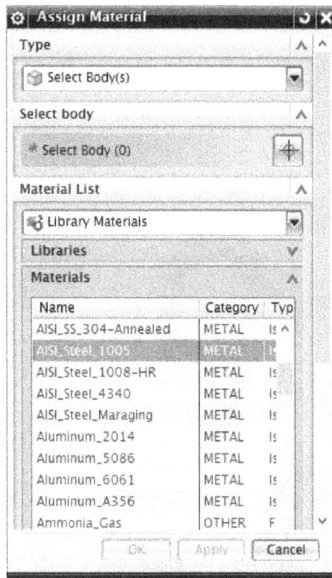

Figure 11-73 The *Assign Material* dialog box

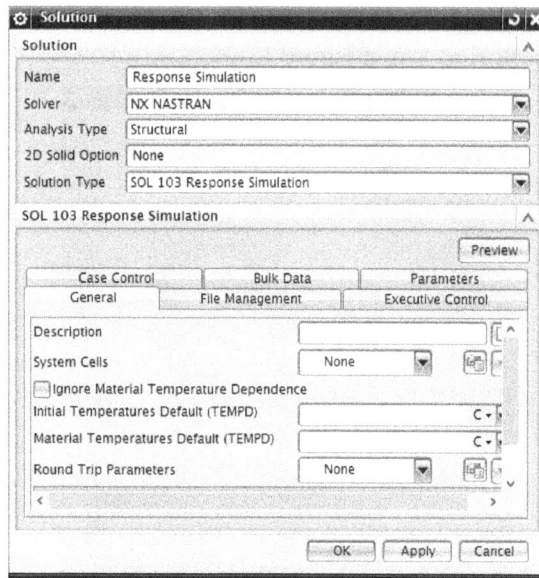

Figure 11-74 The *Solution* dialog box

3. Retain the rest of the settings in the dialog box and choose the **OK** button; you enter in Simulation Environment.

Applying Constraint to the Model

Next, you need to apply constraint to the hook. Since the hook also hangs on other rod, so you need to constrain it.

1. Choose the **User Defined Constraint** tool from the **Constraint Type** drop-down list of the **Loads and Conditions** group of the **Home** tab; the **User Defined Constraint** dialog box is displayed, refer to Figure 11-75.

2. Select the nodes which will be affected while hanging, refer to Figure 11-76. You can use the **Lasso** tool for this selection.

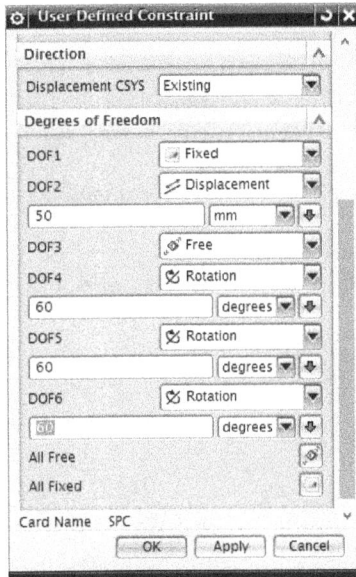

*Figure 11-75 The **User Defined Constraint** dialog box*

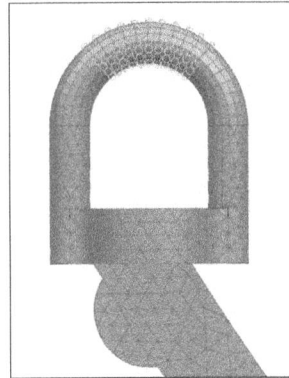

Figure 11-76 Nodes selected for applying constraint

3. Specify the values for different degrees of freedom as given below:
   DOF1 : Fixed,
   DOF2 : Displacement 50 mm
   DOF3 : Free
   DOF4 : Rotation 60 degree
   DOF5 : Rotation 60 degree
   DOF6 : Rotation 60 degree

4. Retain the rest of the settings in the dialog box and choose the **OK** button; the selected nodes are constrained.

Applying the Force Load on the Model

Next, you will apply force on the model. The force load applied on the center position of hook will have more affect on the hook at that point.

1. Choose the **Force** tool from the **Load Type** drop-down list of the **Loads and Conditions** group of the **Home** tab; the **Force** dialog box is displayed.

2. Select the **Magnitude and Direction** option from the drop-down list of the **Type** rollout, refer to Figure 11-77.

3. Select the center node on the inner face of hook, refer to Figure 11-78.

Tip
*To select the centre node, choose the **Node** option from the **Type Filter** drop-down list of the **Selection Group** in the **Top Border Bar**.*

4. Specify **10000** N force load in the edit box when you choose the **Expression** option from the drop-down list of the **Force** area of the **Magnitude** rollout.

5. Specify the load direction **-YC-axis** by using the options of the **Direction** rollout, refer to Figure 11-77.

6. Retain the rest of the settings in the dialog box and choose the **OK** button; the force load is applied on the model, refer to Figure 11-78.

*Figure 11-77 The **Force** dialog box*

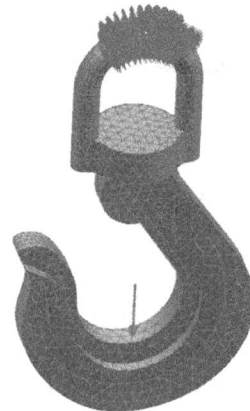

Figure 11-78 Face selected for applying load

Solving and Loading the Result

Next, you need to solve the model.

1. Choose the **Solve** tool from the **Solution** group of the **Home** tab; the **Solve** dialog box is displayed.

2. Make sure the **Solve** option is chosen in the **Submit** drop-down list and choose the **OK** button; the solving process is started. Once the solution process is complete, the **'END OF JOB'** message is displayed in the **Solution Monitor** dialog box.

3. Close all the opened dialog boxes and windows.

4. Right-click on the **Result** node in the **Simulation Navigator**, refer to Figure 11-79 and choose the **Open** option from the shortcut menu; all the results are loaded in the **Post Processing Navigator**, refer to Figure 11-80.

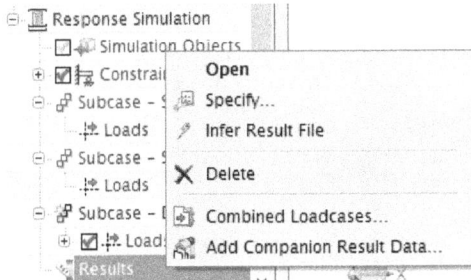

Figure 11-79 Shortcut menu for loading results

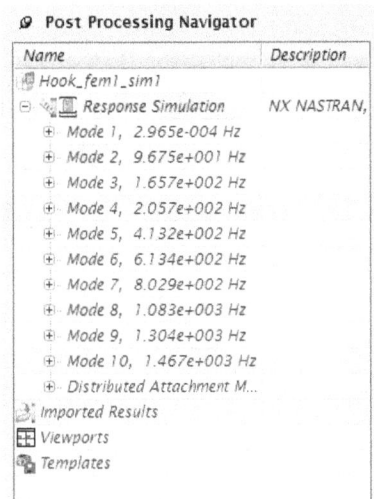

Figure 11-80 The uploaded result

Note
The result contains the 10 modes along which the model is deformed under the different frequencies. You can create the viewport for comparing the deformation results along different modes. Figure 11-81 shows the Nine Views viewport in which different views of modes are plotted.

5. After creating the views, create the **Report** node in the **Simulation Navigator** and export it.

Saving and Closing the Model and Report

1. Choose **File > Save** from the **Ribbon** to save the model.

2. Choose **File > Close > All Parts** from the **Ribbon**; all displayed parts get closed.

3. Also, save and close the report that was opened in the browser.

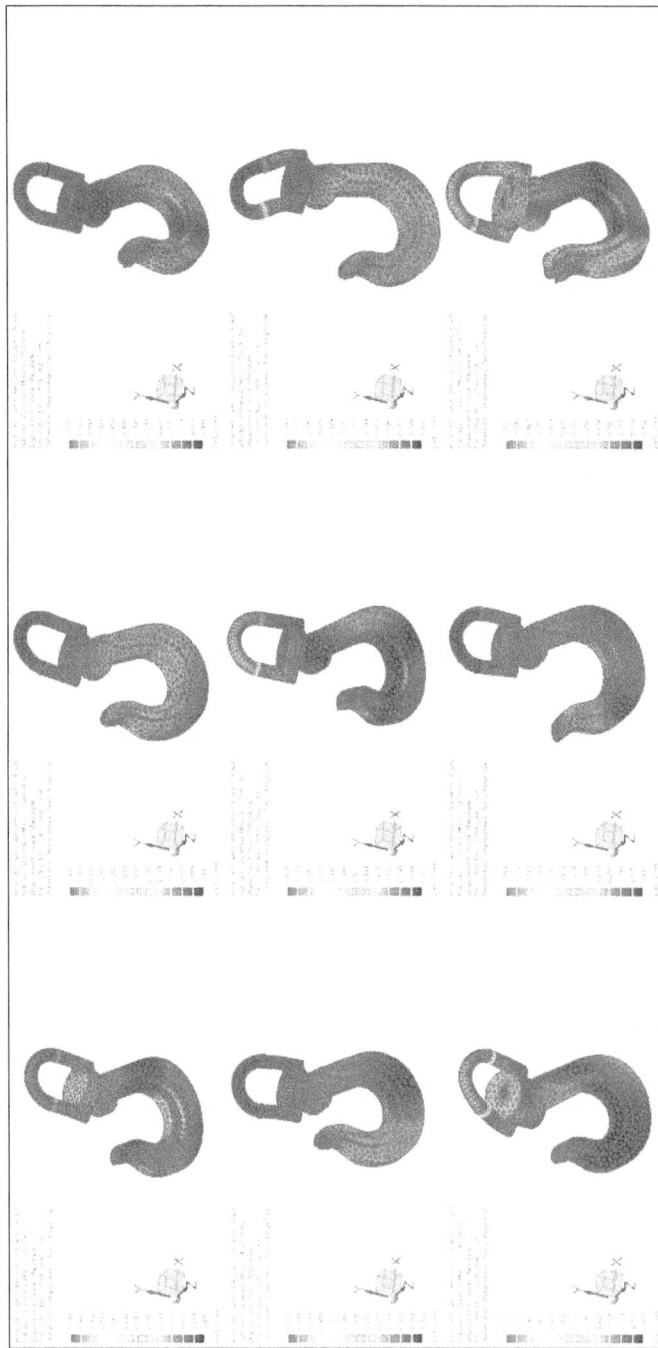

Figure 11-81 Hook deformation for different modes at different frequencies

Index

Other Publications by CADCIM Technologies

The following is the list of some of the publications by CADCIM Technologies. Please visit *www.cadcim.com* for the complete listing.

AutoCAD Textbooks
- AutoCAD 2017: A Problem-Solving Approach, Basic and Intermediate, 23nd Edition
- AutoCAD 2017: A Problem-Solving Approach, 3D and Advanced, 23rd Edition
- AutoCAD 2016: A Problem-Solving Approach, Basic and Intermediate, 22nd Edition
- AutoCAD 2016: A Problem-Solving Approach, 3D and Advanced, 22nd Edition
- AutoCAD 2015: A Problem-Solving Approach, Basic and Intermediate, 21st Edition
- AutoCAD 2015: A Problem-Solving Approach, 3D and Advanced, 21st Edition
- AutoCAD 2014: A Problem-Solving Approach

Autodesk Inventor Textbooks
- Autodesk Inventor 2017 for Designers, 17th Edition
- Autodesk Inventor 2016 for Designers, 16th Edition
- Autodesk Inventor 2015 for Designers, 15th Edition
- Autodesk Inventor 2014 for Designers
- Autodesk Inventor 2013 for Designers
- Autodesk Inventor 2012 for Designers
- Autodesk Inventor 2011 for Designers

AutoCAD MEP Textbooks
- AutoCAD MEP 2016 for Designers, 3rd Edition
- AutoCAD MEP 2015 for Designers
- AutoCAD MEP 2014 for Designers

Solid Edge Textbooks
- Solid Edge ST8 for Designers, 13th Edition
- Solid Edge ST7 for Designers, 12th Edition
- Solid Edge ST6 for Designers
- Solid Edge ST5 for Designers
- Solid Edge ST4 for Designers
- Solid Edge ST3 for Designers
- Solid Edge ST2 for Designers

NX Textbooks
- NX 10.0 for Designers, 9th Edition
- NX 9.0 for Designers, 8th Edition
- NX 8.5 for Designers
- NX 8 for Designers
- NX 7 for Designers
- NX 6 for Designers

SolidWorks Textbooks
- SOLIDWORKS 2016 for Designers, 14th Edition
- SOLIDWORKS 2015 for Designers, 13th Edition
- SolidWorks 2014 for Designers
- SolidWorks 2013 for Designers
- SolidWorks 2012 for Designers
- SolidWorks 2014: A Tutorial Approach
- SolidWorks 2012: A Tutorial Approach
- Learning SolidWorks 2011: A Project Based Approach
- SolidWorks 2011 for Designers

CATIA Textbooks
- CATIA V5-6R2015 for Designers, 13th Edition
- CATIA V5-6R2014 for Designers, 12th Edition
- CATIA V5-6R2013 for Designers
- CATIA V5-6R2012 for Designers
- CATIA V5R21 for Designers
- CATIA V5R20 for Designers
- CATIA V5R19 for Designers

Creo Parametric and Pro/ENGINEER Textbooks
- PTC Creo Parametric 3.0 for Designers, 3rd Edition
- Creo Parametric 2.0 for Designers
- Creo Parametric 1.0 for Designers
- Pro/Engineer Wildfire 5.0 for Designers
- Pro/ENGINEER Wildfire 4.0 for Designers
- Pro/ENGINEER Wildfire 3.0 for Designers

ANSYS Textbooks
- ANSYS Workbench 14.0: A Tutorial Approach
- ANSYS 11.0 for Designers

Creo Direct Textbook
- Creo Direct 2.0 and Beyond for Designers

Autodesk Alias Textbooks
- Learning Autodesk Alias Design 2016, 5th Edition
- Learning Autodesk Alias Design 2015, 4th Edition
- Learning Autodesk Alias Design 2012
- Learning Autodesk Alias Design 2010
- AliasStudio 2009 for Designers

AutoCAD LT Textbooks
- AutoCAD LT 2016 for Designers, 11th Edition
- AutoCAD LT 2015 for Designers, 10th Edition
- AutoCAD LT 2014 for Designers

- AutoCAD LT 2013 for Designers
- AutoCAD LT 2012 for Designers
- AutoCAD LT 2011 for Designers

EdgeCAM Textbooks
- EdgeCAM 11.0 for Manufacturers
- EdgeCAM 10.0 for Manufacturers

AutoCAD Electrical Textbooks
- AutoCAD Electrical 2017 for Electrical Control Designers, 8th Edition
- AutoCAD Electrical 2016 for Electrical Control Designers, 7th Edition
- AutoCAD Electrical 2015 for Electrical Control Designers, 6th Edition
- AutoCAD Electrical 2014 for Electrical Control Designers
- AutoCAD Electrical 2013 for Electrical Control Designers
- AutoCAD Electrical 2012 for Electrical Control Designers
- AutoCAD Electrical 2011 for Electrical Control Designers
- AutoCAD Electrical 2010 for Electrical Control Designers

Autodesk Revit Architecture Textbooks
- Exploring Autodesk Revit 2017 for Architecture, 13th Edition
- Autodesk Revit Architecture 2016 for Architects and Designers, 12th Edition
- Autodesk Revit Architecture 2015 for Architects and Designers, 11th Edition
- Autodesk Revit Architecture 2014 for Architects and Designers
- Autodesk Revit Architecture 2013 for Architects and Designers
- Autodesk Revit Architecture 2012 for Architects and Designers
- Autodesk Revit Architecture 2011 for Architects & Designers

Autodesk Revit Structure Textbooks
- Exploring Autodesk Revit 2017 for Structure, 7th Edition
- Exploring Autodesk Revit Structure 2016, 6th Edition
- Exploring Autodesk Revit Structure 2015, 5th Edition
- Exploring Autodesk Revit Structure 2014
- Exploring Autodesk Revit Structure 2013
- Exploring Autodesk Revit Structure 2012

AutoCAD Civil 3D Textbooks
- Exploring AutoCAD Civil 3D 2017, 7th Edition
- Exploring AutoCAD Civil 3D 2016, 6th Edition
- Exploring AutoCAD Civil 3D 2015, 5th Edition
- Exploring AutoCAD Civil 3D 2014
- Exploring AutoCAD Civil 3D 2013
- Exploring AutoCAD Civil 3D 2012

AutoCAD Map 3D Textbooks
- Exploring AutoCAD Map 3D 2017, 7th Edition
- Exploring AutoCAD Map 3D 2016, 6th Edition
- Exploring AutoCAD Map 3D 2015, 5th Edition

- Exploring AutoCAD Map 3D 2014
- Exploring AutoCAD Map 3D 2013
- Exploring AutoCAD Map 3D 2012
- Exploring AutoCAD Map 3D 2011

3ds Max Design Textbooks
- Autodesk 3ds Max Design 2015: A Tutorial Approach, 15th Edition
- Autodesk 3ds Max Design 2014: A Tutorial Approach
- Autodesk 3ds Max Design 2013: A Tutorial Approach
- Autodesk 3ds Max Design 2012: A Tutorial Approach
- Autodesk 3ds Max Design 2011: A Tutorial Approach
- Autodesk 3ds Max Design 2010: A Tutorial Approach

3ds Max Textbooks
- Autodesk 3ds Max 2017 for Beginners: A Tutorial Approach, 17th Edition
- Autodesk 3ds Max 2016 for Beginners: A Tutorial Approach, 16th Edition
- Autodesk 3ds Max 2017: A Comprehensive Guide, 17th Edition
- Autodesk 3ds Max 2016: A Comprehensive Guide, 16th Edition
- Autodesk 3ds Max 2016 for Beginners: A Tutorial Approach, 16th Edition
- Autodesk 3ds Max 2015: A Comprehensive Guide, 15th Edition
- Autodesk 3ds Max 2014: A Comprehensive Guide
- Autodesk 3ds Max 2013: A Comprehensive Guide
- Autodesk 3ds Max 2012: A Comprehensive Guide
- Autodesk 3ds Max 2011: A Comprehensive Guide

Autodesk Maya Textbooks
- Autodesk Maya 2016: A Comprehensive Guide, 8th Edition
- Autodesk Maya 2015: A Comprehensive Guide, 7th Edition
- Character Animation: A Tutorial Approach
- Autodesk Maya 2014: A Comprehensive Guide
- Autodesk Maya 2013: A Comprehensive Guide
- Autodesk Maya 2012: A Comprehensive Guide

ZBrush Textbooks
- Pixologic ZBrush 4R7: A Comprehensive Guide
- Pixologic ZBrush 4R6: A Comprehensive Guide

Fusion Textbooks
- Blackmagic Design Fusion 7 Studio: A Tutorial Approach
- The eyeon Fusion 6.3: A Tutorial Approach

Flash Textbooks
- Adobe Flash Professional CC2015: A Tutorial Approach
- Adobe Flash Professional CC: A Tutorial Approach
- Adobe Flash Professional CS6: A Tutorial Approach

Computer Programming Textbooks

- Introduction to C++ programming
- Learning Oracle 11g
- Learning ASP.NET AJAX
- Introduction to Java Programming
- Learning Java Programming
- Learning Visual Basic.NET 2008
- Introduction to C++ Programming Concepts
- Learning C++ Programming Concepts
- Introduction to VB.NET Programming Concepts
- Learning VB.NET Programming Concepts

AutoCAD Textbooks Authored by Prof. Sham Tickoo and Published by Autodesk Press

- AutoCAD: A Problem-Solving Approach: 2013 and Beyond
- AutoCAD 2012: A Problem-Solving Approach
- AutoCAD 2011: A Problem-Solving Approach
- AutoCAD 2010: A Problem-Solving Approach
- Customizing AutoCAD 2010
- AutoCAD 2009: A Problem-Solving Approach

Textbooks Authored by CADCIM Technologies and Published by Other Publishers

3D Studio MAX and VIZ Textbooks

- Learning 3DS Max: A Tutorial Approach, Release 4
  Goodheart-Wilcox Publishers (USA)
- Learning 3D Studio VIZ: A Tutorial Approach
  Goodheart-Wilcox Publishers (USA)

CADCIM Technologies Textbooks Translated in Other Languages

SolidWorks Textbooks

- SolidWorks 2008 for Designers (Serbian Edition)
  Mikro Knjiga Publishing Company, Serbia
- SolidWorks 2006 for Designers (Russian Edition)
  Piter Publishing Press, Russia
- SolidWorks 2006 for Designers (Serbian Edition)
  Mikro Knjiga Publishing Company, Serbia

NX Textbooks

- NX 6 for Designers (Korean Edition)
  Onsolutions, South Korea
- NX 5 for Designers (Korean Edition)
  Onsolutions, South Korea

Pro/ENGINEER Textbooks
- Pro/ENGINEER Wildfire 4.0 for Designers (Korean Edition)
  HongReung Science Publishing Company, South Korea
- Pro/ENGINEER Wildfire 3.0 for Designers (Korean Edition)
  HongReung Science Publishing Company, South Korea

Autodesk 3ds Max Textbook
- 3ds Max 2008: A Comprehensive Guide (Serbian Edition)
  Mikro Knjiga Publishing Company, Serbia

AutoCAD Textbooks
- AutoCAD 2006 (Russian Edition)
  Piter Publishing Press, Russia
- AutoCAD 2005 (Russian Edition)
  Piter Publishing Press, Russia
- AutoCAD 2000 Fondamenti (Italian Edition)

Coming Soon from CADCIM Technologies
- SOLIDWORKS Simulation 2016 for Designers
- Exploring RISA 3D 12.0

Online Training Program Offered by CADCIM Technologies
CADCIM Technologies provides effective and affordable virtual online training on animation, architecture, and GIS softwares, computer programming languages, and Computer Aided Design, Manufacturing, and Engineering (CAD/CAM/CAE) software packages. The training will be delivered 'live' via Internet at any time, any place, and at any pace to individuals, students of colleges, universities, and CAD/CAM/CAE training centers. For more information, please visit the following link: *http://www.cadcim.com*.

www.ingramcontent.com/pod-product-compliance
Lightning Source LLC
Chambersburg PA
CBHW080133220326
41598CB00032B/5048

* 9 7 8 1 9 4 2 6 8 9 1 6 4 *